POLARIZATION

POLARIZATION

By

Dr. R.K. Verma

DISCOVERY PUBLISHING HOUSE
NEW DELHI-110002

Reprinted - 2019

First Published - 2006

ISBN: 978-81-8356-123-5

Polarization

Published by:

DISCOVERY PUBLISHING HOUSE PVT. LT

4383/4B, Ansari Road, Darya Ganj

New Delhi-110 002 (India)

Phone: +91-11-23279245, 23253475; 43596065

E-mail: discoverybooksindia@gmail.com

discoverypublishinghouse@gmail.com

web: www.discoverypublishinggroup.com

Printed at:

Infinity Imaging Systems

Delhi

PREFACE

This book "Polarization" covers the course in Geometrical and Physical optics for most of Universities in India. This book was planned to covers Polarization (Polarization by Reflection, Polarization by refraction, Double refraction, the Polariods, Nicol Prism. Double Image Prisms, Analysis of Polarization in a given beam of light). The language of the book has been kept as simple as could be consistent with precision and brevity.

I wish to express my sincere thanks to all of them who have helped me in one way or the other in the preparation of this enlarged edition. My thanks are due to M/s Discovery Publishing House for their keen interest in bringing out this separate text.

Criticism and suggestion for further improvement shall be gratefully acknowledge.

Author

Contents

1

POLARIZATION

INTRODUCTION

The phenomena of interference and diffraction would occur with light whether the waves are transverse or longitudinal. In this chapter we will deal with some special results arising from the transverse character of light. Interference and diffraction phenomena proved that light is a wave motion and enabled the determination of the wavelength.

However, they do not give any indication regarding the character of the waves. Whether the light waves are longitudinal or transverse, or whether the vibrations are linear or circular cannot be deduced from the above two phenomena, as all kinds of waves under suitable conditions exhibit interference and diffraction. In 1861 Arago and Fresnel showed that light waves vibrating in mutually perpendicular planes do not interfere.

In 1817 Thomas Young explained the absence of interference by postulating that light waves are *transverse waves*. About fifty years later. Maxwell developed electromagnetic theory and suggested that light waves are electromagnetic waves. As electromagnetic waves are transverse waves, it is obvious that light waves too are transverse waves. The concept of transverse nature leads to the concept of polarisation. Light coming from common light sources is unpolarized.

The state of polarization cannot be detected by unaided human eye. An understanding of polarisation is essential for understanding the propagation of electromagnetic waves guided through wave-guides and optical fibres. Polarised light has many important applications in industry and engineering. One of the most important applications is in liquid crystal displays (LCDs) which are widely used in wristwatches, calculators, TV screens etc.

EVIDENCE THAT WAVES OF LIGHT ARE TRANSVERSE

For electromagnetic waves in the microwave region (λ + 1 cm) the transverse character can be demonstrated using a wire-grid polarizer. It has number of straight parallel wires (Fig. 1.1) of an electric conductor. Let the plane of the grid bexy plane with length of wires along JC-direction, and let the waves travel along the z-direction. If one more wire-grid (not shown) is placed parallel to the first and is rotated in its own plane, then no waves pass if the wires in the two grids are mutually perpendicular, while waves pass if the wires in the two grids are parallel.

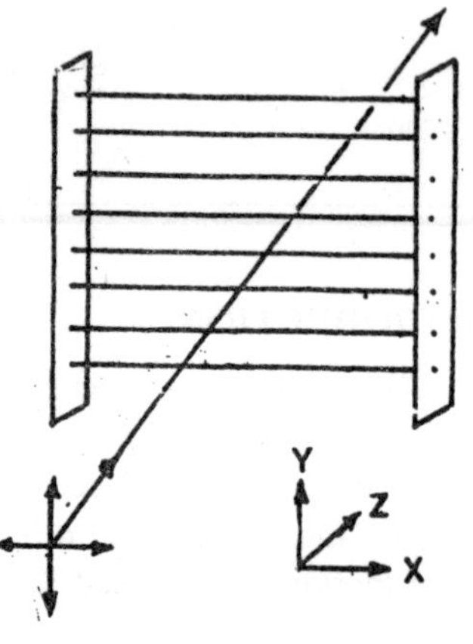

Fig. 1.1 : The wire-grid polarizer.

The explanation is simple. The x-oscillations of electric field E in the e.m. waves make the electrons in the wires oscillate and the energy is thus dissipated as heat They-osallations of E field are unable to do this (assuming wire diameter << λ) and hence pass through.

POLARIZATION OF LIGHT

We have seen in an earlier chapter that there are two possible ways of wave propagation:

(1) Longitudinal waves in which the motion of individual particles is along the line of propagation, and

(2) Transverse waves in which the motion is perpendicular to the line of propagation. Which kind of wave motion do we have in the case of light?

The important difference between the longitudinal and transverse waves is that the latter can be "polarized." To understand this important notion let us look at a wave in the direction of its propagation as shown in Fig. 1.2. In the case of longitudinal waves (a), the motion of the particles takes place perpendicular to the surface of the paper and will not be noticeable from the direction we are observing it. In the case of transverse waves (b) and (c), the motion of the particles is in the plane of the paper and easily observable in that projection. We call the transverse wave "natural" or "nonpolarized" if the motion of the particles takes place in all possible directions (b); if the motion is only in one direction (c), the wave is "polarized." The notion of polarization can be clarified by the analogy give in Fig. 1.3. Suppose we have a sieve, made of a set of parallel wires without cross wires, and drop matches on it in such a way that the falling matches remain horizontal but may have different orientations in the horizontal plane. It is clear that only those matches that are parallel to the sieve wires will pass through. If under this "match polarizer" we place a similar "match analyzer," the "beam of matches" will pass through only if the wires in the lower sieve run parallel to those of the upper sieve. We can extend this analogy to the case of longitudinal waves by dropping the matches in a vertical position. In this case, all the matches will pass through, regardless of the relative positions of the two sieves.

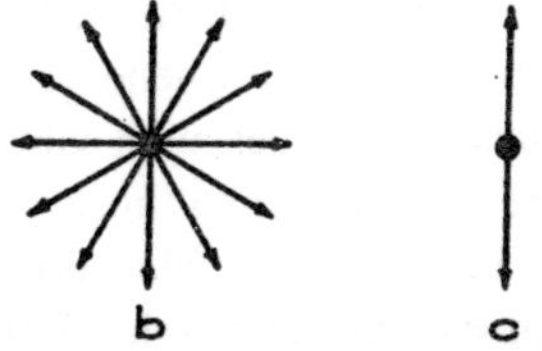

Fig. 12 : Cross-section of a wave in the direction of its propagation.

To our eyes, the plane of vibration of light waves makes no difference—light that has been polarized so the vibration is all in a single plane (Fig. 1.2c) looks exactly the same as ordinary light that consists of myriads of photons whose planes of vibration are aligned in every

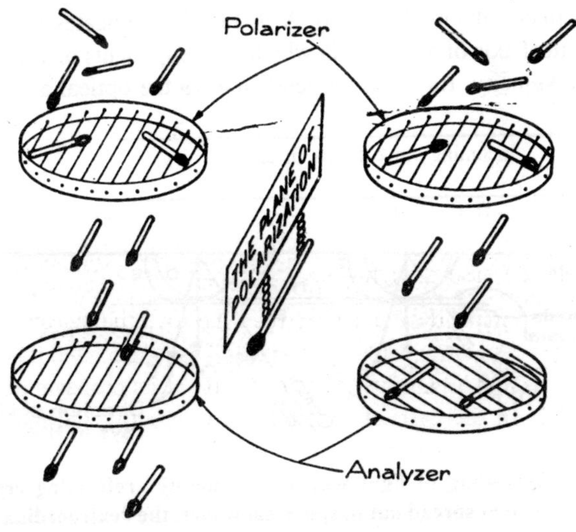

Fig. 1.3 : "Polarized matches."

possible direction (Fig. 1.2b). Much of the light we see is at least partially polluted, because whenever light is reflected from a smooth *non-metallic* surface the waves with vibrations parallel to the surface are reflected more intensely than those with vibrations perpendicular to the surface. The light from the blue sky is also partially polarized. The, white light from the sun, in passing through the molecules of the air, has its blue component scattered, or reflected, from the molecules more than are the longer red waves, and this scattered blue light is partially polarized. The eyes of bees can not only distinguish polarized light from non-polarized, but they can determine the plane of polarization. Bees apparently use this ability to tell directions and navigate back to the hive, and can do so even when the sun is covered by clouds as long as there is a patch of blue sky visible.

DOUBLY REFRACTING CRYSTALS

Certain crystals, such as quartz and calcite, as well as many others (including ice, to a very slight extent) have different properties in different directions. Electrical properties, heat conductivity, and optical properties depend on the direction in which they are measured in the crystal. Fig. 1.4 shows the spreading of light waves in such a crystal. In one particular direction, called the *optical axis* ("axis" here means a direction, and not a particular line), all light travels at the same speed, no matter what its plane of vibration. Ray A in Fig. 1.5 indicates a ray parallel to the optical axis of a crystal, in which all components of vibration have the same velocity. Ray B is perpendicular to the optical axis direction,

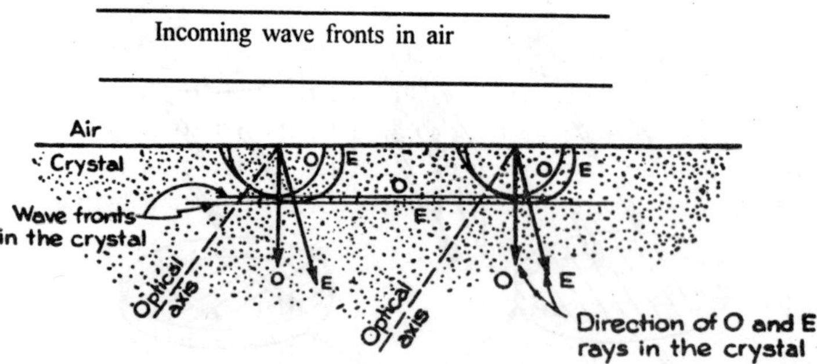

Fig. 1.4 : Spreading of light waves in a doubly refracting crystal. The "ordinary" waves spread out in spherical waves, the "extraordinary" waves in elliptical waves. In the direction called the "optical axis," the speed of both ordinary and extraordinary waves are the same.

and all of the vibrations in B can be resolved into components that are either parallel to the optical axis (E) or perpendicular to it (O). The O components (called the "ordinary" ray) travel at the same speed as A; the E components (called the "extraordinary" ray) travel at a different speed. (In Fig. 1.5 they are shown with a higher speed than O, but in some crystals the speed may be slower.)

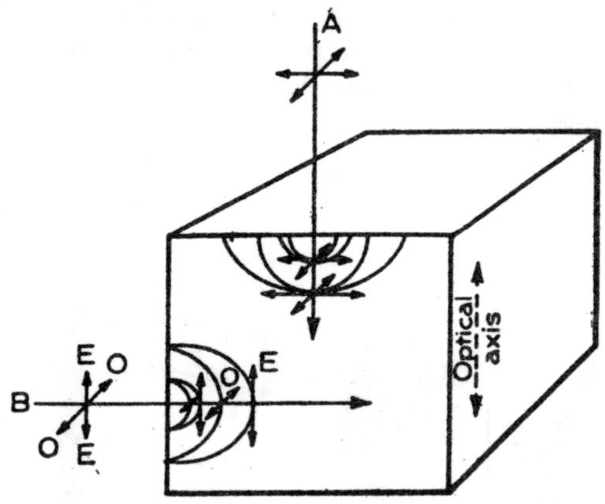

Fig. 1.5 : In ray A, parallel to the optical axis, vibrations ore all perpendicular to the optical axis, and all travel at the same speed. Ray B is perpendicular to the optical axis, and its vibrations will hove components both perpendicular and parallel to the optical axis. The components perpendicular to the optical axis travel at the normal speed, and comprise the "ordinary" ray. The components vibrating parallel 10 the optical axis (the "extraordinary" ray) travel at a different speed.

We have defined the index of refraction, v, to be the ratio of the speeds of light in air (or vacuum) and in the material. A doubly refracting crystal will have two indices of refraction: re for the extraordinary ray and v_0 for the ordinary ray. Indices of refraction for the ordinary and the extraordinary ray are given here for a few commonly used materials:

	v_0	v_E
Calcite	1.658	1.486
Quartz	1.544	1.553
Tourmaline	1.637	1.619

Tourmaline has the property of being relatively opaque to the ordinary ray, so that only the extraordinary is transmitted through it. Thus, if a ray of unpolarized light-falls on a tourmaline crystal in a direction perpendicular to its optical axis, the light coming through the crystal will emerge completely polarized—*i.e.*, with all the vibrations in a single plane that is parallel to the optical axis. Iodo-quinine sulfate crystals have this same property of absorbing one ray and being transparent to the other. "Polaroid" is composed of billions of tiny needle-shaped iodoquinine sulfate crystals lined up with their optical axes parallel and imbedded in a sheet of plastic. Light passing through a sheet of Polaroid is almost completely polarized, with the vibrations being parallel to the length of the imbedded crystals.

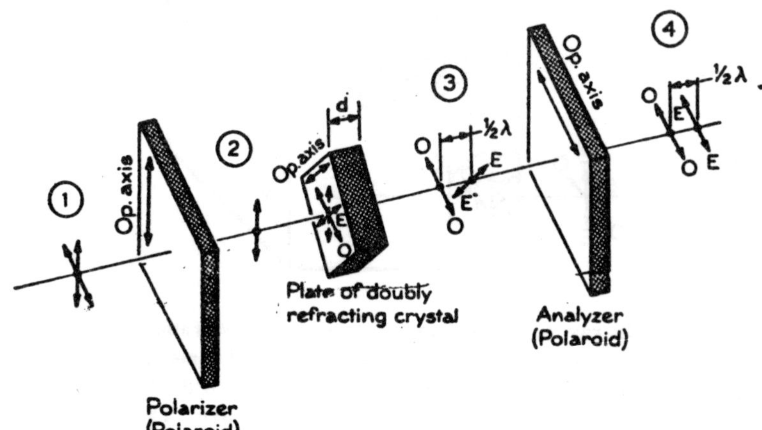

Fig. 1.6 : The production of colors by meant of the interference of polarized light.

The production of colors by means of doubly refracting material gives an interesting insight into the behaviour of polarized light. Fig. 1.6 shows a narrow beam of unpolarized monochromatic light:

(1) Falling on a piece of polaroid with a vertical axis. The light emerging from this polarizer is all vibrating in a vertical plane,

(2). The beam next falls on a slab of doubly refracting crystal with its optical axis 45° from the vertical, The crystal resolves die incoming vertical vibrations into two components, E and O, which are parallel and perpendicular to the optical axis of the crystal. These vibrations will be transmitted through the crystal at different speed and, if the thickness of the crystal is Just right, the extraordinary ray will gain a half wave length on the ordinary ray by the time they emerge into the air again

(3). Since the vibrations of E and O are at right angles to one another, there can be no interference between them. However, if we now have the beam pass through another piece of polaroid (the analyzer) with an axis 45° to the plane of both E and O, only the components parallel to the analyzer axis can come through.

(4) These components now lie in the same plane of vibration and are able to interfere with one another. Since E and O are a half wave length out or phase and of equal amplitude, they will annul one another and the light of the beam will be extinguished.

We have considered the beam to be monochromatic, and it is plain that if we let white light fall on our apparatus the thickness of the plate may be suitable to extinguish light of some wave lengths (if the retardation is any odd number of half wave lengths) and reinforce others (if the retardation is any integral number of wave lengths). Ordinary cellophane is doubly refracting, and if a wad of crumpled cellophane is placed between two sheets of polaroid, so that the light passes through different thicknesses of cellophane, beautiful colored patterns can be seen.

The production of colors by the above means is used in the study of mineral crystals. Plate III shows a photograph of crystals taken in this way.

Velocity of Light

The first attempt to measure the propagation of light was undertaken by Galileo in a very primitive way. One evening, he and his assistant placed themselves on two distant hills in the neighbourhood of Florence, each of them carrying a lantern with a shutter. Galileo's assistant was instructed to open his lantern as soon as he noticed the flash from the one carried by his master. If light was propagating with a finite speed, the flash from the assistant's lantern would have been observed by Galileo with a certain delay. The result of this experiment was, however, completely negative, and we know now very well why. Light propagates so fast that the expected delay in Galileo's experiment must have been about one hundred-thousandth of a second, which is quite unnoticeable to human senses.

The first successful measurement of the velocity of light was carried out in 1675 by the German astronomer, Roemer, who replaced Galileo's assistant by the moons of the planet Jupiter, thus increasing the distance to be covered by light by a factor of hundreds of millions. Roemer's method is illustrated in Fig. 1.7, which shows the orbits of the earth, Jupiter, and one of its moons. Moving around the planet, the moons are periodically eclipsed as they enter the broad cone of shadow cast by Jupiter. Studying these eclipses, Roemer noticed that sometimes they took place as much as eight minutes ahead of schedule and sometimes with a delay of eight minutes. He also noticed that the eclipses were early when the earth and Jupiter were on the same side of the sun (1st position) and delayed in the opposite case (2nd position). Ascribing correctly the observed irregularities to the difference of time taken by light to cover the changing distance between the earth and Jupiter, Roemer calculated that light must be propagating through space at a speed of about 300,000 kilometers per second (3×10^{10} cm/sec).

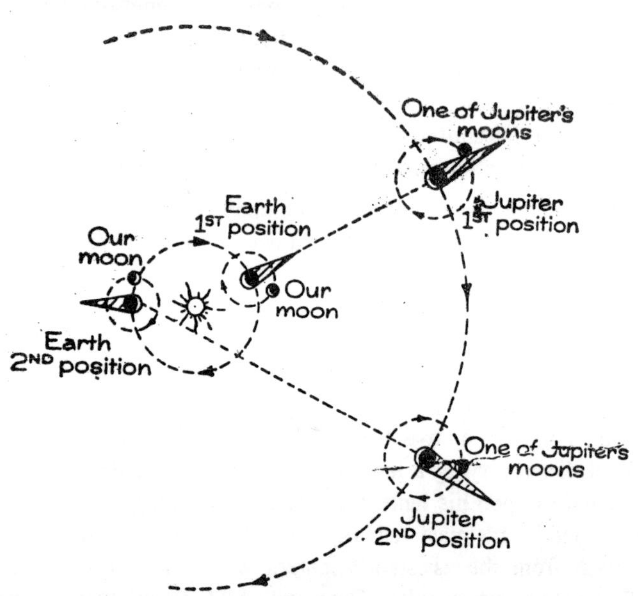

Fig. 1.7 : Roemer's method of measuring the velocity of light by observing the eclipses of Jupiter's moons.

The first laboratory measurement of the speed of light was carried out in 1849 by the French physicist, H. L. Fizeau (1819-1896), whose apparatus is shown in Fig. 1.8. It consists essentially of a pair of cogwheels set at the opposite ends of a long axle. The wheels were positioned in such a way that the cogs of one were opposite the intercog openings of the other, so that a light beam from the source could not be seen by the eye no matter what the position of the wheels. However, if the wheels were set in fast rotation and the speed of this rotation was such that the wheels moved by half the distance between the neighbouring cogs during time taken by light to propagate from one wheel to the other, was exported to pass through without being stopped. In order to observe the effect at the speed of a few thousand revolutions per minute, which was about the maximum that Fizeau could achieve, he had to lengthen the path of the light beam by using four mirrors as shown in Fig. 1.8. This direct laboratory measurement gave a value for the speed of light that stood in reasonably good agreement with that obtained by Roemer's astronomical method.

If, for example, Fizeau used wheels with 100 teeth spinning at 3,000 rev/min, we can compute the necessary lengthening of the path of the light rays. While the light travels to the distant mirrors and back again, the wheels must have turned through the angle between a tooth and a space, in 1/200 of a revolution. At 3,000 rev/min this would require:

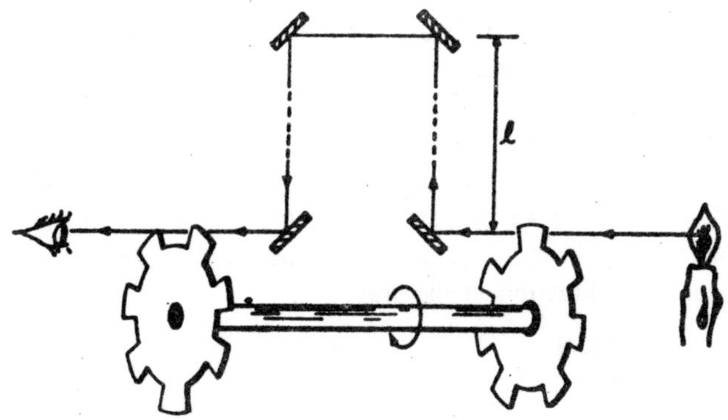

Fig. 1.8 : Fizeau's method for measuring the velocity of light.

$$\frac{1}{3{,}000 \times 200} = \frac{1}{600{,}000} \text{ min, or } 10^{-4} \text{ sec}$$

We know that the speed of light is almost exactly 3×10^{10} cm/sec, so in this time interval of 10^{-4} sec it would have had to travel 3×10^{6}cm or 30 km. Thus, I, the distance to the far mirrors, would have to be about 15 km or approximately 9 min. Later more accurate determinations of c, as the speed of light is universally called, were made by the American physicist, A. A. Michelson, by a similar scheme that used a rotating mirror instead of a toothed cogwheel.

More recently, a polarization phenomenon has been employed to serve as a more accurately timed shutter than rotating wheels or mirrors. Certain substances, such as glass and many organic liquids, become doubly refracting when they are in a strong electric field. The Kerr cell is a glass container, generally filled with nitrobenzene, that is placed where the doubly refracting material is used in Fig. 1.6. The orientation of polarizer and analyzer are adjusted so that no light normally passes through. But when an electronic device creates an electric field in the cell, the nitrobenzene becomes doubly refracting, and a flash of light can pass through the analyzer. Modem electronics can time these flashes to an accuracy of less than 10^{-6} sec, and very accurate values of c can be determined without having to use long light paths.

Work is still constantly being done by new and increasingly refined methods, and we are now fairly sure that the velocity of light in a vacuum is not far from 2.99776×10^{10}cm/sec.

PREFERENTIAL DIRECTION IN A WAVE

Waves are basically of two types, namely longitudinal waves and transverse waves. A wave in which particles of medium oscillate to and fro along the direction of wave propagation is called a *longitudinal wave.* Sound waves and waves produced on a spring are examples of longitudinal waves. The longitudinal waves consist of alternate compressions and rare factions. In these waves, particle displacement (vibration) and wave propagation occur in the *same* direction. Therefore, there does not exist any preferential direction in the wave.

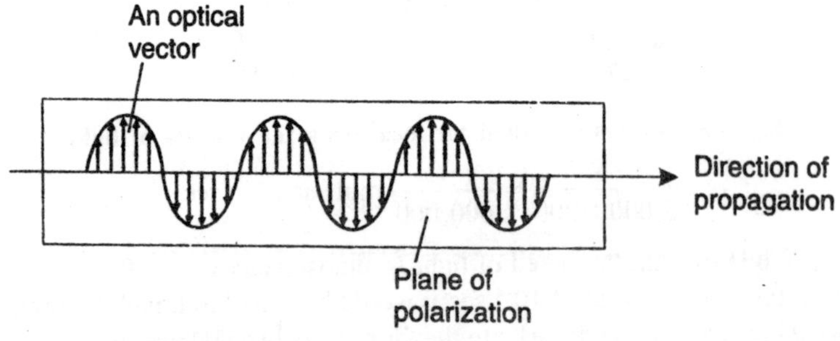

Fig. 1.9

A wave in which every particle of the medium oscillates *up* and *down* at right angles to the direction of wave propagation is called a transverse wave. Ripples on water surface and waves on a rope are examples of transverse waves. The wave propagation takes in the form of alternating crests and troughs. In a transverse wave, the direction of particle displacement occurs perpendicular to the wave propagation. Hence, the direction normal to the wave propagation is the *preferential direction* in a transverse wave.

The existence of a preferential direction leads to the characteristic phenomenon known as polarisation of transverse waves. Polarisation is not found with longitudinal waves as they do not possess a directional property. Polarisation is specific to transverse waves.

An electromagnetic wave is a transverse wave consisting of electric and magnetic fields vibrating perpendicular to each other and to the direction of propagation. The vibrating electric vector E and the direction of wave propagation form a plane. In olden days this plane was called the *plane of vibration.* However, it has become common now days to refer to it as the *plane of polarisation.*

We adopt the same notation and define the *plane of polarisation of the wave as the plane*, which *contains the optical vector E and the direction of propagation.*

POLARISED LIGHT

The simplest type of an electromagnetic wave is a wave in which the direction of vibrations of electric vector E is strictly confined to a single direction in the plane perpendicular to the direction of propagation of the wave. Such a light is said to be *plane polarised light.* If the wave is coming towards the eye, the electric vector appears executing a linear vibration normal to the ray direction. Hence, a plane-polarised wave is also known as a linearly polarised wave. We therefore offer the following definitions:

A *plane-polarised* light wave is a wave in which the electric vector is everywhere confined to a single plane.

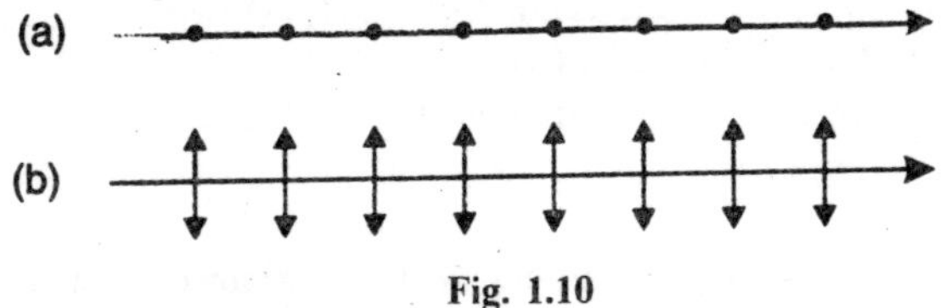

Fig. 1.10

A *linearly polarised* light wave is a wave in which the electric vector oscillates in a given, constant orientation.

Now-a-days the adjective "linearly polarised" is more preferred and frequently used than "plane polarised". Ray diagrams, as shown in Fig. 1.10, represent the linearly polarised light.

NATURAL LIGHT

An ordinary light source consists of a very large number of atomic emitters. Each atom radiates a plane polarised wave train for about 10^{-8}s. As time progresses the orientation of the plane of polarisation changes randomly and rapidly at the rate of 10^8 times per second. Our eyes cannot respond to such rapid changes. In the time that the eye responds to light, millions of wave trains are emitted by the source. Further, the frequencies of the wave trains are not exactly identical. Thus, ordinary light comprises of a heterogeneous group of wave trains having different wavelengths and vibrating in different planes perpendicular to the direction in which the ray is propagating (Fig. 1.11a). The random orientation of the planes of polarisation leads to equal probability of all directions about the propagation direction, as shown in (Fig. 1.11b). Therefore, in natural light the optical vector has no specific favoured orientation. *Light in which the planes of vibration are symmetrically distributed about the propagation direction of the wave is known as the unpolarized light.*

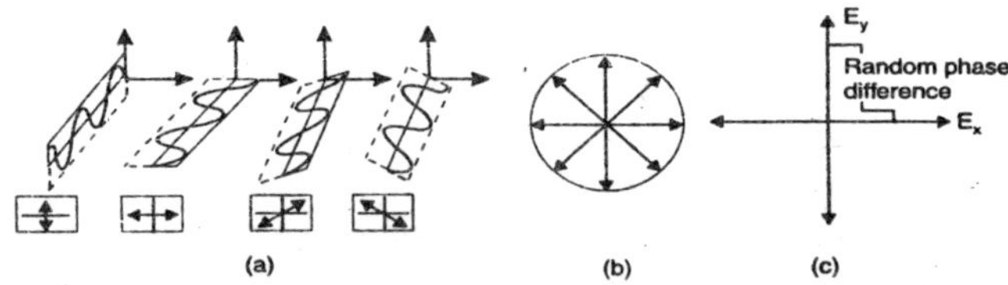

Fig. 1.11 : Unpolarized light (a) component waves having different planes of polarization (b) Schematic representation, (c) Representation of unpolarised light.

We may resolve each optical vector in unpolarized light into two rectangular components lying parallel and perpendicular to a chosen direction (Fig. 1.11c). Instead of showing a large number of optical vectors as in (Fig. 1.11b), we use only two vectors, as in (Fig. 1.11c), for representing unpolarized light. Due to the random distribution of the optical vectors, the amplitude of the component vectors will be equal. Therefore, if the intensity of the incident unpolarized light is I_0 the intensity of each component will be 1/2 I_0. Thus, the intensity of unpolarized light is equally distributed between the two components.

PRODUCTION OF LINEARLY POLARIZED LIGHT

Linearly polarised light may be produced from unpolarized light using of the following five optical phenomena:

(i) Reflection,

(ii) Refraction,

(iii) Scattering,

(iv) Selective absorption (dichroism) and

(v) Double refraction.

Polarization by Reflection

E.L. Malus, the French engineer and scientist discovered in 1808 polarisation of natural light by reflection from the surface of glass. He noticed that when natural light is incident on a plane sheet of glass at a certain angle, the reflected beam is plane polarized.

Fig. 1.12 shows an unpolarized light beam AB incident on a glass surface. The incident ray AB and the normal NBN' define the *plane of incidence.* The electric vector E of the ray AB can be resolved into two components, one perpendicular to the plane of incidence and the other lying in the plane of incidence. The perpendicular component is represented by dots and is called the *s-component*. The parallel component is represented by the arrows and is called the *p-component.*

In case of completely unpolarized light the two components are of equal magnitude. When natural light is incident on the surface, the two components are reflected to different extents. The reflected ray BC has a predominance of s-component (Fig. 1.12a) and as such it is partially polarised. As the angle of the incidence of light is varied, the extent of polarisation of the reflected ray varies. At a particular angle θ_p the reflection coefficient for p-component goes zero (Fig. 1.13) and the reflected beam does not contain any p-component. It contains only s-component and is totally *linearly polarised.* The angle θ_p is called the *polarizing angle.*

This particular method of obtaining linearly polarised light is not advantageous. The intensity of the reflected beam is very weak as only 15% of the s-component is reflected. In contrast, the refracted beam is strong and consists of a mixture of p-component, which is transmitted 100% and the balance 85% of s-component.

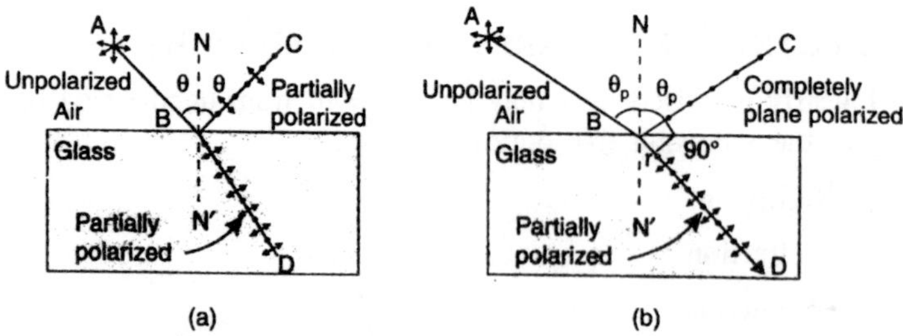

Fig. 1.12 : Production of plane polarized light by the method of reflection– (a) When an unpolarized light is incident on a reflecting surface, the reflected and refracted beams are partially polarized, (b) The reflected beam is completely polarized when the angle of incidence equals the polarizing angle θ_P which satisfies Brester's law.

Brewster's Law

Sir David Brewster performed a series of experiments on the polarization of light by reflection at a number of surfaces. He found that the polarising angle depends upon the refractive index of the medium. In 1892, Brewster proved that *the tangent of the angle at which polarisation is obtained by reflection is numerically equal to the refractive index of the medium.* If θ_P is the angle and u. is the refractive index of the medium, then

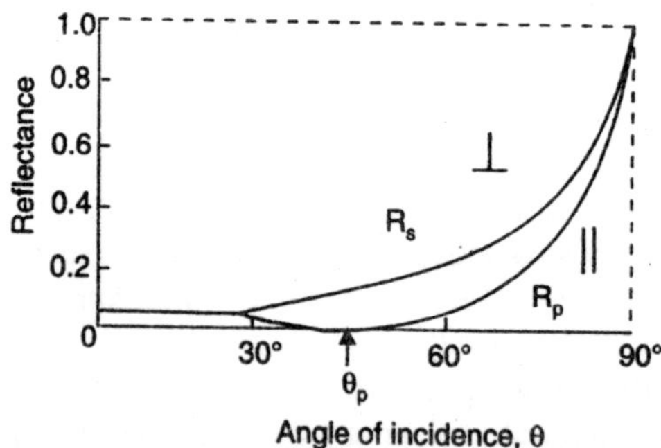

Fig. 1.13 : Variation of reflectance of perpendicular and parallel components of a wave incident on air-to-glass boundary with the angle of incidence. Note that the reflectance for p-component becomes zero at Brewster's angle.

$$\mu = \tan \theta_P \qquad ...(1)$$

This is known as Brewster's law.

If natural light is incident on a smooth surface at the polarizing angle, it is reflected along BC and refracted along BD, as shown in

Fig. 1.12 (b). Brewster found that the maximum polarisation of reflected ray occurs when it is at right angles to the refracted ray. It means that $\theta_P + r = 90°$

$$\therefore \quad r = 90° - \theta_P \qquad ...(2)$$

According to Snell's law,

$$\frac{\sin \theta_P}{\sin r} = \frac{\mu_2}{\mu_1} \qquad ...(3)$$

where, μ_2 is the absolute refractive index of reflecting surface and μ_1 is the refractive index of the surrounding medium. It follows from eqn. (2) and eqn. (3) that,

$$\frac{\sin \theta_P}{\sin (90° - \theta_P)} = \frac{\mu_2}{\mu_1}$$

or

$$\frac{\sin \theta_P}{\sin \theta_P} = \frac{\mu_2}{\mu_1}$$

$$\therefore \quad \tan \theta_P = \frac{\mu_2}{\mu_1} \qquad ...(4)$$

Eqn. (4) shows that the polarising angle depends on the refractive index of the reflecting surface. The polarising angle θ_P is also known as *Brewster angle* and denoted as θ_B. Light reflected from any angle other than Brewster angle is partially polarised.

Application of Brewster's Law

(i) Brewster's law can be used to determine the refractive indices of opaque materials.

(ii) It helps us in calculating the polarizing angle necessary for total polarization of reflected light for any material if its refractive index is known. However, the law is not applicable for metallic surfaces.

(iii) Two windows known as *Brewster windows* are used in gas lasers. They are arranged at Brewster angle at the two ends of the laser tube. Every time light passes through the windows on its way to the reflecting mirrors of the optical resonator, s-component is removed. At the end, light emerging from the laser consists of linearly polarised light of p-component.

(iv) Another application utilises the Brewster angle for transmitting a light beam into or out of an optical fibre without reflection losses.

Polarization by Refraction—Pile of Plates

When unpolarized light is incident at Brewster angle on a smooth glass surface, the reflected light is totally polarized while the refracted light is partially polarized. If natural light is transmitted through a single plate, the transmitted beam is only partially polarized. If a stack of glass plates is used instead of a single plate, reflections from successive surfaces occur leading to the filtering of the s-component in the transmitted ray. Ultimately, the transmitted ray consists of p-component alone. If I_P and I_S are the intensities of the parallel and perpendicular components in the refracted light, the degree of polarisation of the transmitted light is given by

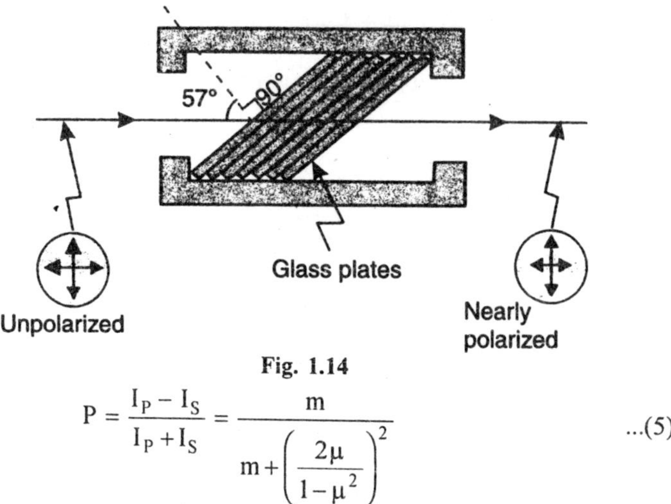

Fig. 1.14

$$P = \frac{I_P - I_S}{I_P + I_S} = \frac{m}{m + \left(\frac{2\mu}{1-\mu^2}\right)^2} \quad ...(5)$$

where m is the number of plates required and 4 is the refractive index of the material.

It is found that a stack of about 15 glass plates is required for this purpose. The glass plates are supported in a tube of suitable size and are held inclined at an angle of about 33° to the axis of the tube, as shown in Fig. 1.14. Such an arrangement is called a *pile of plates*. Unpolarized light enters the tube and is incident on the plates at Brewster angle and the transmitted light will be polarised parallel to the plane of incidence.

However, this method of polarising light is also not efficient because a good portion of light is lost in reflections.

Polarization by Scattering

If a narrow beam of natural light is incident on a transparent medium containing a suspension of ultramicroscopic particles, the light scattered is partially polarized. The degree of polarisation depends on the angle of scattering. The beam scattered at 90° with respect to the incident direction is linearly polarized. The direction of vibration of E vector in the scattered light will be perpendicular to the plane defined by the direction of propagation and the direction of observation, as illustrated in Fig. 1.15. Sunlight scattered by air molecules is polarised. The maximum effect is observed on a clear day when the Sun is near the horizon. The light reaching an observer on the ground from directly overhead is polarised to the extent of 70% to 80%.

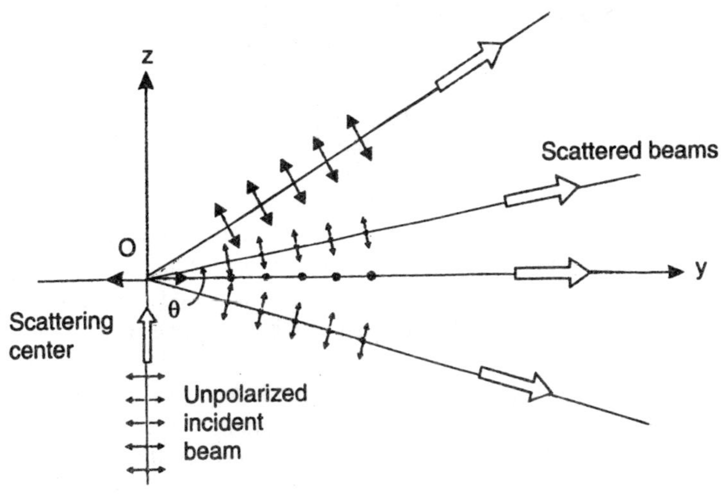

Fig. 1.15

Polarization by Selective Absorption

In 1815 Biot discovered that certain mineral crystals absorb light selectively. When natural light passes through a crystal such as tourmaline, it is split into two components, which are polarized in mutually perpendicular planes. The crystal strongly absorbs light that is polarized in a direction parallel to a particular plane in the crystal but freely transmits the light component polarized in a perpendicular direction. This difference in the absorption for the rays is known as *selective absorption* or *dichroism*. If the crystal is of proper thickness, one of the components is totally absorbed and the other component emerging from the crystal is linearly polarized. Selective absorption is illustrated in Fig. 1.16. A good example of dichroic crystal is a tourmaline crystal.

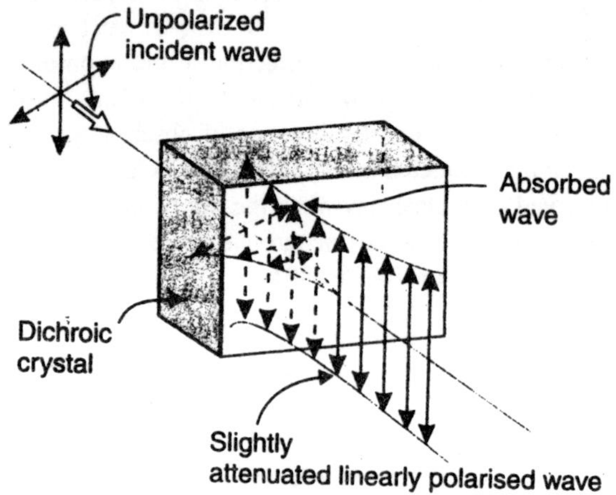

Fig. 1.16

The difference in absorption in different directions may be understood from the electron theory. When the frequency of the incident light wave is close to the natural frequency of the electron cloud, the light waves are absorbed strongly. Crystals that exhibit selective absorption are *anisotropic*. The crystal splits the incident wave into two waves. If the light wave frequency is close to the natural frequency of the electron cloud, then the component having its vibrations perpendicular to the principal plane of the crystal gets absorbed. The component with parallel vibrations is less absorbed and transmitted. The transmitted light is linearly polarised. However, the difficulty is that the crystals cannot be grown to sufficiently bigger

Polarization by Double Refraction

The double refraction phenomenon was discovered in 1669 by Erasmus Bartholinus during his studies on calcite ($CaCO_3$) crystals, which are known more as Iceland spar. When light is incident on a calcite crystal, it is split into two refracted rays (Fig. 1.17) differing in then-properties. The phenomenon of causing two refracted rays by a crystal is called *birefringence* or *double refraction*. The crystals are said to be birefringent.

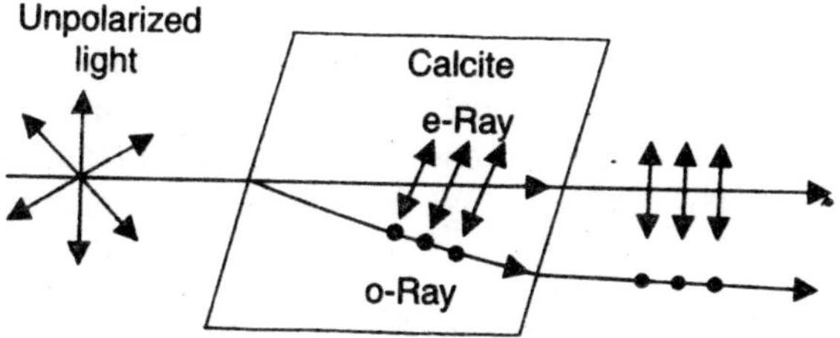

Fig. 1.17

The two rays produced in double refraction are linearly polarised in mutually perpendicular directions. One of the rays obeys Snell's law of refraction and hence is known as an ordinary ray or o-ray. The other ray does not obey Snell's law and is called an *extraordinary ray* or *e-ray*. Hence, the two rays can be distinguished from each other. If one of the rays is eliminated, the light transmitted by the crystal will be a linearly polarised light.

POLARIZER AND ANALYSER

A polarizer is an optical device that transforms unpolarized light into polarized light (Fig. 1.18). If it produces linearly polarised light, it is called a linear polarizer. A linear polarizer is associated with a specific direction called the transmission axis of the polarizer. If natural light is incident on a linear polarizer, only that vibration that is parallel to the transmission axis is allowed through the polarizer while the vibration that is in a perpendicular direction is totally blocked. An analyser is a device, which is used to identify the direction of vibration of linearly polarised light. A polarizer and an analyser are fabricated in the same way and have the same effect on the incident light.

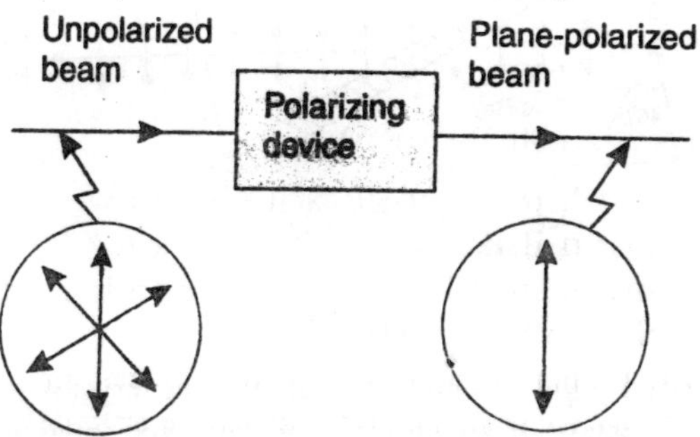

Fig. 1.18 : A polarizer transforms unpolarized light into polarized light.

Fabrication of Linear Polarizer

Practical linear polarizers are fabricated using birefringent crystals or dichroic crystals. Out of the polarisers based on double refraction, Nicol prism is the widely used one. Polarizer sheets made using dichroism are known as Polaroid sheets. We describe here the preparation of Nicol prism and Polaroid sheets.

(i) *Nicol Prism :* A Nicol prism is made from calcite crystal. It was designed by William Nicol in 1820. A rhomb of calcite crystal about three times as long as it is thick, is obtained by cleavage from the original crystal. The ends of the rhombohedron are ground until they make an angle of 68° instead of 71° with the longitudinal edges. This piece is then cut into two along a plane perpendicular both to the principal axis and to the new end surfaces. The two parts of the crystal are then cemented together with Canada balsam, whose refractive index lies between the refractive indices of calcite for the o-ray and e-ray. $\mu_o = 1.66$, $\mu_e = 1.486$ and $\mu_{canada\ baisam} = 1.55$. The position of the optic axis AB is as shown in Fig. 1.19. The refractive index for e-ray depends upon the direction in which e-ray is propagating in the crystal. The difference between the refractive index of o-ray and that for e-ray goes on increasing with the angle between the two rays in the crystal. When this angle is 90°, the difference is a maximum. Thus, for a fixed value for μ_o, the μ_e has its maximum or minimum value in perpendicular direction. In the above case $\mu_e = 1.486$ represents the minimum value.

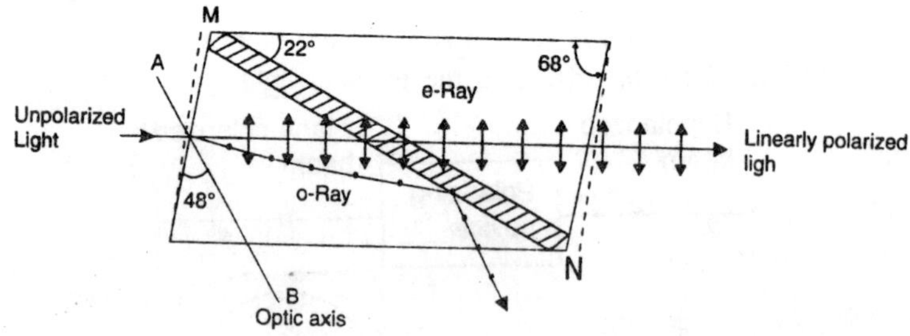

Fig. 1.19

Unpolarized light is made to fall on the crystal as shown in Fig. 1.19 at an angle of about 15°. The ray after entering the crystal suffers double refraction and splits up into o-ray and e-ray. The two rays with their directions of vibrations are as shown in the Fig. 1.19. The values of the refractive indices and the angles of incidence at the Canada balsam layer are such that the e-ray is transmitted while the o-ray is internally reflected. The face where the o-ray is incident is blackened so that the o-ray is completely absorbed. Then we get only the linearly polarised e-ray coming out of the Nicol with direction of vibration as shown in the Fig. 1.19. Thus, the Nicol works as a polarizer.

The Nicol prism is the most widely used polarizing device. Nicols are good polarizers but they are expensive and have a limited field of view of about 28°.

For studying the optical properties of transparent substances, two Nicols are used–one as polarizer and the other as analyser.

(ii) *Polaroid Sheets* : In 1928 E.H. Land invented a method of aligning small crystals to obtain large polarising sheets. The sheets are called Polaroid sheets. They are fabricated as follows. A clear plastic sheet of long chain molecules of PVA (polyvinyl alcohol) is heated and then stretched in a given direction to many times its original length.

During the stretching process the PVA molecules become aligned along the direction of stretching. The sheet is then laminated to a rigid sheet of plastic to stabilise its size. It is then exposed to iodine vapour. The iodine atoms attach themselves to the straight long chain PVA molecules and consequently form long parallel conducting chains. The conduction electrons associated with iodine can move along the chains.

When natural light is incident on the sheet, the component that is parallel to the chains of iodine atoms induces current in the conducting chains and is therefore strongly absorbed. Consequently, the light transmitted contains only the component that is perpendicular to the direction of molecular chains. The direction of E-vector in the transmitted beam corresponds to the *transmission axis* of the Polaroid sheet. These sheet polarisers are inexpensive and can be made in large sizes.

Polaroid sheets are widely used in sunglasses, camera filters etc. to eliminate the unwanted glare from objects. They are used respectively as polarisers and analysers to produce polarised light and to examine the state of polarisation of light.

Working

Nicol prisms and Polaroid sheets are widely used for the production and detection of linearly polarised light. These components are appropriately mounted in metal ring structures such that the transmission axis can be rotated and the amount of rotation may be read from the graduations marked on the ring.

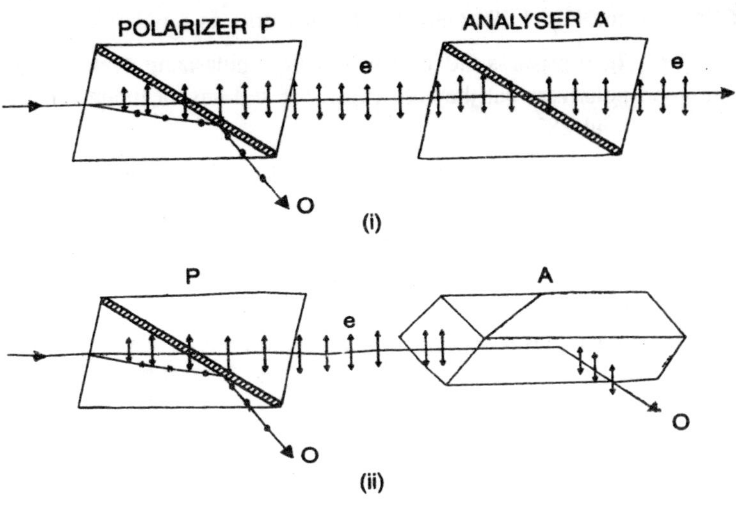

Fig. 1.20

When two Nicol prisms P and A are placed adjacent to each other as shown in Fig. 1.20, one of them acts as a polarizer and the other acts as an analyser. If an unpolarized ray of light is incident on the Nicol prism P, a linearly polarised e-ray emerges from P with its vibration direction lying in the principal section of P. The state of the polarisation of the light emerging from polariser P can be examined with another polarizer A, which for convenience is called an analyser. Let now this ray be incident on the second Nicol prism A, whose principal section is parallel to that to P. The vibration direction of the ray will be in the principal section of A and hence it is transmitted unhindered through the analyser A.

If the Nicol prism A is gradually rotated, the intensity of the e-ray decreases in accordance. When its principal section becomes perpendicular to that of the Nicol prism P, the vibrations of the ray, emerging from

P and incident on A, will be perpendicular to the principal section of A. In this position the ray behaves as o-ray inside the prism A and is totally internally reflected by the Canada balsam layer. Hence no light is transmitted by the prism A. In this configuration, the two Nicol prisms P and A are said to be *crossed.* If the Nicol prism A is further rotated through another 90°, the intensity of light emerging from A will go on increasing. The intensity will become a maximum when its principal section is again parallel to that of the prism P. Thus, the prism P produces linearly polarised light while the prism A detects it. Hence the prism P is called a polarizer and .the prism A an analyser.

Fig. 1.21 shows the Polaroid sheets in three different configurations. When the transmission axis of the analyser A is parallel to that of polarizer P, the light passes unhindered through the analyser, as shown in Fig. 1.21(a). If the transmission axes are at an angle θ, light is partially transmitted (Fig. 1.21). When the axes are perpendicular to each other, no light is transmitted (Fig. 1.21).

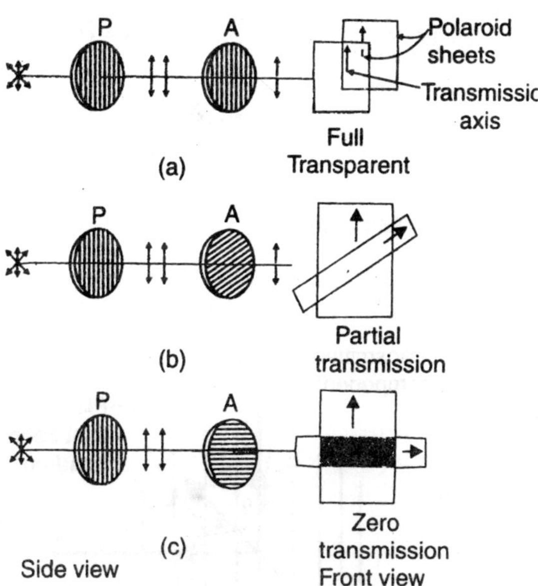

Fig. 1.21 : Polarizer sheets with transmission axes–(a) Parallel allowing transmission of light, (b) Oriented at an angle, allowing partial transmission, (c) Crossed (at 90°) causing extinction of light.

Effect of Polarizer on Natural Light

When unpolarized light passes through a polarizer, the intensity of the transmitted light will be exactly half that of the incident light. We can prove this as follows.

Let E_0 be the amplitude of vibration of one of the waves of the unpolarized light incident on the polarizer. Let E_0 make an angle θ with the transmission axis of the polarizer. E_0 may be resolved into its rectangular components E_x and E_y as shown in Fig. 1.22.

The polarizer blocks the component E_x and transmits the component E_y. The intensity of the transmitted light is then

$$I = E^2_y \ (\cos\theta)^2 = E^2_o \cos^2\theta = I_o \cos^2\theta \qquad ...(6)$$

In unpolarized light, all values of Q starting from 0 to In are equally probable.

$$\therefore \qquad I = I_0 \{\cos^2\theta\} \qquad ...(7)$$

$$= \frac{I_0}{2\pi}\int_0^{2\pi}\cos^2\theta\, d\theta = \frac{I_0}{2\pi}\int_0^{2\pi}\left(\frac{1+\text{cis}\, 2\theta}{2}\right) d\theta$$

$$= \frac{I_0}{4\pi}\left[(\theta)_0^{2\pi} + \left\{\frac{\sin 2\theta}{2}\right\}_0^{2\pi}\right]$$

$$= \frac{I_0}{4\pi}(2\pi = 0)$$

$$\therefore \qquad I = \frac{1}{2} I_0$$

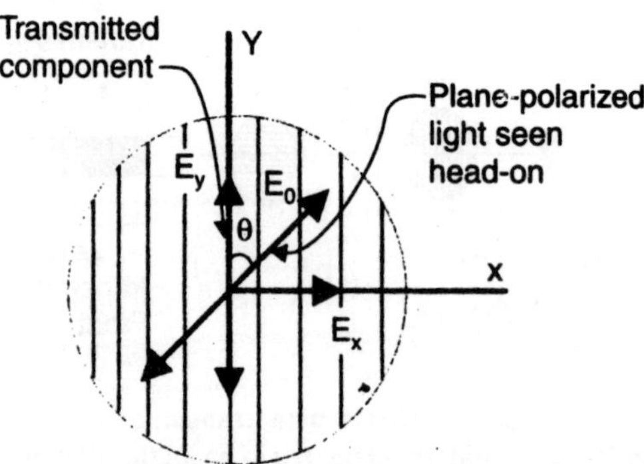

Fig. 1.22 : Action of a polarizing sheet on a linearly polarized wave oriented at an angle θ w.r.t. the transmission axis. The component E_X is absorbed and E_Y component is transmitted.

Thus, if unpolarized light of intensity I_0 is incident on a polarizer, the intensity of light transmitted through the polarizer is $\frac{I_0}{2}$.

Effect of Analyser on Plane Polarized Light—Malus' Law

When natural (unpolarized) light is incident on a polarizer, the transmitted light is linearly polarized. If this light further passes through an analyser, the intensity varies with the angle between the transmission axes of the polarizer and analyser. Malus studied the phenomenon in 1809 and formulated the law that bears his name. It states that the *intensity of the polarized light transmitted through the analyser is proportional to cosine square of the angle between the plane of transmission of the analyser and the plane of transmission of the polarizer. This is known as Malus' law.*

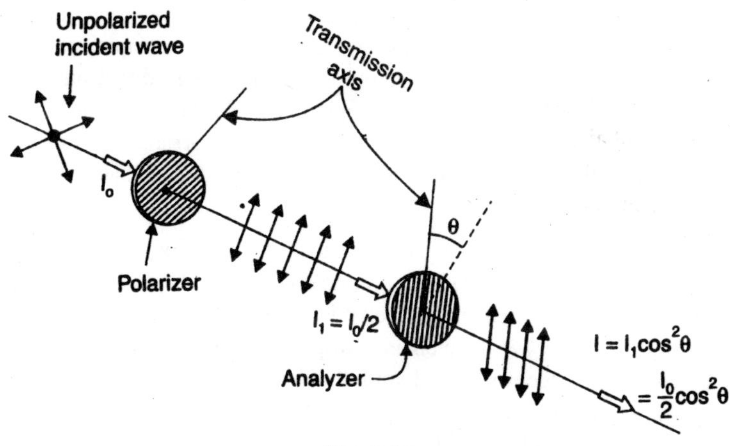

Fig. 1.23

If unpolarized light of intensity I_o is incident on a polarizer, plane polarized light of intensity $\frac{I_o}{2}$ is transmitted by it. Let us denote $\frac{I_o}{2}$ by I_1. This plane polarized light then passes through the analyser. Let E be the amplitude of vibration and θ be the angle that this vibration makes with the axis of the analyser. E can be resolved into two rectangular components: (i) E_y, parallel to the plane of transmission of the analyser and (ii) E_x, perpendicular to the plane of analyser. It is only the parallel component E that is transmitted by the analyser.

$$E_y = E\cos\theta$$

The intensity corresponding to this component is

$$I = E^2 \cos^2 \theta = I_1 \cos^2 \theta = \left(\frac{I_0}{2}\right) \cos^2 \theta$$

Case (i): If $\theta = 0°$	axes parallel	$I = I_1$
Case (ii): If $\theta = 90°$	axes perpendicular	$I = 0$
Case (i): If $\theta = 180°$	axes parallel	$I = I_1$
Case (i): If $\theta = 270°$	axes perpendicular	$I = 0$

Thus, we obtain two positions of maximum intensity and two positions of zero intensity when we rotate the axis of the analyser with respect to that of the polarizer.

ANISOTROPIC CRYSTALS

When a light beam is incident on an *isotropic medium* such as a glass slab, it refracts as a single ray. An optically isotropic material is one in which the index of refraction is the same in all directions. Glass, water and air are examples of isotropic materials. The atoms in a crystal are arranged in a regular periodic manner. If the arrangement of atoms differs in different directions within a crystal, then the physical properties vary with the direction.

The thermal conductivity, electrical conductivity, velocity of light and hence refractive index etc., properties depend on the crystallographic direction along which the property is measured. Then we say that the crystal is *anisotropic*. In such anisotropic crystals the force of interaction between the electron cloud and the lattice is different in different crystallographic directions. The natural frequency of the electron cloud is likewise dependent on the direction in which the electrons are caused to vibrate by the incident light wave. This results in different velocities in different directions and the index of refraction is different in different directions within the crystal.

The anisotropic crysstals are divided into two classes: *uniaxial* and *biaxial* crystals. In case of uniaxial crystals, one of the refracted rays is an ordinary ray and the other is an extraordinary ray. In biaxial crystals both the refracted rays are extraordinary rays. Calcite, tourmaline and quartz are examples of uniaxial crystals whereas mica, topaz and aragonite are examples of biaxial crystals.

We study here only about the uniaxial crystals.

CALCITE CRYSTAL

The most common example of uniaxial crystals is calcite.

Calcite crystals are rhombohedral, as seen from Fig. 1.24. It is bounded by six faces, each of which is a parallelogram with angles equal to 102° and 78°. The angles between the edges meeting at each of the two diametrically opposite corners A and H are obtuse (larger than 90°). Hence, these two comers are called *blunt comers* of the crystal.

Optic Axis

A line bisecting any of the blunt comers is the optic axis (Fig. 1.24). In fact any line parallel to this line is also an optic axis. Thus, the optic axis is a direction and not a specific line in the crystal. It is to be noted that the optic axis is not obtained by joining the two blunt comers. Only in a special case, when the three edges of the crystal are equal, the line joining the two blunt comers A and H coincide with the crystallographic axis of the crystal and it gives the direction of the optic axis. The optic axis is actually the axis of symmetry of the crystal. A ray of light propagating along optic axis does not suffer double refraction, because the structure of the crystal is symmetric about that direction.

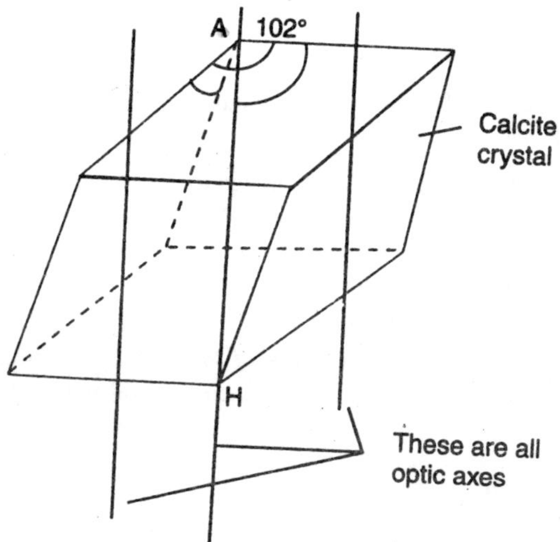

Fig. 1.24

The optic axis is the direction in a uniaxial crystal along which the e-ray and the o-ray travel with the *same* speed and consequently double refraction does not take place along this direction. The corresponding refractive index is the refractive index for ordinary light, say μ_0.

Principal Section

A plane containing the optic axis and perpendicular to a pair of opposite faces of the crystal is called the *principal section* of the crystal for that pair of faces. Thus, there are three principal sections passing through any point within the crystal, one corresponding to each pair of opposite faces. A principal section always cuts surfaces of calcite crystal in a parallelogram having angles 71° and 109° (Fig. 1.25a). (Fig. 1.25b) shows a face of the crystal in which the end-view of the principal section AB of (Fig. 1.25a) is shown by the dotted line AB. The lines parallel to AB represent the end-views of other principal sections parallel to AB with in the crystal.

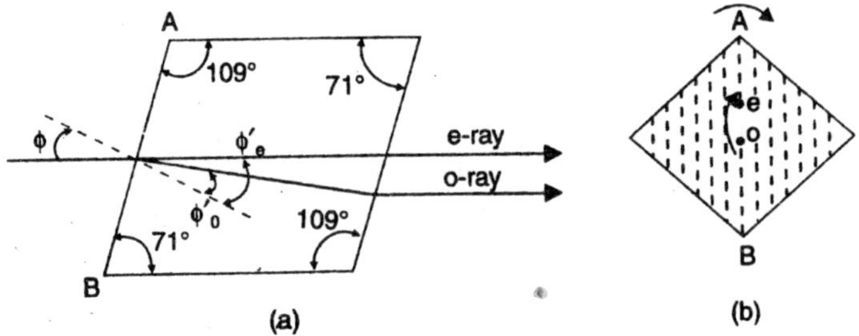

Fig. 1.25 : When the calcite crystal is rotated around the o-ray, e-ray describes a right circular cone around it.

Defining principal section is not enough to understand the directions of vibrations for the o-ray and e-rays. Hence, two more planes are defined as principal plane for the o-ray and the principal plane for the e-ray. The plane containing the optic axis and the o-ray is called the principal plane of the o-ray and the plane containing the optic axis and the e-ray is called the *principal plane of the e-ray*. The directions of vibrations in the o-ray and e-ray can be understood with reference to these planes.

Double Refraction

Fig. 1.25 (a) shows a principal section of calcite crystal. A ray of light is incident on the face AB of the crystal and it travels along the principal section. The, ray is split into two rays, namely o-and e-rays. The o-ray travels through the crystal without deviation while the e-ray is refracted at some angle. As the opposite faces of the crystal are parallel, the rays emerge out parallel to the incident ray. Within the crystal the o-ray *always* lies in the plane of incidence whereas e-ray does not lie in the plane of incidence, e-ray lies in the plane of incidence only when the plane of incidence is a principal section.

If a mark (dot or cross) is made on a paper and then the calcite crystal (AB face) is placed on it, two images are seen through the crystal, as illustrated in Fig. 1.25 (b). The images are produced by the o-ray and e-ray. The intensities of the images are lesser than that of the original mark. The line joining them lies in the principal section. If now the crystal is rotated slowly about an axis passing through the o-image, the e-image moves round in a circle while the o-image remains stationary. It shows that the velocity of propagation of o-ray is the same in all directions, while that of e-ray changes with direction.

The e-ray and o-ray are linearly polarised. The e-ray has its vibrations (*i.e.*, the optical vector) *parallel* to the principal section whereas the vibrations (optical vector) in o-ray are *perpendicular* to the principal section, as indicated in Fig. 1.17. The vibration directions can be established by examining the rays through a polarizer. As the polarizer is held in the path of the rays and rotated slowly, the intensity of one of the images, say the o-image, increase while that of the e-image decreases. In one position, the intensity of the o-image will be a maximum while the e-image is extinguished. Further rotation through 90° from this particular position, causes the o-image to disappear and e-image intensity to become a maximum. It proves that the e- and o-rays are linearly polarised in mutually perpendicular directions.

HUYGENS' EXPLANATION OF DOUBLE REFRACTION

The propagation of light may be treated in terms of wave surfaces. If a point source of natural light is imagined to be embedded within an isotropic substance such as glass, it gives rise to one wave surface, which is spherical in shape. The wavefront stimulates atoms, which then act as sources of spherical wavelets, all of which are in phase. They expand in all directions with the same velocity.

In case of a double refracting crystal, two wave surfaces will be formed simultaneously, with the result that the beam will split into two rays. The wave surface corresponding to o-ray propagates with the same velocity in all directions and is therefore spherical. The wave surface corresponding to e-ray is an ellipsoid of revolution about the optic axis, since e-ray travels with different velocities in different directions within the crystal. *The two wave surfaces touch each other at two points and the line joining these points defines the direction of the optic axis.* As light propagates through the crystal, the two wave surfaces travel in different directions in the crystal. Ultimately, two refracted rays emerge from the crystal.

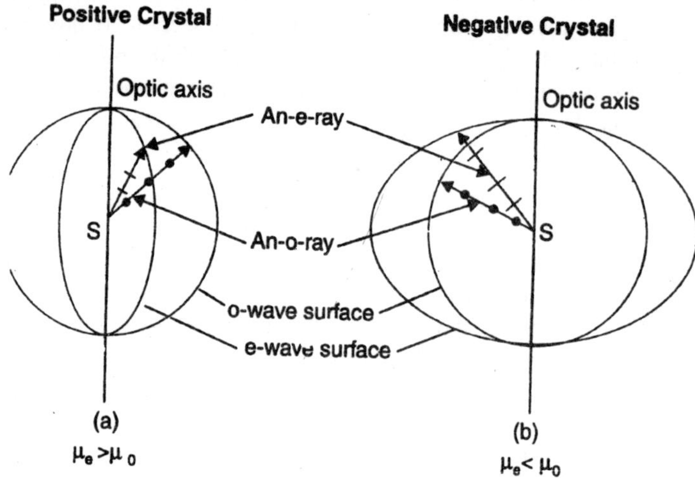

Fig. 1.26 : Huygens wave surfaces produced by a point source S-embedded in the birefringent crystal; (a) a positive crystal (b) a negative crystal.

Because of two different wave fronts, two different types of situations arise. The spherical wave front of o-ray may be enclosed by the ellipsoidal wave front of e-ray in one type of crystals. Such crystals are known as *negative crystals*. They are called negative crystals because the refractive index corresponding to the e-ray is less than that corresponding to o-ray. Calcite crystal is an example of negative type crystals. In the other case, the extraordinary wave front lies within the ordinary wave front and such crystals are called positive crystals. They are positive because the refractive index for the extraordinary ray is greater than that of o-ray.

o-Ray and e-Ray

We now compare the properties of o- and e-rays.

(i) Ordinary ray obeys the conventional laws of refraction, whereas the e-ray does not conform to them.

(ii) Both o-ray and e-ray are plane polarized. They are polarized in mutually perpendicular planes. The electric vector of o-ray vibrates perpendicular to the principal section of o-ray while the vibrations of e-ray take place parallel to the principal section of e-ray.

(iii) O-ray travels with the same speed in all directions within the crystal. The e-ray travels with different speeds along different directions in the crystal. However, the speed of e-ray will be equal to that of o-ray along the optic axis direction.

(iv) Because o-ray travels with the same velocity in all directions, the refractive index corresponding to it has a constant value. On the other hand, the refractive index for e-ray varies from direction to direction. The principal refractive index for o-ray is defined as follows:

$$\mu_o = \frac{c}{\upsilon_o} = \frac{\text{velocity of light in a vacumn}}{\text{Velocity of r - ray in the crystal}} \quad ...(8)$$

The principal refractive index for e-ray in positive crystals is defined as follows:

$$\mu_e = \frac{c}{(\upsilon_e)_{min}} = \frac{\text{velocity of light in a vacumn}}{\text{Minimum velocity of e - ray in the crystal}} \quad ...(9)$$

The principal refractive index for e-ray in negative crystals is defined as follows:

$$\mu_e = \frac{c}{(\upsilon_e)_{max}} = \frac{\text{velocity of light in a vacumn}}{\text{Maximum velocity of e - ray in the crystal}} \quad ...(10)$$

(v) When natural light is incident on an anisotropic crystal at an angle to the optic axis, it splits into o-and e-rays, which travel in different directions with different velocities (Fig. 1.27).

When natural light is incident in a direction normal to the optic axis, o-ray and e-ray propagate in the *same direction* in the crystal but with *different velocities*, as shown in Fig. 1.27 (a). In a negative crystal e-ray leads o-ray and in case of a positive crystal o-ray leads e-ray.

When natural light is incident on the crystal in a direction parallel to the optic axis, it does not split into two rays, as shown in Fig. 1.27(b). We may say that the o-and e-rays travel in the same direction with the same velocity.

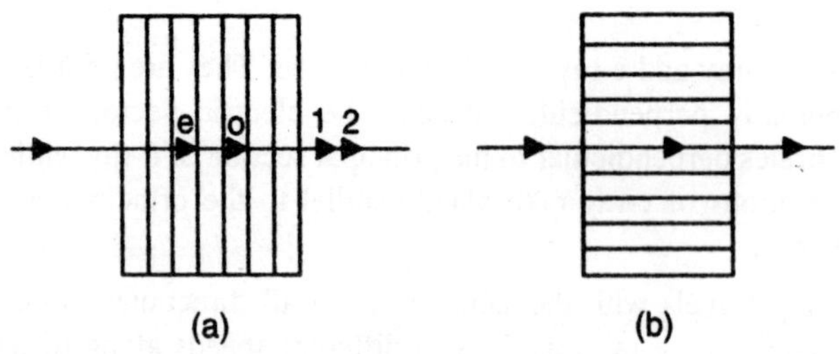

Fig. 1.27

(a) Double refraction, the o-ray and e-ray travel with different velocities in the same direction, when light is incident normal to the optic axis direction of a birefringent crystal.

(b) No double refraction and e-ray and o-ray travel with same velocity along the optic axis direction. The parallel lines on the crystal face indicate the optic axis direction.

(vi) The distinction of o-ray and e-ray exists only within the crystal. Once they emerge from the crystal, they travel with the same velocity. The rays outside the crystal differ only in their direction of travel and plane of polarization. The designation of o-ray and e-ray has no meaning outside the crystal.

Positive Crystals and Negative Crystals

We compare here the characteristics of the positive and negative crystals.

(i) In positive uniaxial crystals, the ellipsoid of revolution corn-spending to the e-ray is totally contained within the sphere corresponding to the o-ray. In negative uniaxial crystals, the ellipsoid of revolution for e-ray lies completely outside the sphere corresponding to o-ray. The two cases are depicted in Fig. 1.26.

(ii) In positive crystals the e-ray velocity has a maximum value along the optic axis and a minimum value in a direction perpendicular to the optic axis. On the other hand, in negative crystals the velocity of e-ray has a minimum value parallel to the optic axis and a maximum value in a direction perpendicular to the optic axis.

(iii) In positive crystals, e-ray travels slower than o-ray in all directions except along the optic axis.

$\upsilon_e = \upsilon_o$ – parallel to optic axis

$\upsilon_e < \upsilon_o$ – other directions

In negative crystals, o-ray travels slower than e-ray in all directions except along the optic axis.

$\upsilon_e = \upsilon_o$ – parallel to optic axis

$\upsilon_e > \upsilon_o$ – other directions

(v) In positive crystals the principal refractive index for e-ray is larger than the principal refractive index for o-ray.

$$\mu_e > \mu_o$$

In negative crystals the principal refractive index for o-ray is larger than the principal refractive index for e-ray.

$$\mu_e < \mu_o$$

(v) *Birefringence* or amount of double refraction of a crystal is defined as

$$\Delta\mu = \mu_e - \mu_o \quad \text{...(11)}$$

$\Delta\mu$, is a positive quantity for positive crystals as $\mu_e > \mu_o$ in these crystals.

$\Delta\mu$ is a negative quantity for negative crystals as $\mu_e < \mu_o$ in these crystals.

HUYGENS' CONSTRUCTION OF WAVEFRONTS

A thinner rectangular cross-section of a crystal can be cut from a bigger crystal in three different ways. It can be cut in such a way that the optic axis lies :

(i) inclined to the refracting face

(ii) parallel to the refracting face

(iii) perpendicular to the refracting face. The path of o-ray and e-ray within a uniaxial crystal can be determined using Huygens' principle of secondary wavelets. We take the example of a negative crystal for the purpose of tracing the paths of o-ray and e-ray.

Case. 1. Optic axis inclined to the refracting edge

Let MNN' M' represent the calcite (negative) crystal. Let CD represent the monochromatic plane wave front incident normal to the crystal surface. The optic axis is *inclined to the refracting edge*. According to Huygens the phenomenon of double refraction involves two types of propagation of light waves. The wave known as o-wave has a spherical wave front whereas the e-wave has an ellipsoidal wave front. The two wave fronts touch each other along the optic axis of the crystal. In a negative crystal the e-ray velocity is greater than the o-ray velocity and hence the ellipsoidal surface lies outside the spherical surface.

Let us consider that parallel beam of light CD falls *normally* on the surface of the negative crystal. As soon as the parallel beam strikes the crystal boundary, each point on the wave front becomes a source of secondary disturbance. The points A and B are chosen for the purpose of illustration. According to Huygens principle the points A and B produce elliptical and spherical wavelets. The position of the two wave fronts after a lapse of 't' seconds can be determined as follows. A circle of radius '$\upsilon_0 t$' is drawn taking A as centre. Similarly, a circle of the same radius is drawn with B as centre. The ellipsoidal wave front can be drawn if the major and minor axes are known. In a negative crystal the major axis is equal to '$2\upsilon_e t$' and minor axis is '$2\upsilon_e t$'. The circle and the ellipse touch each other, as required, along the optic axis. If we now draw the common tangents to the secondary wavelets, they represent the plane wave fronts corresponding to the two rays. Thus KL is the tangent to

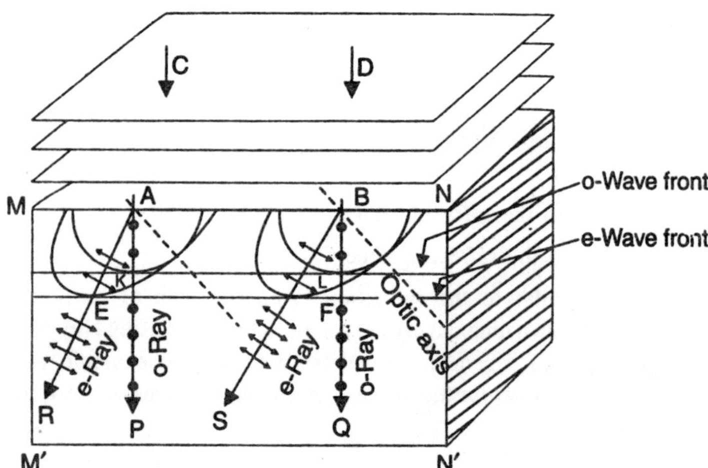

Fig. 1.28 : Unpolarized light beam is incident at normal on a calcite crystal slice. Optic axis is inclined to the incident ray. Huygens wavelets showing double refraction, and difference in velocities of o-ray and e-ray.

the spherical wavelets and EF is the tangent to the ellipsoidal wave fronts. If now we join the point of origin of wavelets to the points of tangency, the direction of propagation of the rays will be known. Thus, if we join A to K and B to L, we obtain the direction of propagation of o-ray. Similarly, the lines AE and BF show the direction of propagation of e-ray. We find that in this case the o-ray and e-ray travel along different directions with different velocities.

Case 2(a) : Optic axis in the plane of incidence and parallel to the refracting edge

Let MNN' M' represent the refracting face of the calcite (negative) crystal. Let CD represent the monochromatic plane wavefront incident normal to the crystal surface. Let the optic axis be *parallel to the refracting edge MN and lie in the plane of incidence.* Fig. 1.29 (a). According to Huygens the phenomenon of double refraction involves propagation of light waves of two types. The wave known as o-wave has a spherical wavefront whereas the e-wave has an ellipsoidal wavefront. The two wavefronts touch each other along the optic axis of the crystal. In a negative crystal the e-ray velocity is greater than the o-ray velocity and hence the ellipsoidal surface lies outside the spherical surface.

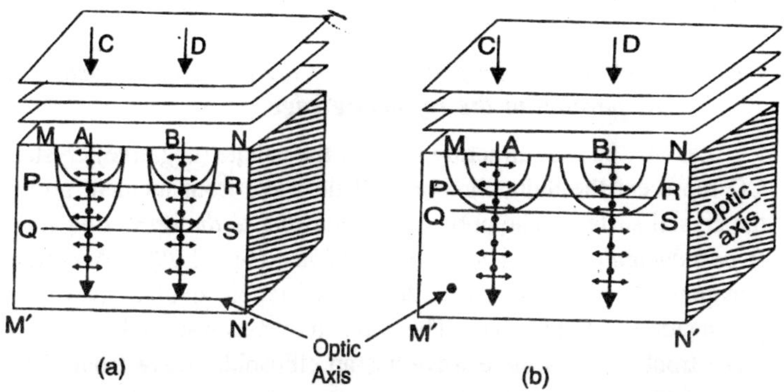

Fig. 1.29 : : Unpolarized light beam incident normally on a calcite slab at right angles to the optic axis. Double refraction does not occur but o-ray and e-ray travel with different velocities in the crystal, (a) Optic axis in the plane of incidence and parallel to the crystal surface; (b) Optic axis perpendicular to the plane of incidence and parallel to the crystal surface.

Let us consider that parallel beam of light CD falls *normally* on the surface of the negative crystal. As soon as the parallel beam strikes the crystal boundary, each point on the wavefront becomes a source of secondary disturbance. The points A and B are chosen for the purpose of illustration. According to Huygens' principle the points A and B produce elliptical and spherical wavelets. The two wavelets touch each other along AM and BN, which is the direction of optic axis, as required.

The wave fronts at any instant can be drawn as tangential surfaces to the two wavelets. Note that the two wave fronts here are parallel to each other. The position of the two wavefronts after a lapse of 't' seconds can be determined as follows. A circle of radius '$\upsilon_0 t$' is drawn taking A as centre. Similarly, a circle of the same radius is drawn with B as centre. The ellipsoidal wavefront can be drawn if the major and minor axes are known. In a negative crystal the major axis is equal to '$2\upsilon_e t$' and minor axis is '$2\upsilon_0$'. PR is the tangent to the spherical wavelet and therefore represents the refracted wavefront for the o-ray. Similarly, QS represents the refracted wavefront for e-ray. If now we join the point' of origin of wavelets to the points of tangency, the direction of propagation of the rays will be known. Thus, if we join A to P and B to R, we obtain the direction of propagation of o-ray. By joining A to Q and B to S we get the direction of propagation of e-ray. We find that in this case the o- ray and e-ray travel along the same direction but have different velocities.

Case 2(b): Optic axis perpendicular to the plane of incidence and parallel to the refracting edge

Let MN be the refracting edge of the calcite (negative) crystal. Let CD represent the monochromatic plane wave front incident normal to the crystal surface. The optic axis is *parallel to the refracting edge but is perpendicular to the plane of incidence* (Fig. 1.29b). According to Huygens the phenomenon of double refraction involves two types of propagation of light waves. The wave known as o-wave has a spherical wave front whereas the e-wave has an ellipsoidal wave front. The two wave fronts touch each other along the optic axis of the crystal. In a negative crystal the e-ray velocity is greater than the o-ray velocity and hence the ellipsoidal surface lies outside the spherical surface.

Let us consider that parallel beam of light CD falls *normally* on the surface of the negative crystal. As soon as the parallel beam strikes the crystal boundary, each point on the wave front becomes a source of secondary disturbance. The points A and B are chosen for the purpose of illustration. According to Huygens principle the points A and B produce elliptical and spherical wavelets. Since the spherical and ellipsoidal wavelets touch each other along a line perpendicular to the plane of incidence, the wavelets appear spherical in the plane of incidence. The position of the two wave fronts after a lapse of 't' seconds can be determined as follows. A circle of radius '$\upsilon_0 t$' is drawn taking A as centre. Similarly, a circle of the same radius is drawn with B as centre. The ellipsoidal wave front can be drawn if the major and minor axes are known. In a negative crystal the major axis is equal to '$2\upsilon_e t$' and minor axis is '$2'\upsilon_0 t$'. The circle and the-ellipse touch each other, as required, along the optic axis, which is perpendicular to the surface MN. If we now draw the common tangents to the secondary wavelets, they represent the plane wavefronts corresponding to the two rays. Thus PR is the common tangent to the spherical wavelets and therefore represents the refracted wave front for the o-ray. Similarly, QS represents the refracted wave front for e-ray. If now we join the point of origin of wavelets to the points of tangency, the direction of propagation of the rays will be known. Thus, if we join A to P and B to R, we obtain the direction of propagation of o-ray. By joining A to Q and B to S we get the direction of propagation of e-ray. We find that in this case the o-ray and e-ray travel along the same direction but have different velocities.

Case 3 : Optic axis perpendicular to the refracting edge and lying in the plane of incidence

Let MNN' M' represent the calcite (negative) crystal. Let CD represent the monochromatic plane wave front incident normal to the crystal surface. The optic axis is *perpendicular to the refracting edge and lies in the plane of incidence*. According to Huygens the phenomenon of double refraction involves two types of propagation of light waves. The wave known as o-wave has a spherical wave front whereas the e-wave has an ellipsoidal wave front. The two wave fronts touch each other along the optic axis of the crystal. In a negative crystal the e-ray velocity is greater than the o-ray velocity and hence the ellipsoidal surface lies outside the spherical surface.

Let us consider that parallel beam of light CD falls *normally* on the surface of the negative crystal. As soon as the parallel beam strikes the crystal boundary, each point on the wavefront becomes a new point source of light. The points A and B are chosen for the purpose of illustration. According to Huygens' principle the points A and B produce elliptical and spherical wavelets. Since the two wavelets must touch each other along the optic axis, they touch at points P and Q (Fig. 1.30). Note that the section of the ellipsoid lies outside the section of the circle. The minor axis of the ellipse and the radius of the circle are equal. The position of the two wave fronts after a lapse of 't' seconds can be determined as follows. A circle of radius '$\upsilon_0 t$' is drawn taking A as centre. Similarly, a circle of the same radius is drawn with B as centre. The ellipsoidal wave front can be drawn if the major and minor axes are known. In a negative crystal the major axis is equal to '$2\upsilon_0 t$' and minor axis is '$2\upsilon_0 t$'. The circle and the ellipse touch each other, as required, along the optic axis, which is perpendicular to the refracting edge MN. If we now draw the common tangents to the secondary wavelets, they represent the plane wavefronts corresponding to the two rays.

Thus PQ is the common tangent to the spherical wavelets as well as the ellipsoidal wave fronts. If now we join the point of origin of wavelets to the points of tangency, the direction of propagation of the rays will be known. Thus, if we join A to P and B to Q, we obtain the direction of propagation of o-ray as well as e-ray. We find that in this case the o- ray and e-ray travel along the same direction with the same velocity.

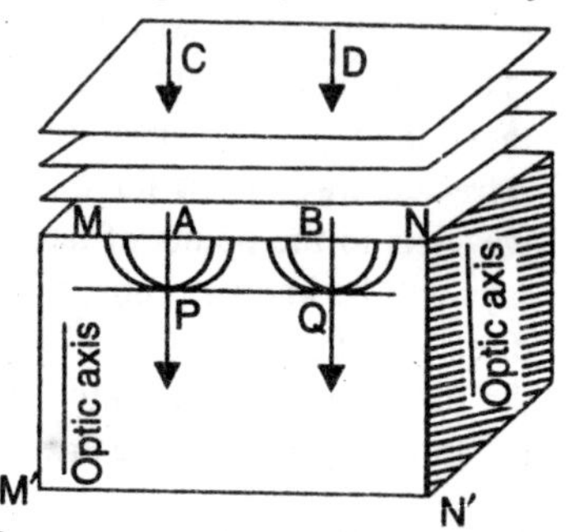

Fig. 1.30 : Unpolarized light beam incident normally on a calcite slab parallel to the optic axis. The o-ray and e-ray travel along the same velocity.

EXPERIMENTAL DETERMINATION OF PRINCIPAL REFRACTIVE INDICES

For determining the refractive index for the extraordinary ray a calcite crystal is cut in the form of a prism with the optic axis perpendicular to the refracting edge of the prism and perpendicular to the base BC. It can also be cut with the optic axis parallel to the refracting edge of the prism. The prism is placed on the spectrometer table and is adjusted for the minimum deviation position for the extraordinary rays. The angle of minimum deviation δ_m is determined and the principal refractive index for the extraordinary ray is calculated from the relation (12a).

$$\mu_e = \frac{\left(\frac{A+\delta_m}{2}\right)}{\sin\frac{A}{2}} \qquad \text{...(12a)}$$

For a given wavelength, the ordinary and the extraordinary rays are separated while passing through the prism. Therefore, the angle of minimum deviation δ_m' for the ordinary ray can be measured and thus its refractive index can be calculated from the relation (12b).

$$\mu_0 = \frac{\sin\left(\frac{A + \delta'_m}{2}\right)}{\sin\frac{A}{2}} \qquad ...(12b)$$

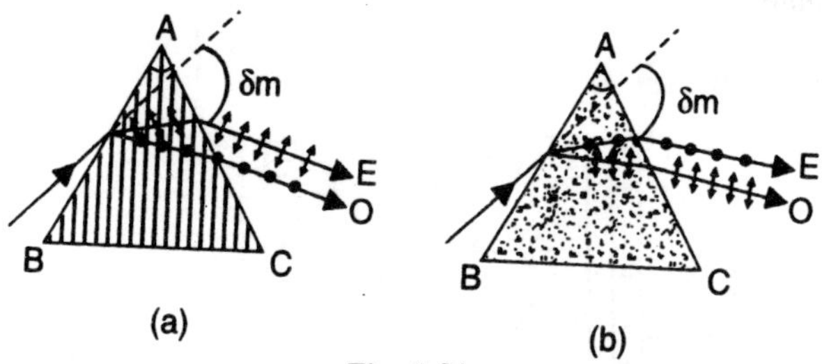

(a)

(b)

Fig. 1.31

ELECTROMAGNETIC THEORY OF DOUBLE REFRACTION

The birefringent crystals belong to the group of non-conducting materials. They are known, as dielectric materials. If a dielectric crystal is placed between the plates of a parallel plate capacitor and a voltage is applied, the crystal becomes polarized. This polarization is dielectric polarization and is different from the optical polarization. The voltage produces an electric field E and under the action of the electric field, electron clouds in the atoms of the crystal distort with the result that the centres of action of negative charge and positive charge get separated by a small distance. It means that electrical dipoles are produced throughout the crystal. Negative charges of the dipoles lie on the side of the positively charged plate of the capacitor and positive charges lie on the side of the negatively charged plate. Thus electric charges of opposite nature are induced on the opposite faces of the crystal in response to the applied electric field (Fig. 1.32). This is known as *dielectric polarization.*

The electric field, polarization and displacement in the crystal are related by

$$D = \varepsilon_0 + P \qquad ...(13)$$

where $$D = \varepsilon_0\varepsilon_r E \qquad ...(14)$$

and $$P = \chi \varepsilon_0 E \qquad \text{...(15)}$$

ε_r and χ are known as *relative permittivity* and *electric susceptibility* of the crystal. The relative permittivity is related to the refractive index of the material through the relation

$$\mu = \sqrt{\varepsilon_r} = \sqrt{\frac{\varepsilon}{\varepsilon_0}} \qquad \text{...(16)}$$

We know that refractive index is related to the velocity of light in the medium.

$$\mu = \frac{c}{\upsilon} \qquad \text{...(17)}$$

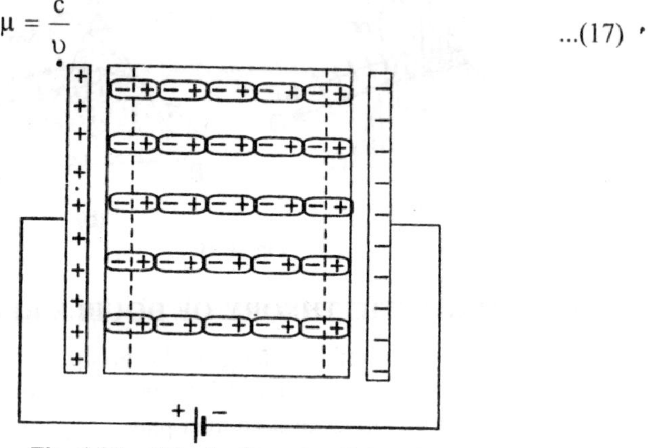

Fig. 1.32 : Polarization of a dielectric

It is not necessary for dielectric polarization that the crystal be placed in between the plates of a capacitor. The electric field of a light wave propagating in the crystal also can cause dielectric polarization. According to the above relations the dielectric polarization caused by the light wave is related to the velocity (direction of propagation) of light wave with in the crystal.

In case of isotropic crystals, the vectors D, E and P are parallel and ε_r, μ and χ are scalars. In anisotropic crystals the induced polarization is in a direction different from that of the field. As a result, ε_r, μ and χ are tensor quantities. The components of polarization are given by

$$\begin{aligned} &P_x \ \varepsilon_0 (\chi_{11} E_x + \chi_{12} E_y + \chi_{13} E_z) \\ &P_y \ \varepsilon_0 (\chi_{21} E_x + \chi_{22} E_y + \chi_{23} E_z) \\ &P_z \ \varepsilon_0 (\chi_{31} E_x + \chi_{32} E_y + \chi_{33} E_z) \end{aligned} \qquad \text{...(18)}$$

The above equations can be simplified by choosing the co-ordinate axes in such a way that the off-diagonal elements vanish. Then

$$P_x \ \varepsilon_0 \ \chi_{11} \ E_x$$
$$P_y \ \varepsilon_0 \ \chi_{22} \ E_y \qquad ...(19)$$
$$P_z \ \varepsilon_0 \ \chi_{33} \ E_z$$

Similarly, we can express $D_x = \varepsilon_{11} \ E_x$,

$$D_y = \varepsilon_{22} \ E_y \qquad ...(20)$$
$$D_z = \varepsilon_{33} \ E_z$$

These directions are called the *principal axes* of the crystal and the corresponding diagonal terms ε_{11}, ε_{22} and ε_{33} the *principal permittivities.* The variation in the permittivity and the corresponding variation in refractive index and hence in wave velocity is the origin of double refraction. μ_x, μ_y and μ_z are the three principal *refractive indices* corresponding to the three principal permittivities. *The subscripts x, y and z relate to the direction of polarization of the light waves and not to their direction of propagation.* The principal permittivities and refractive indices are related as

$$\mu_x = \sqrt{\frac{\varepsilon_{11}}{\varepsilon_0}}, \ \mu_y = \sqrt{\frac{\varepsilon_{22}}{\varepsilon_0}} \ \text{ and } \ \mu_z = \sqrt{\frac{\varepsilon_{33}}{\varepsilon_0}} \qquad ...(21)$$

Index Ellipsoid or Optical Indicatrix

The different speeds of light in different directions in a crystal are conveniently represented by a three-dimensional figure that shows the refractive index for light waves in their direction of polarization. The representation is known as the *indicatrix. The optical indicatrix is spherical or ellipsoidal surface having the three different refractive indices as axes.*

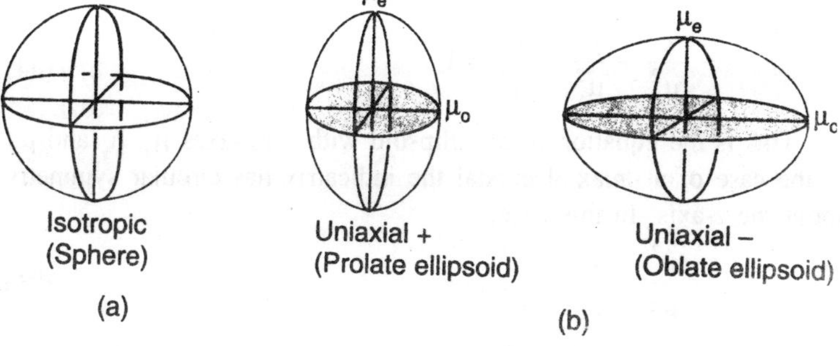

Fig. 1.33

In case of isotropic crystals, the axes are all of the same length and consequently the refractive indices in three directions are the same. Therefore, the indicatrix is a *sphere*, as shown in Fig. 1.33 (a). In case of uniaxial crystals, a unique axis exists normal to the plane of two equal axes. The velocity of propagation of light along this unique axis (z-direction) is independent of the wave polarisation and this direction is the *optic axis* direction. The refractive index along this direction is μ_e. In a plane perpendicular to the optic axis, the refractive index is independent of direction, that is $\mu_x = \mu_y$ denoted by μ_0. Therefore, in uniaxial crystals the indicatrix is an ellipsoid having a circular cross-section in the xy-plane. The ellipsoid is prolate if $\mu_e > \mu_0$ and oblate if $\mu_e < \mu_0$ (Fig. 1.33b).

The form of optical indicatrix can be obtained as follows. The energy density in a dielectric medium is given by

$$W = \frac{1}{2} D.E \qquad ...(22)$$

$$W = \frac{1}{2}\left[\frac{D_x^2}{\varepsilon_{11}} + \frac{D_y^2}{\varepsilon_{22}} + \frac{D_z^2}{\varepsilon_{33}}\right] \qquad ...(23)$$

or

$$\frac{1}{2}\left[\frac{D_x^2}{W\varepsilon_{11}} + \frac{D_y^2}{W\varepsilon_{22}} + \frac{D_z^2}{W\varepsilon_{33}}\right] = 1$$

or

$$\left[\frac{D_x^2}{2W\mu_x^2\varepsilon_0} + \frac{D_y^2}{2W\mu_y^2\varepsilon_0} + \frac{D_z^2}{2W\mu_z^2\varepsilon_0}\right] = 1$$

Writing $x^2 = \frac{D_x^2}{2\varepsilon_0 W}$, $y^2 = \frac{D_y^2}{2\varepsilon_0 W}$ and $z^2 = \frac{D_z^2}{2\varepsilon_0 W}$, we get

$$\frac{x^2}{\mu_x^2} + \frac{y^2}{\mu_y^2} + \frac{z^2}{\mu_z^2} = 1 \qquad ...(24)$$

This is the equation of an ellipsoid with semi-axes μ_x, μ_y and μ_z. In the case of a uniaxial crystal the indicatrix has circular symmetry about the z-axis. In this case,

$$\frac{x^2}{\mu_0^2} + \frac{y^2}{\mu_0^2} + \frac{z^2}{\mu_e^2} = 1 \qquad ...(25)$$

Let us now consider light propagating in a direction r, at an angle θ to the optic axis. Because of the circular symmetry, we can choose that the y-axis should coincide with the projection of r on the xy-plane. The plane normal to r intersects the ellipsoid in the shaded ellipse. The two allowed directions of polarization are parallel to the axes of the ellipse and thus correspond to OP and OQ (Fig. 1.34). They are thus perpendicular to r as well as to each other. The two waves polarized along these directions have refractive indices given by OP = μ_o and OQ = $\mu_e(\theta)$. In case of the e-ray, the plane of polarization varies with θ as does the refractive index.

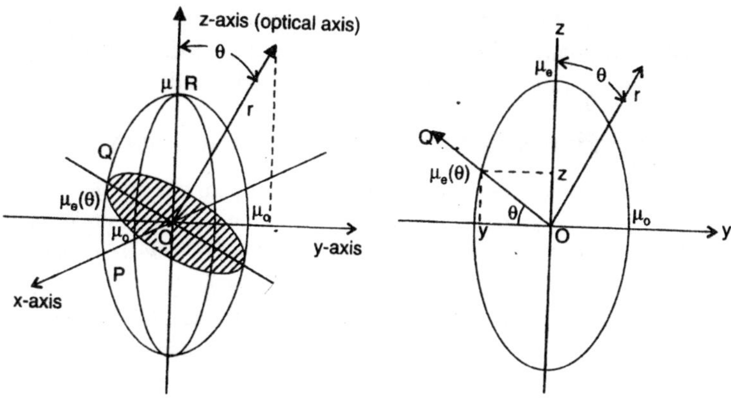

Fig. 1.34 **Fig. 1.35**

We can determine the relationship between μ_e (θ), μ_e and μ_o with the help of the Fig. 1.35. The figure shows the intersection of the indicatrix with the yz-plane. From the diagram, it is seen that

$$\mu_e^2(\theta) = z^2 + y^2 \quad ...(26)$$

and

$$z = \mu_e(\theta)\sin\theta \quad ...(27)$$

$$\therefore \quad \mu_e^2(\theta) = \mu_e^2(\theta)\sin^2\theta + y^2$$

or

$$y^2 \mu_e^2(\theta)\cos^2\theta \quad ...(28)$$

The equation of the ellipse shown in Fig. 1.33 is

$$\frac{y^2}{\mu_o^2} + \frac{z^2}{\mu_e^2} = 1 \quad ...(29)$$

Substituting eqn. (27) and eqn. (28) into eqn. (29), we obtain

$$\frac{1}{\mu_e^2(\theta)} = \frac{\cos^2\theta}{\mu_o^2} + \frac{\sin^2\theta}{\mu_e^2} \qquad ...(30)$$

Thus, for $\theta = 0°$, that is propagation along the optic axis, $\mu_e\,(0°) = \mu_o$ while for $\theta = 90°$, $\mu_e\,(90°) = \mu_e$. The two polarizations which can be propagated correspond to the maximum and minimum refractive indices given by the index ellipsoid. For propagation parallel to the optic axis (that is z-direction), there is no birefringence as the section of the ellipsoid perpendicular to this direction is a circle. For propagation perpendicular to the optic axis, the birefringence will be a maximum, the permitted polarizations will be parallel to the y-axis with refractive index μ_0 and parallel to the z-axis with refractive index μ_e.

PHASE DIFFERENCE BETWEEN e-RAY AND o-RAY

We have seen that natural light incident on the surface of an anisotropic crystal undergoes double refraction and produces two plane polarized waves. *Even when a plane polarized light wave is incident on a birefringent crystal such that its electric vector makes an angle with the optic axis, then the polarized wave splits into two polarized waves namely e-ray and o-ray*. Let us consider the particular case of a slice of a positive crystal where the optic axis is parallel to refracting face of the crystal (Fig. 1.26a). The two waves travel along the same direction in the crystal but with different velocities. As a result, when the waves emerge from the rear face of the crystal, an optical path difference would have developed between them. The optical path difference can be calculated as follows:

Let d be the thickness of the crystal.

The optical path for o - ray within the crystal $= \mu_0 d$

The optical path for e - ray within the crystal $= \mu_e d$

The optical path difference between e - ray and o - ray $\Delta = (\mu_e - \mu_o)d$ ...(31)

Consequently, a phase difference arises between the two waves. It is given by

$$\delta = \frac{2\pi}{\lambda}(\mu_e - \mu_o)d \qquad ...(32)$$

As the two component waves are derived from the same incident wave, the two waves are in phase at the front face and have emerged from the crystal with a constant phase difference (Fig. 1.36) and hence it may be expected that the waves are in a position to interfere with each other. However, as the planes of polarization of o-ray and e-ray are perpendicular to each other, interference cannot take place between e- and o-rays. The waves instead combine with each other to give an elliptically polarized wave.

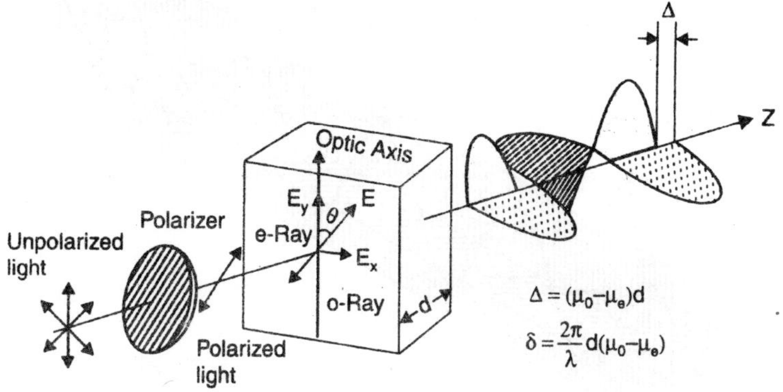

Fig. 1.36

TYPES OF POLARISED LIGHT

We can now sum up the various types of polarised light as follows.

(i) *Unpolarized light*, which consists of sequence of wave trains, all oriented at random. It is considered as the resultant of two optical vector components, which are incoherent.

(ii) *Linearly polarised light*, which can be regarded as a resultant of two coherent linearly polarised waves.

(ii) *Partially polarised light*, which is a mixture of linearly polarised light and unpolarised light. Partially polarized light is represented as shown in Fig. 1.37.

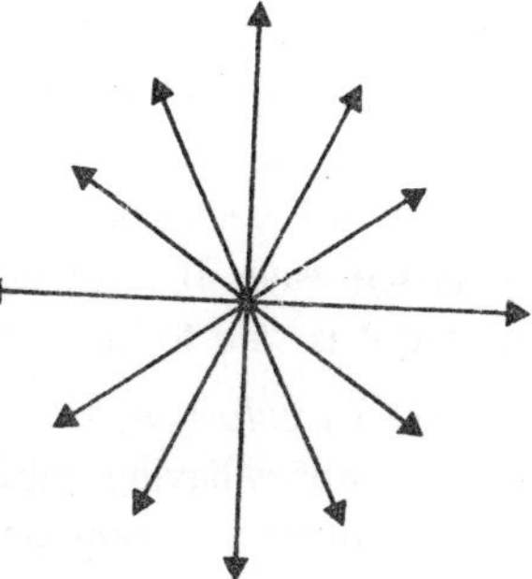

Fig. 1.37 : Schematic representation of partially polarized light.

(iv) *Elliptically polarised light*, which is the resultant of two coherent waves having different ampli-tudes and a constant phase difference of 90° (Fig. 1.38). In elliptically polarized light, the magnitude of electric vector E changes with time and the vector E rotates about the direction of propagation.

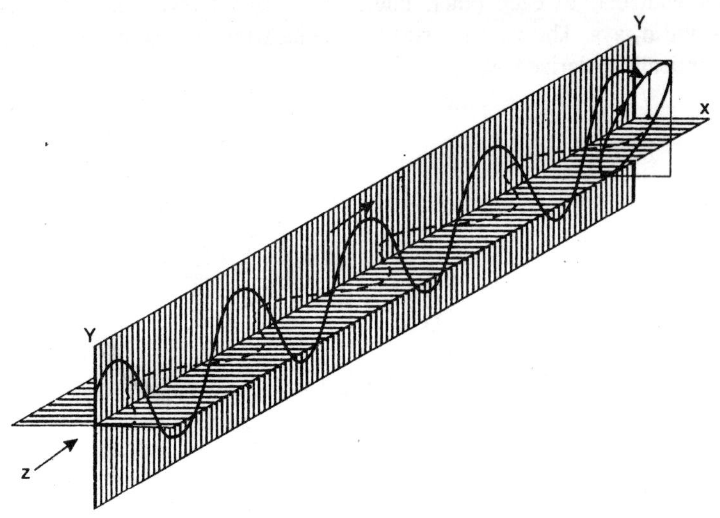

Fig. 1.38 : Elliptically polarised light is produced when two orthogonal coherent waves having different amplitudes and a phase difference of 90°, superpose on each other.

If we imagine that we are looking at the light wave advancing towards us, we would observe that the tip of the E vector traces an ellipse. If we look from sides, we would find that the tip of E sweeps a flattened helix in space. When we are looking back towards the source, if the rotation of E vector occurs clockwise, it is said to be a right-*elliptically*-polarised wave. If it rotates anticlockwise, as we look back toward the source it is said to be a *left-elliptically polarised* wave.

(v) *Circularly polarised light*, which is the resultant of two coherent waves having same amplitudes and a constant phase difference of 90°. A light wave is said to be *circularly polarised*, if the magnitude of the electric vector E stays constant but the vector rotates about the direction of propagation such that it goes on sweeping a circular helix in space (Fig. 1.39). If we imagine that the wave is advancing toward our eyes, we would find that the tip of the E vector of the wave traces a circle. If the rotation of the tip of E is clockwise, as seen by an observer looking back towards the source, then the wave is said to be right-circularly polarised. If the tip of E rotates anticlockwise, as seen by an observer looking back toward the source, the wave is said to be left-circularly polarised.

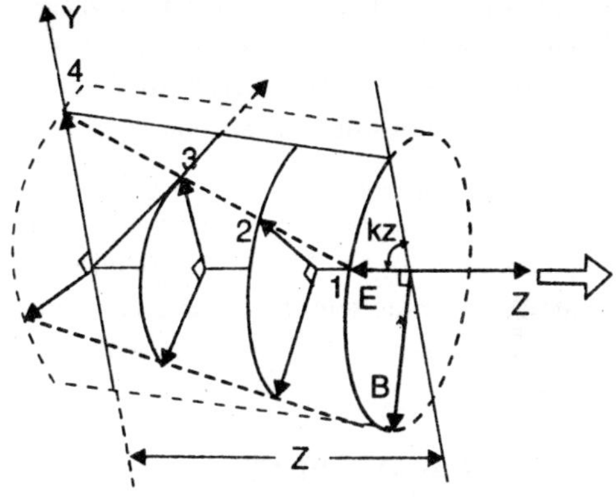

Fig. 1.39

EFFECT OF POLARIZER ON TRANSMISSION OF POLARISED LIGHT

(i) If unpolarized light is incident on a polarizer, the transmitted light will be linearly polarised light. The intensity of the transmitted polarised light will be half the intensity of unpolarized light incident on the polarizer. The intensity of the transmitted light does not change on rotation of the polarizer.

(ii) If partially polarised light is incident on a polarizer, the intensity of the transmitted light will be dependent on the direction of the transmission axis of the polarizer. The intensity of the transmitted light will vary from a maximum value I_{max} to a minimum value I_{min} in one full rotation of the polarizer. Two positions of I_{max} and two positions of I_{min} occur in one complete rotation,

(iii) If plane polarised light is incident on the polarizer, the intensity of transmitted light varies from zero to a maximum value. Two positions of zero intensity and two positions of full intensity I occur in one complete rotation of the polarizer.

(iv) When circularly polarised light is incident on a polarizer, the intensity of the transmitted light stays constant in any position of the polarizer. The circular vibrations may be resolved into two mutually perpendicular linear vibrations of equal amplitude. When the circularly polarised light is incident on the polarizer, the vibrations parallel to its transmission axis pass through the polarizer while the perpendicular component is obstructed. When the polarizer is rotated, there is always a component of constant intensity parallel to the axis of the polarizer, which is freely transmitted. Hence, the intensity of the transmitted light is the same for all positions of the polarizer.

(v) In case of elliptically polarised light, the intensity of the light transmitted through the polarizer varies with the rotation of the polarizer from I_{max} to I_{min}. I_{max} is found when the polarizer axis coincides with the semi-major axis of the ellipse and I_{min} occurs when the polarizer axis coincides with the semi-minor axis of the ellipse.

RETARDERS OR WAVE PLATES

Retarders are a class of optical elements that serve to change the state of polarization of an incident wave. The operation of a retarder is very simple. When plane polarized light is incident on a retarder, it splits the light into two plane polarized light waves and one of the waves lags behind the other by a known amount. Upon emerging from the retarder, the two waves superpose on each other to produce a wave, which is of a different state of polarization. A quarter wave plate and a half wave plate are two important retarders. As calcite is brittle and difficult to handle in the form of thin slices, it is not generally used to make retardation plates. Retarders are frequently made from quartz but more often they are made using the biaxial crystal mica.

Quarter Wave Plate

A *quarter wave* plate is a thin plate of birefringent crystal having the optic axis parallel to its refracting faces and its thickness adjusted such that it introduces a quarter-wave ($\lambda/4$) path difference (or a phase difference of 90°) between the e-ray and o-ray propagating through it.

When a plane polarized light wave is incident on a birefringent crystal having the optic axis parallel to its refracting face, the wave splits into e-wave and o-wave. The two waves travel along the same direction but with different velocities. As a result, when they emerge from the rear face of the crystal, an optical path difference would be developed between them. Thus, for a quartz wave plate,

$$(\mu_e - \mu_o)\, d = \frac{\lambda}{4} \qquad ...(33)$$

$$d = \frac{\lambda}{4[\mu_e - \mu_o]} \qquad ...(34)$$

A quarter wave plate introduces between e-ray and o-ray a phase difference δ given by

$$d = \frac{2\pi}{\lambda}\Delta = \frac{\pi}{2} = 90^\circ$$

A quarter-wave plate is used in producing elliptically or circularly polarised light. It converts plane polarised light into elliptically or circularly polarised light depending upon the angle that the incident light vector makes with the optic axis of the quarter wave plate.

Action of quarter wave plate on elliptically and circularly polarised light:

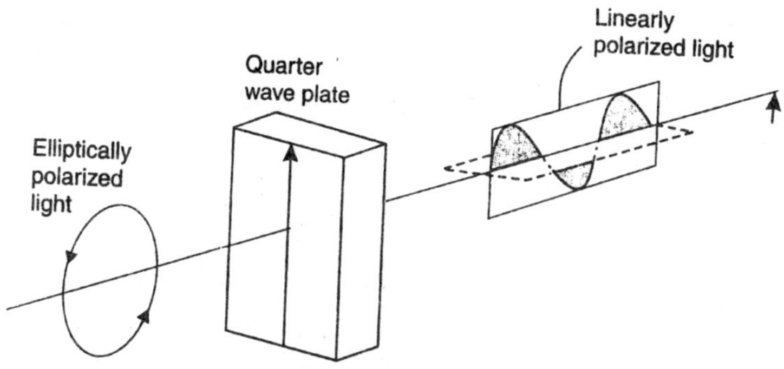

Fig. 1.40 : Action of a quarter wave plate. Linearly polarized light is produced in either case of elliptically or circularly polarized light incident on the plate.

Let us consider elliptically polarised light being incident on a quarter wave plate. Elliptically polarised light may be viewed as made up of two coherent plane polarised waves of different amplitudes, and differing in phase by 90°. the quarter wave plate introduces an additional phase difference of 90° leading to a total phase difference of 180° between the two component waves. When they emerge out of the plate, they combine to form linearly polarised wave, as shown in Fig. 1.40. The action of the quarter wave plate on circularly polarised light wave is similar. Circularly polarised light incident on a quarter wave plate is converted into linearly polarised light.

Half Wave Plate

A half wave plate is a thin plate of birefringent crystal having the optic axis parallel to its refracting faces and its thickness chosen such that it introduces a half-wave ($\lambda/2$) path difference (or a phase difference of 180°) between e-ray and o-ray.

When a plane polarized light wave is incident on a quartz crystal having the optic axis parallel to its refracting faces, it splits into two waves: o-and e-waves. The two waves travel along the same direction inside the crystal but with different velocities. As a result, when they emerge from the rear face of the crystal, an optical path difference would be developed between them.

$$(\mu_e - \mu_o)\, d = \frac{\lambda}{2} \qquad \text{...(35)}$$

$$d = \frac{\lambda}{2\,(\mu_e - \mu_o)} \qquad \text{...(36)}$$

A half wave plate introduces between e-ray and o-ray a phase difference δ given by

$$\delta = \left(\frac{2\pi}{\lambda}\right) \Delta = \pi = 180°$$

Rotation of the plane of polarisation of linearly polarised light by a half wave plate

Now let a plane polarized light be incident normally on the half-wave plate. Let the electric vector E make an angle θ with the optic axis of the half wave plate (Fig. 1.41). The incident wave splits into two waves, e-and o-waves. The waves progressively develop path difference as they travel through the crystal and they emerge with a phase difference of 180°. When the two waves combine, they yield a plane-polarised wave, which has its plane of polarization rotated through an angle of 2θ. Therefore, *a half-wave plate rotates the plane of polarization of the incident plane polarised light through an angle* 2θ. The half wave plate will invert the handedness of elliptical or circular polarised light, changing right to left and *vice versa*.

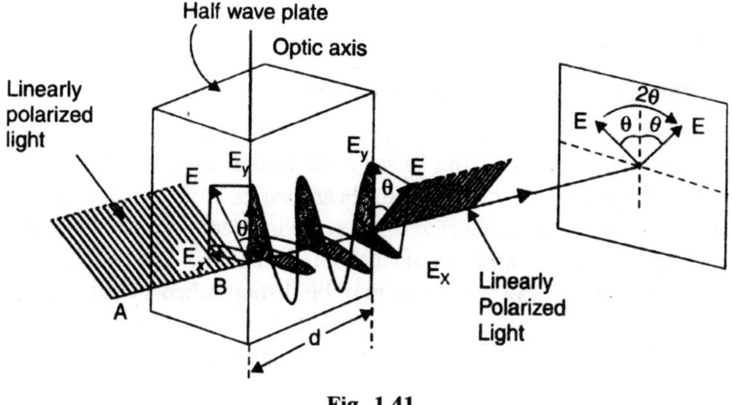

Fig. 1.41

PRODUCTION OF ELLIPTICALLY POLARIZED LIGHT

A quarter wave plate and a polarizer are the optical devices necessary to produce elliptically polarized light from unpolarized light.

Unpolarized light is first converted to plane polarized light by allowing it to pass through a polarizer (a polaroid sheet or a Nicol prism). The plane polarized light is then made incident on a quarter wave plate (Fig. 1.42). The quarter wave plate or the polarizer is rotated such that the electric vector E of plane polarized light wave makes an angle θ (≠ 45°) with the optic axis of the quarter wave plate. The incident ray divides into o-ray and e-ray of amplitudes E sin θ and E cos θ. The rays travel along the same direction in the crystal with different velocities. The two rays are polarized in orthogonal planes. They are in phase at the front face but progressively get out of phase as they travel through the crystal. When they emerge out of the crystal they will have a path

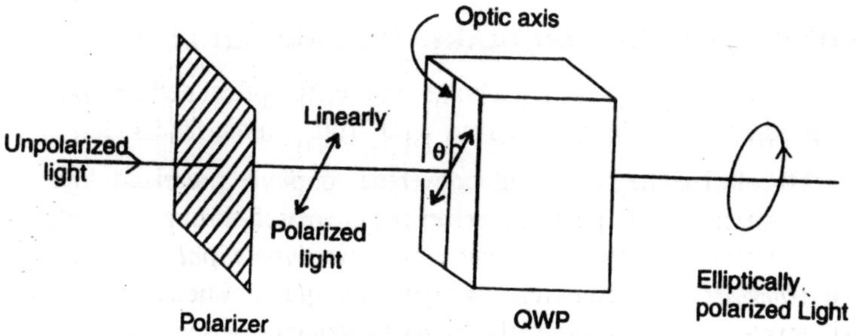

Fig. 1.42

difference of $\lambda/4$ or a phase difference of 90°. When they combine, they produce elliptically polarized light.

Detection of Elliptically Polarized Light

The light beam is allowed to pass through an analyser (a polaroid sheet or a Nicol prism). If on rotating the analysing polaroid sheet or Nicol, the intensity of the emerging beam varies from a maximum to a minimum value, but is never zero, then the incident light is elliptically polarized. A similar result would be obtained if the incident light is partially polarized. The two cases may be distinguished by inserting a quarter wave plate in the path of light before it falls on the analyser. If the original light is elliptically polarized, it may be considered as resultant of two coherent plane

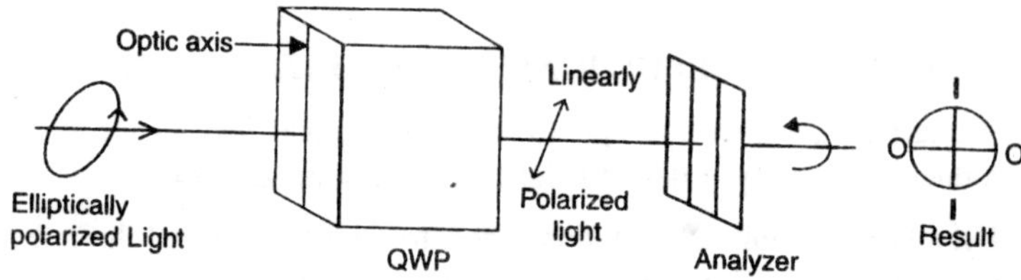

Fig. 1.43

polarized waves, that is e-ray and o-ray, which are out of phase by 90°. If the light passes through the quarter wave plate, an additional phase difference of 90° is introduced between the e-ray and o-ray. Therefore, the total phase difference becomes 180° between the e-ray and o-ray. On emerging from the quarter plate, the e-and o-rays combine to produce plane polarized light. If the light coming out of quarter wave plate is examined with an analyzer, light will be extinguished twice in one full rotation of the polarizer as shown in Fig. 1.43.

PRODUCTION OF CIRCUIARLY POLARIZED LIGHT

A quarter wave plate and a polarizer are the optical devices required for producing circularly polarized light from unpolarized light.

Unpolarized light is first converted to plane polarized light by allowing it to pass through a polarizer (a polaroid sheet or a Nicol prism). Plane polarized light is then made to be incident on a quarter wave plate. The polarizer and the quarter wave plate are rotated such that the electric vector E of the plane-polarized wave makes an angle of 45° with the optic axis of the quarter wave plate. The plane polarized wave incident on the quarter wave plate splits into two rays, o-ray and e-ray of equal amplitude ($E_1 \cos 45° = E_2 \sin 45°$). The two rays travel in the same direction inside the crystal but with different velocities (Fig. 1.44). The two rays are in phase at the front face of the crystal but progressively get out of phase as they travel through the crystal. As they emerge from the rear face of the crystal, they will have a path difference of $\lambda/4$ or phase difference of 90°. The two rays are linearly polarized in mutually perpendicular directions. When they combine, they produce circularly polarized light.

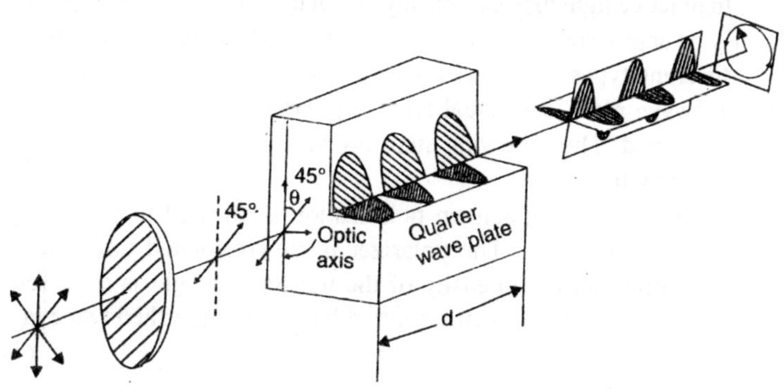

Fig. 1.44

Detraction of Circularly Polarized Light

The light beam is allowed to pass through an analyser (a polaroid sheet or a Nicol prism). If on rotating the analysing polaroid sheet or Nicol, the intensity of the emerging beam remains uniform, then the incident light is circularly polarized. A similar result would be obtained if the incident light is ordinary unpolarized light. The two cases may be distinguished by inserting a quarter wave plate in the path of light before it falls on the analyser. If the given light is circularly polarized, it may be considered as resultant of two coherent plane polarized waves, that is e-ray and o-ray, which are out of phase by 90°. If the light passes through the quarter wave plate, an additional phase difference of 90° is introduced between the e-ray and o-ray. Therefore, the total phase difference becomes 180° between the e-ray and o-ray. On emerging from the quarter plate, the e-and o-rays combine to produce plane polarized light. Therefore, if the light coming out of quarter wave plate is examined with an analyser, light will be extinguished twice in one full rotation of the polarizer as shown in Fig. 1.45.

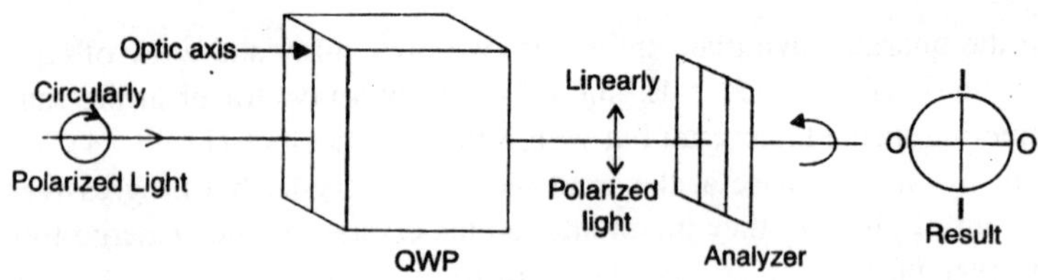

Fig. 1.45

ANALYSIS OF POLARIZED LIGHT

In practice light may exhibit any one of the three types of polarization, or may be unpolarized or a mixed type. The unaided eye cannot distinguish the different types of polarization. However, using a polarizer and a quarter wave plate, the actual type of polarization of a light beam can be ascertained. The following steps are used in the analysis of the type of polarization.

(i) The light of unknown polarization is allowed to fall normally on a polarizer. The polarizer is slowly rotated through a full circle and the intensity of the transmitted light is observed. If the intensity of the transmitted light is extinguished twice in one full rotation of the polarizer, then the incident light is plane *polarized.*

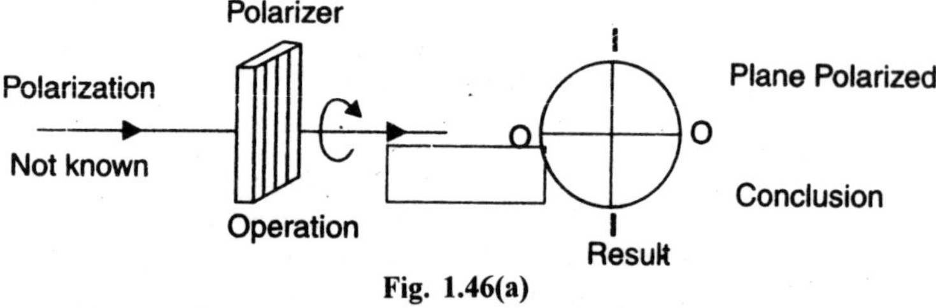

Fig. 1.46(a)

(ii) If the intensity of the transmitted light varies between a maximum and a minimum value but does not become extinguished in any position of the polarizer, then the incident light is either elliptically polarized or partially polarized.

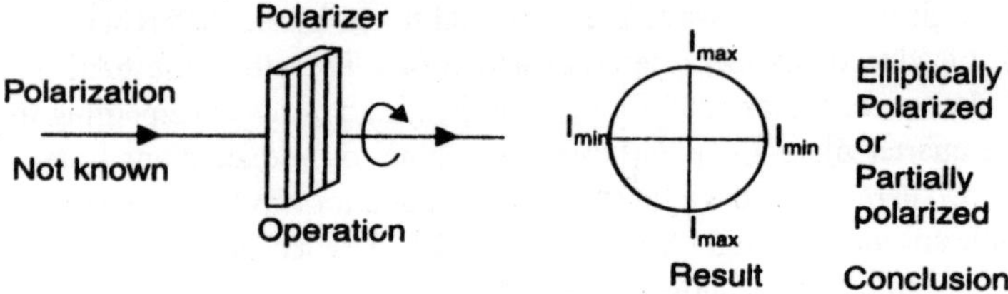

Fig. 1.46(b)

(iii) If the intensity of the transmitted light remains constant on rotation of the polarizer, then the incident light is either circularly polarized or unpolarized.

To distinguish between elliptically polarized and partially polarized or between the circularly polarized and unpolarized light, we take the help of a quarter wave plate. The light is first made to be incident on the quarter wave plate and then it passes through the polarizer.

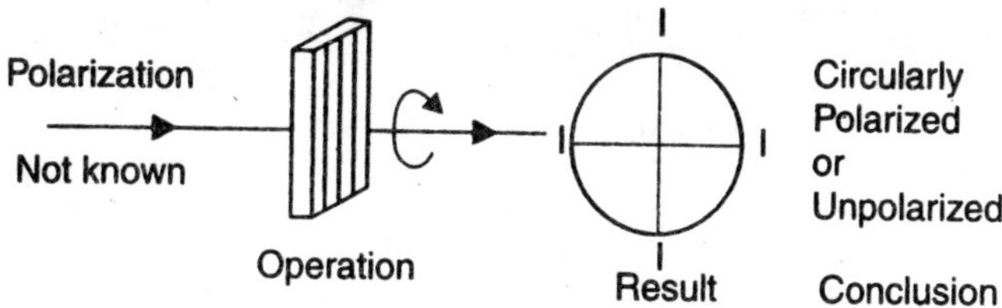

Fig. 1.46(c)

(iv) If the incident light is elliptically polarized, the quarter wave plate converts it into a plane polarized beam. When this linearly polarized light passes through the polarizer, it would be extinguished twice in one full rotation of the polarizer.

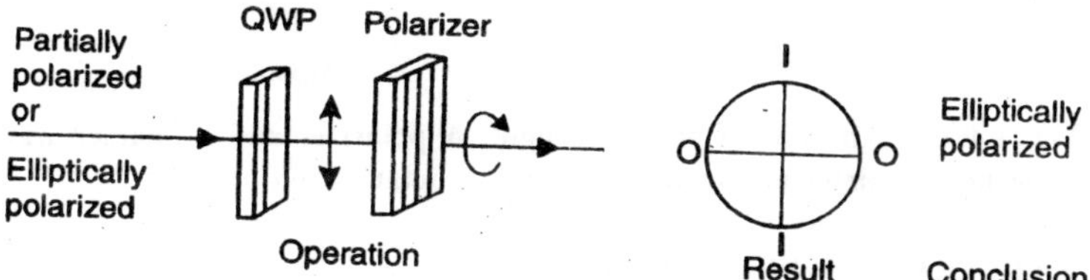

Fig. 1.46(d)

On the other hand, if the transmitted light intensity varies between a maximum and a minimum without becoming zero, then the incident light is partially polarized.

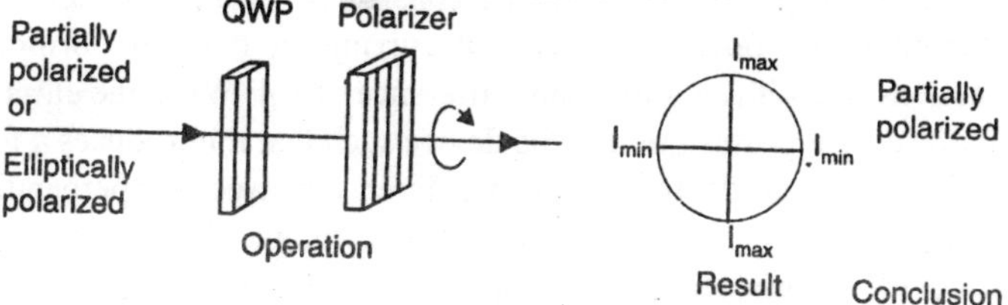

Fig. 1.46(e)

(v) If the incident light is circularly polarized, the quarter wave plate converts it into plane polarized light. When this linearly plane polarized light passes through the polarizer, it would be completely extinguished twice in one full rotation of the polarizer.

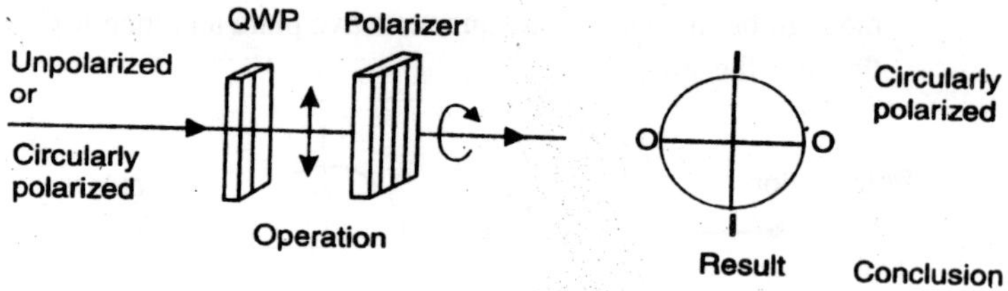

Fig. 1.46(f)

On the other hand, if the intensity of the transmitted light stays constant, then the incident light is unpolarized.

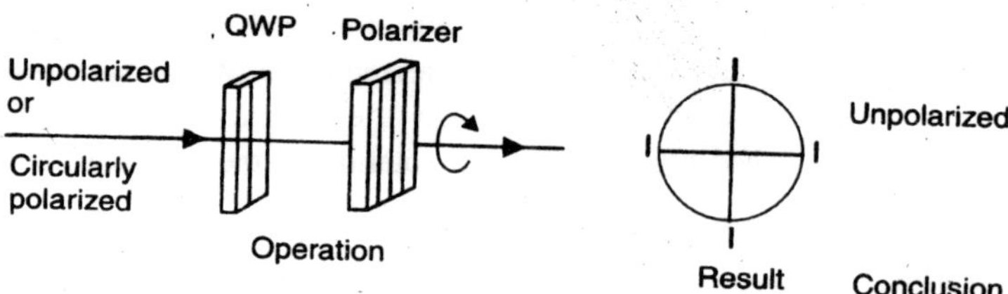

Fig. 1.46(g) :Analysis of polarized light. A polarizer and a quarter wave plate help in determining the type of the polarization of light.

BABINET COMPENSATOR

A *compensator* is an optical device whose function is to compensate a path difference. It is used in conjunction with a polarizer and analyzer combination to investigate elliptically polarized light. The compensator helps in determining the axis of the ellipse and the ratio of their lengths. Elliptically polarized light may be considered as the resultant of two coherent plane polarized waves occurring in mutually orthogonal directions and with an initial path difference of $\lambda/4$. When the elliptically polarized light passes through the device such that it introduces a further path difference of $\lambda/4$, the total path difference between the perpendicular waves becomes $\lambda/2$ and the vibrations recombine after emerging from the device to produce plane polarized light. From the analysis of this plane polarized light, the information regarding the incident elliptically polarized light can be obtained.

Construction

The Babinet compensator is made of two wedge-shaped quartz sections, ABC and ADC, having equal acute angles. The wedges are placed against each other such that they form a small rectangular block as shown in Fig. 1.47(a). One of the quartz wedges is fixed and the other can be displaced along their plane of contact with the help of a micrometer screw arrangement. Thus, the combination acts as a plate of variable thickness.

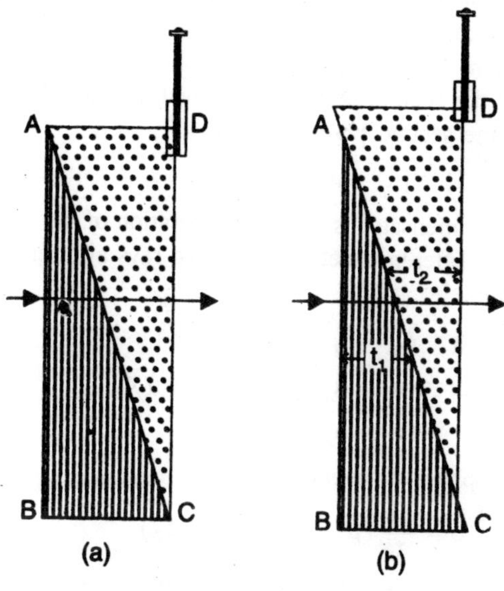

Fig. 1.47

The optic axis of the first section is parallel to its refracting edge AB and the optic axis of the second section is in a direction perpendicular to the edge. The two optic axes are perpendicular to each other and also perpendicular to the incident beam.

Production of Polarized Light

Let plane polarized light be incident normally on the face AB of the compensator. It splits into e-ray and o-ray parallel and perpendicular to AB respectively. The e-ray travels slower than o-ray in the first section, since quartz is a positive uniaxial crystal. When these rays enter the second section, the e-ray becomes o-ray since the optic axis in the second section is in a direction normal to that in the first prism. Similarly, o-ray becomes e-ray. Thus, the two rays exchange their velocities in passing from one section to the other section. The net effect is that one section cancels the effect of the other.

If d_1 is the thickness of the first section, and μ_e and μ_o are the refractive indices of quartz for e-and o-rays respectively, the path difference between the e- and o-rays in the first section will be

$$\Delta_1 = [\mu_e - \mu_o]\, d_1$$

As the principal planes of the two sections are at right angles, the e-and o-rays change their roles in going from the first section to the second section. The velocities of e-ray and o-ray interchange and if the thickness of the second section is d_2, then the path difference between the rays in the second prism will be

$$\Delta_2 = [\mu_e - \mu_o]\, d_2$$

As the compensator is thin, the separation of the rays is negligible. The net path difference between the two rays after emerging from the crystal will be

$$\Delta = \Delta_1 + \Delta_2$$

$$\Delta = (\mu_e - \mu_o)\, d_1 + (\mu_o - \mu_e)\, d_2$$

$$= (\mu_e - \mu_o)\,(d_1 - d_2) \quad ...(37)$$

The net phase difference is

$$\delta = \frac{2p}{\lambda}(\mu_e - \mu_o)(d_1 - d_2) \quad ...(38)$$

For a ray passing through the centre of the compensator where $d_1 = d_2$, the net path difference and hence the phase difference is zero. It means that the effect of one wedge is exactly cancelled by the other. This is true for all wavelengths and the incident wave is transmitted as such. Plane polarized light incident on the compensator will emerge as plane polarized light with its plane of vibration parallel to that of the incident light.

Any desired thickness difference $(d_1 - d_2)$ can be achieved at the centre of the compensator by moving the second section relative to the first section. Thus, any desired value of phase difference can be obtained between the e and o-rays. Therefore, the light emerging will be either plane or circular or elliptically polarized light, depending on the phase difference.

Thus, the compensator has the same effect as that of a wave plate of varying thickness. The advantage of compensator is that it can be arranged to suit any wavelength where as a quarter wave plate is designed to suit only one particular wavelength.

Analysis of Elliptically Polarized Light

Using the compensator, one can determine the characteristics of elliptically polarized light.

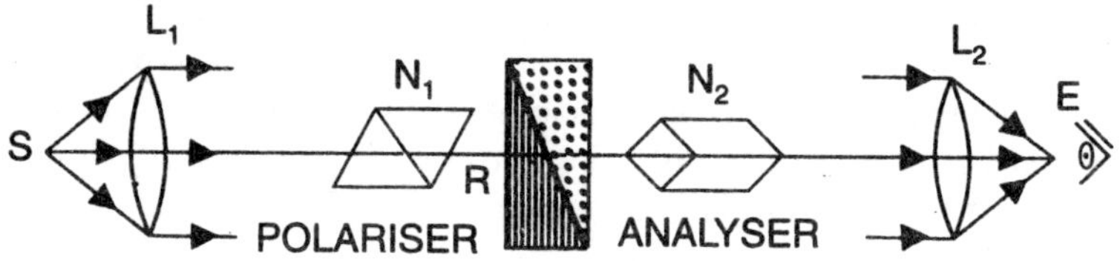

Fig. 1.48

Let the compensator be placed between crossed polarizer N_1 and analyser N_2 as shown in Fig. 1.48. Let the transmission axis of polarizer be oriented at 45° with respect to the optic axis of wedge ABC of the compensator. At midpoint R the light emergent from the compensator is plane polarized in the same plane as transmitted by N_1 and therefore it will be extinguished by the analyser N_2. Similarly, at distances from the midpoint for which the retardation is 1λ, 2λ, 3λ,.....wλ, and the emergent light is plane polarized in the same plane as transmitted by N_1 and hence will be extinguished by the analyser. So the field of view is crossed by a series of equidistant parallel dark bands.

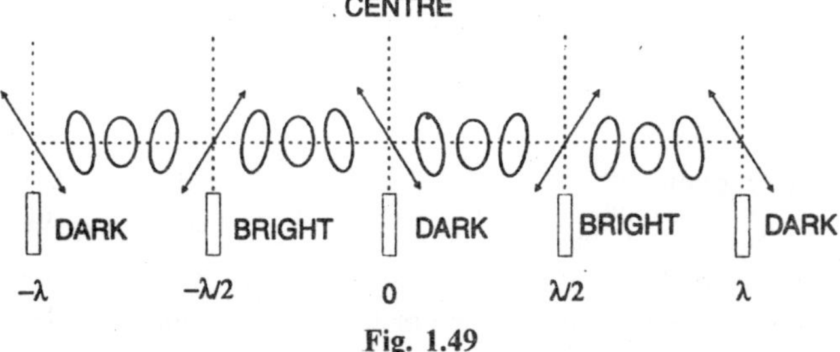

Fig. 1.49

At positions between them, where the path difference corresponds to an odd multiple of $\frac{\lambda}{2}$, *i.e.*, $\frac{\lambda}{2}, \frac{3\lambda}{2}, \frac{5\lambda}{2}, \ldots \frac{(2m+1)\lambda}{2}$, the transmitted light is plane polarized. The analyser transmits the light completely and those regions will be bright. In all other cases, the emerging light is elliptically polarized with varying parameters of the ellipse, as shown in Fig. 1.49. If white light is used, the central band will be dark while others will be coloured.

By using white light source, the compensator is adjusted such that the central dark band is under cross wire and the micrometer reading is noted. The micrometer screw is turned through an angle such that the compensator introduces a phase difference of $\pi/2$ at cross wire. Then elliptically polarized light is made to be incident on the compensator. The central dark band undergoes a shift with respect to the cross wire. The compensator is rotated through an angle a in its own plane until the central dark band is on the cross wire. The axes of the incident elliptically polarized light are parallel to the optic axes of the wedges of the compensator.

Phase Difference

The elliptical vibration can be regarded as made of two mutually perpendicular linear vibrations, which are having a phase difference, δ. δ can be determined as follows.

First the compensator is illuminated with white plane-polarised light and the micrometer is adjusted to bring the central dark band on the cross-wires. The white light is then replaced by elliptically polarised light. The central band shifts to a point where the original phase difference δ between the two component vibrations of elliptical polarised light is exactly balanced by the phase difference introduced by the compensator. This phase difference is determined by rotating the screw until the central dark band is again on the cross-wires. If this rotation is ϕ, then

$$\frac{\delta}{2\pi} = \frac{\phi}{\alpha} \text{ or } \delta = \frac{2\pi\phi}{\alpha} \quad ...(39)$$

Position of Axes

The position of the major and minor axes of the given elliptical vibration can be found as follows. The compensator is illuminated with white polarised light and the micrometer screw is adjusted to bring the central dark band on the cross-wires. The screw is then turned through an angle $\alpha/2$ so that the compensator introduces a phase difference of 90°. The central dark band now is not on the cross wires. The elliptically polarised light is made incident on the compensator. Then the compensator is rotated in its own plane until the central dark band again comes on to the crosswires. The axes of the incident light are parallel to the optic axes of the wedges of the compensator.

Ratio of the Axes

Referring to the Fig. 1.50, OA and OB represent the optic axes of the two wedges of the compensator. OC is the direction of the principal section of the analyser. DD' is the direction of vibration of light emerging from the compensator at the crosswires. The tangent of the angle θ that the principal section of the analyser makes with the optic axes of the wedge gives the ratio of the axes. Thus,

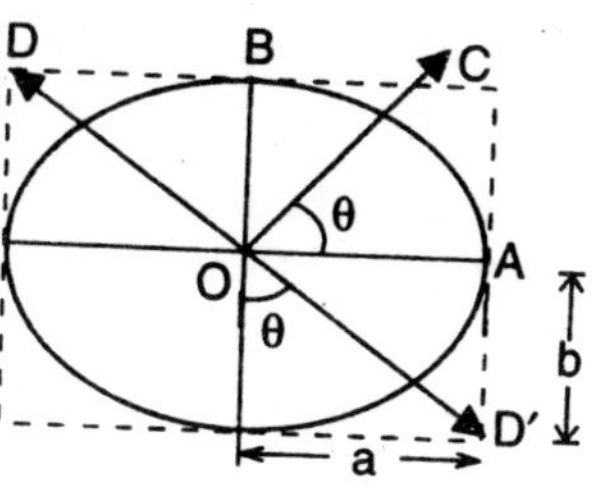

Fig. 1.50

$$\tan \theta = a/b$$

θ is determined by rotating the analyser until the bands disappear and the field becomes uniformly illuminated. The angle of rotation is θ.

Advantages : A quarter wave plate produces a fixed phase difference between o-ray and e-ray and can be used only for monochromatic light of one particular wavelength. In case of a compensator, the phase difference between the rays can be varied continuously and hence a compensator, made of a combination of wedges, can be used for light of any given wavelength.

FRESNEL'S RHOMB

Fresnel designed a rhomb of glass whose angles are 54° and 126° as shown in Fig. 1.51. Its functioning is based on the fact that a phase difference of n $\pi/4$ is introduced between the component vibrations when light is totally internally reflected back at glass - air- interface when the angle of incidence is 54°.

A ray of light enters normally at one end of the rhomb and is totally internally reflected at the point B along BC. The angle of incidence at B is 54°, which is more than the critical angle of glass. Let the incident light be plane polarised and let the vibrations make an angle of 45° with the plane of incidence. Its components (i) parallel to the plane of incidence and (ii) perpendicular to the plane of incidence are equal. These components after reflection at the point B undergo a phase difference of $\pi/4$ or a path difference of $\lambda/8$. A further phase difference of $\pi/4$ or a path difference of $\lambda/8$ is introduced between the components when the ray BC is totally internally reflected back along CD. Therefore, the final emergent ray DE has two components, vibrating at right angles to each other and they have a path difference of $\lambda/4$. Therefore, the emergent ray is circularly polarised.

If the light entering the Fresnel's rhomb is circularly polarised, a further path difference of $\lambda/4$ is introduced between the component vibrations. The total path difference between the component vibrations is $\lambda/2$. Therefore, the emergent light is plane polarised and its vibrations make an angle of 45° with the plane of incidence.

When an elliptically polarised light is passed through a Fresnel's rhomb, a further path difference of $\lambda/4$ is introduced between the component vibrations. The total path difference between the component vibrations is $\lambda/2$ and the emergent light is plane polarised.

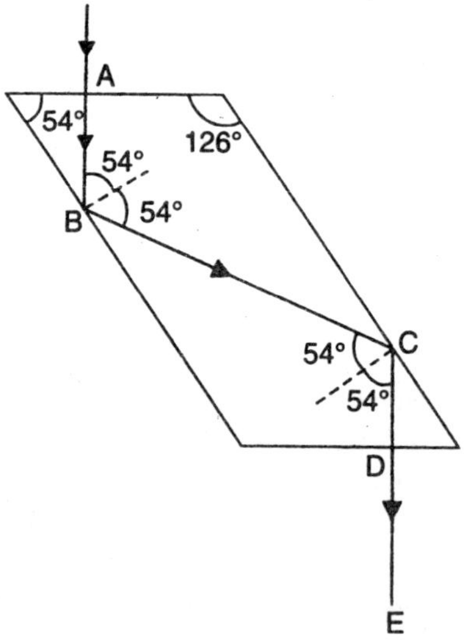

Fig. 1.51

Thus, Fresnel's rhomb works very much similar to a quarter wave plate. A quarter wave plate is used for a particular wavelength light whereas a Fresnel's rhomb can be used for light of all wavelengths.

DOUBLE IMAGE POLARIZING PRISMS

Nicol prism cannot be used with ultraviolet light on account of the canada balsam layer which absorbs UV rays. Sometimes it is also desirable to have both the ordinary and extraordinary rays widely separated. For this purpose, two prisms, namely Rochon prism and Wollaston prism, are used.

1. Rochon Prism

It consists of two prisms ABC and BCD (of quartz or calcite) cut with their optic axes as shown in Fig. 1.52. The prism ABC is cut such that the optic axis is parallel to the face AB and the incident light. The prism BCD has the optic axis perpendicular to the plane of incidence. Light incident normally on the face AC of the prism passes without deviation up to the boundary BC. In the prism BCD, the ordinary ray passes without deviation. If the prisms are made of quartz, the extraordinary ray is deviated as shown in Fig. 1.52.

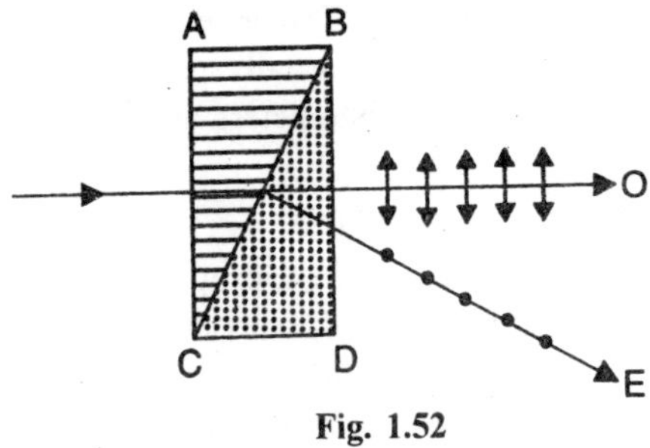

Fig. 1.52

In the case of calcite, the extraordinary ray will be deviated to the other side. The prisms ABC and BCD are cemented together by glycerine or castor oil. Here, the ordinary emergent beam is achromatic whereas the extraordinary beam is chromatic.

2. Wollaston Prism

It consists of two prisms ABC and BCD of quartz or calcite cut with their optic axes as shown in Fig. 1.53. They are cemented together by glycerine or castor oil.

A ray of light is incident normally on the face AC of the prism ABC. The ordinary and the extraordinary rays travel along the same direction but with different speeds. After passing BC the ordinary ray behaves as the extraordinary ray and the extraordinary ray behaves as the ordinary

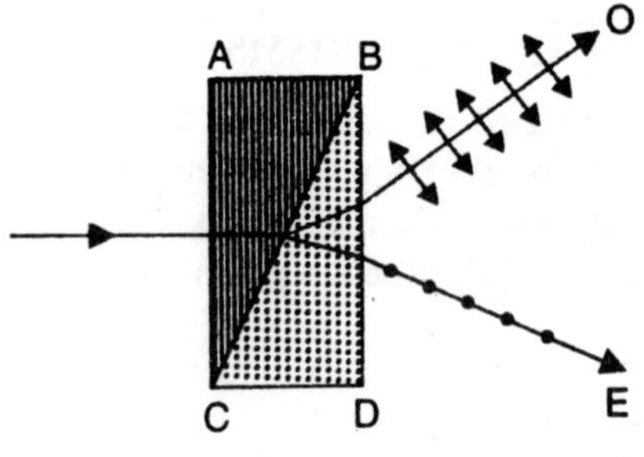

Fig. 1.53

ray while passing through the prism BCD. One ray is bent towards the normal while the other is bent away from the normal. In quartz $\mu_e > \mu_o$. Therefore, the ordinary ray while passing the boundary BC is refracted towards the normal as an extraordinary ray while the extraordinary ray is refracted away from the normal as ordinary ray, as shown in Fig. 20.48. If the prisms are made from calcite, the directions of the o-ray and e-ray are interchanged. While coming out of the face BD of the prism, the o-ray and e-ray are diverged. The prism is useful in determining the percentage of polarisation in a partially polarised beam. Double image prisms are used in spectrophotometers and pyrometers.

OPTICAL ACTIVITY

When a beam of plane polarised light propagates through a quartz crystal along the optic axis, the plane of polarisation steadily turns about the direction of the beam, as shown in Fig. 1.54. The optical rotation can be detected as follows. If two Polaroid sheets or Nicol prisms are held in crossed configuration and if a beam of unpolarized light is viewed through them, the field of view appears to be completely dark. Now let a quartz crystal, cut with its faces perpendicular to the optic axis, be inserted between the polarisers such that light is incident normally on the crystal. The field of view now appears lit up indicating that the light is not cut off by the analyser. In order to cut off the transmitted light, we find that the analyser is to be rotated through a certain angle. The experiment establishes that the plane polarised light produced by the polarizer remains plane polarised while passing through the quartz crystal but the plane of polarisation is rotated through an angle. This angle is the angle through which the analyser is rotated in order to cut off the light totally.

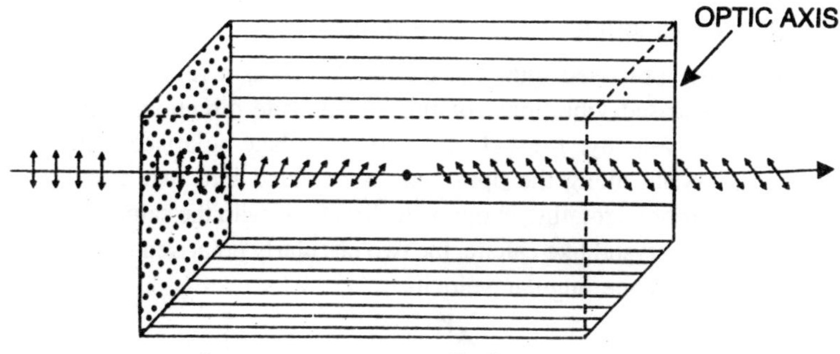

Fig. 154(a)

The ability to rotate the plane of polarisation of plane polarised light by certain substances is called *optical activity*. Substances, which have the ability to rotate the plane of the polarised light passing through them, are called optically active substances. Quartz and cinnabar are examples of *optically active* crystals while aqueous solutions of sugar, tartaric acid are optically active solution and liquid.

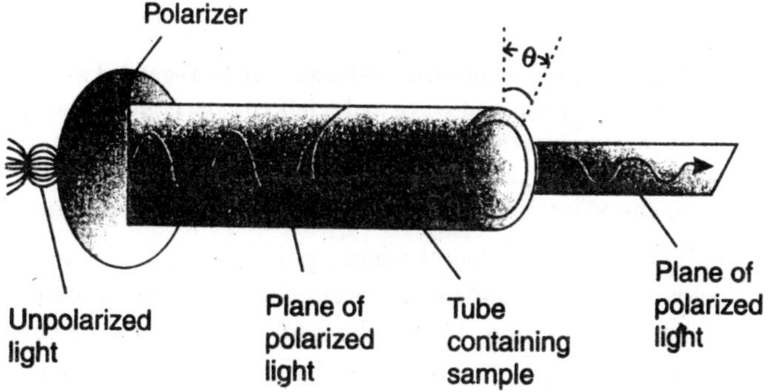

Fig. 1.54(b) : Rotation of the plane-polarized light. The plane is rotated by an angle after the light passes through an optically (a) active crystal or (b) solution.

Optically active substances are classified into two types.

(i) *Dextrorotatory substances* : Substances which rotate the plane of polarization of the light towards the right are known as right-handed or dextrorotatory.

(ii) *Laevorotatory substances* : Substances which rotate the plane of polarization of the light towards the left are known as left-handed or laevorotatory

Fresnel's Explanation of Optical Rotation.

A linearly polarised light can be considered as a resultant of two circularly polarised vibrations rotating in opposite directions with the same angular velocity. Fresnel assumed that plane-polarised light on entering a crystal along the optic axis is resolved into two circularly polarised vibrations rotating in opposite directions with the same angular frequency. In a crystal like calcite, the two circularly polarised vibrations travel with the same angular velocity.

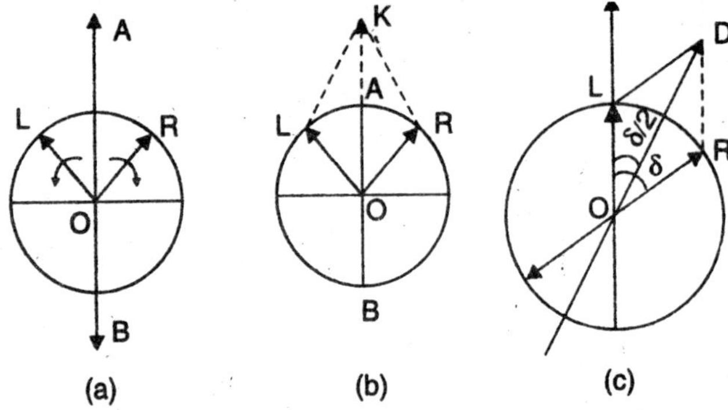

Fig. 1.55 : (a) A linearly polarized light wave can be regarded as a resultant of a right regularly polarized wave and a left circularly polarized wave. (b) In optically inactive materials, the vibrations corresponding to the circularly polarized waves rotate with equal angular velocity, (c) In optically active substances, the vibrations rotate with different angular velocities.

In Fig. 1.55, OL is the circularly polarised vector rotating in the anticlockwise direction and OR is the circularly polarised vector rotating in the clockwise direction. The resultant of OR and OL is the vector OA (Fig. 1.55b). According to Fresnel, when linearly polarised light enters a crystal of calcite along the optic axis, the circularly polarised vibrations rotating in opposite directions have the same velocity. The resultant vibration will be along OK. Therefore, crystals like calcite do not rotate the plane of vibration.

In case of quartz, the linearly polarised light, the component having clockwise rotation travels faster than the anticlockwise component. When the components emerge out of the crystal, they are at an angle 5. The resultant of these two vectors OR and OL is now along OD (Fig. 1.55 c). Before entering the crystal, the plane of vibration is along OA and after emerging from the crystal, it is along OD. The plane of vibration has rotated through an angle $\delta/2$. The angle through which the plane of vibration is rotated depends on the thickness of the crystal.

Analytical Treatment

Let us assume that linearly polarised light is incident normally on a quartz plate cut perpendicular to the optic axis. Let the vibrations in the incident light be represented by

$$y = a \cos \omega t$$

On entering the crystal these vibrations are broken up into two equal and opposite circular vibrations which are represented by

$$x_1 = \frac{a}{2} \sin \omega t$$

and $$y_1 = \frac{a}{2} \cos \omega t \quad \text{–clockwise circular polarisation}$$

$$x_2 = \frac{a}{2} \sin \omega t$$

and $$y_2 = \frac{a}{2} \cos \omega t \quad \text{–anti-clockwise circular polarisation}$$

These circular components travel through the crystal with different velocities. When they emerge from the crystal, there is a phase difference ° between them. In case of quartz crystal, which is a right-handed optically active crystal, the clockwise component travels faster. The circular components are then represented by

$$x_1 = \frac{a}{2} \sin (\omega t + \delta)$$

and $$y_1 = \frac{a}{2} \cos (\omega t + \delta) \quad \text{–clockwise component}$$

$$x_2 = \frac{a}{2} \sin \omega t$$

and $$y_2 = \frac{a}{2} \cos \omega t \qquad \text{–anti-clockwise component}$$

The resultant displacements along the two axes are

$$X = x_1 + x_2 = \frac{a}{2} \left[\sin (\omega t + \delta) - \sin \omega t\right]$$

$$= a \cos \left[\omega t + \frac{\delta}{2}\right] \sin \frac{\delta}{2} \qquad \text{...(40)}$$

and $$Y = y_1 + y_2 = \frac{a}{2} \left[\cos (\omega t + \delta) + \cos \omega t\right]$$

$$= a \cos\left[\omega t + \frac{\delta}{2}\right]\cos\frac{\delta}{2} \qquad ...(41)$$

The resultant vibrations along the X-axis and Y-axis have the same phase. Therefore, the resultant vibration is plane polarised and it makes an angle δ/2 with the original direction. Thus, the plane of polarisation is rotated through an angle δ/2 on passing through the crystal.

Dividing eqn. (39) by eqn. (40), we obtain

$$\frac{X}{Y} = \tan\frac{\delta}{2} \qquad ...(42)$$

If μ_R and μ_L are the refractive indices of quartz in the direction of optic axis for clockwise and anticlockwise circularly polarised light and d is the thickness of the crystal plate, then the phase difference δ is given by

$$\delta = \frac{2\pi}{\lambda}(\mu_L - \mu_R)d \qquad ...(43)$$

Hence the plane of polarisation is rotated through

$$\theta = \frac{\delta}{2} = \frac{\pi d}{\lambda}(\mu_L - \mu_R) \qquad ...(44)$$

Experimental Verification of Fresnel's Theory

Fresnel showed that linearly polarised light on entering an optically active crystal is resolved into two circularly polarised vibrations. He arranged alternatively a number of negative and positive optically active prisms to form a rectangular block, as shown in Fig. 1.56. The optic axis is parallel to the base of each prism. When linearly polarised light is incident normally on the first crystal surface AB, the two component circular vibrations travel along the same direction but with different speeds. When the light beam passes through the first oblique surface BC, the R-component, which is faster in the first prism, becomes slower in the second prism. The opposite is the situation for L-component. It means that the second prism BCD acts as a denser medium for the R-component and rarer medium for the L-component. As a result, the R-component bends towards the base in the second prism while L-component bends away from the base. Therefore, the two components are separated apart while they travel through prism BCD. Again at the boundary of the next prism, the speeds are interchanged and the R-component bends away from the base and the L-component bends towards the base. The net result is that the two beams are separated more and more as they pass through the successive prisms. When they ultimately emerge out, they are widely separated.

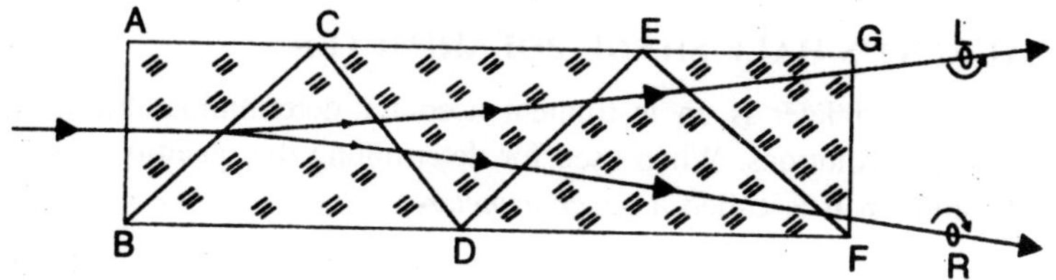

Fig. 1.56

When the two waves are analysed by a quarter wave plate and a Nicol prism, both are found to be circularly polarised in opposite directions. Thus, Fresnel's hypothesis is verified.

SPECIFIC ROTATION

If an optically active material is kept between two crossed polarizers, the field of view becomes bright. In order to get darkness once again, the analyser has to be rotated through an angle. The angle through which the analyser is rotated equals the angle through which the plane of polarization is rotated by the optically active substance. This angle depends on

(a) the thickness of the substance,

(b) density of the material or concentration of the solution,

(c) wavelength of light, and

(d) the temperature.

The amount of rotation θ caused by crystalline materials is given by

$$\theta = al$$

In solutions the amount of rotation θ is given by

$$\theta = scl$$

where c is the concentration and s is called the specific rotation.

The specific rotation for a given wavelength of light at a given temperature is defined conventionally as the rotation produced by one decimetre long column of the solution containing 1 gm of optically active material per c.c .of solution.

$$[S]_\lambda^t = \frac{\theta}{l \times C} = \frac{\text{Rotation in degrees}}{\text{Length in decimeters} \times \text{conc. in gm/c.c.}}$$

$$= \frac{10\theta}{l(\text{cm})C}$$

LAURENT'S HALF SHADE POLARIMETER

A polarimeter is an instrument used for determining the optical rotation of solutions. When used for determining the quantity of sugar in a solution it is called a *saccharimeter*.

Construction : A polarimeter consists of a glass tube for holding the solution under test held between crossed Nicol prisms, N_1 and N_2. Beyond the polarising Nicol prism a half-shade plate is located which is used for accurately adjusting the two Nicol prisms for crossed position. G is a glass tube which contains the optically active solution. Light from a monochromatic source is rendered parallel by the lens L and is incident on the polarizer, N_1. The light transmitted by the polarizer is plane polarised. The polarised beam then passes through the half-shade plate and then through the glass tube G. The light emerging from the solution will be incident on the analyser N,. The light is observed through a telescope T. The analysing Nicol N_2 can be rotated about the axis of the tube and the rotation can be measured with the help of a graduated circular scale.

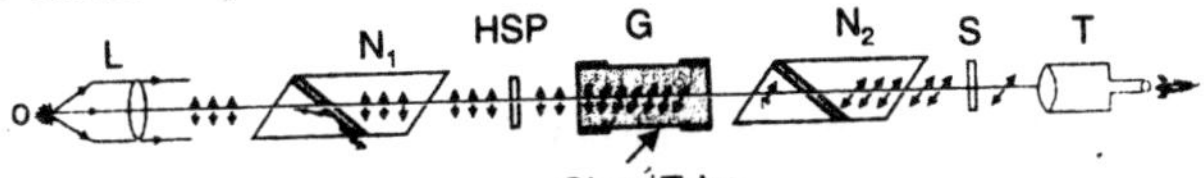

L N1 HSP G N2 S T

Glass Tube Containing Solution

Fig. 1.57

Working : To find the specific rotation of a solution, the analyser is first adjusted such that field of view is completely dark. Then the glass tube is filled with the solution and is held in position. The field of view now becomes illuminated. The field of view can be again be made dark by rotating the analyser through a certain angle which gives the optical rotation of the solution. The practical difficulty in this method is la, determination of the exact position for which complete darkness is achieved. The difficulty is overcome by using what is known as a *Laurent's half-shade device* (Fig. 1.58). It consists of a semicircular half wave plate ACB of quartz cemented to a semicircular plate ADB of glass. The optic axis of the wave plate is parallel to the line of separation AB. The half wave plate introduces a phase difference of 180° between e-ray and o-ray passing through it. The thickness of the glass plate is such that it transmits the same amount of light as done by the quartz half wave plate. One half of the incident light passes through the quartz plate ACB and the other half through the glass plate ADB. The light after passing through the polarizer is incident normally on the half shade plate and has vibrations along OR On passing through the glass, half the vibrations will remain along OP but on passing through the quartz half, the vibrations will split into e-and o-rays. The o-vibrations are along OD and e-vibrations are along OA. The half wave plate introduces a phase difference forward between the two vibrations. The vibrations of o-ray will occur along OC instead of OD on emerging from the plate. Therefore the resultant vibration will be along OQ whereas the vibrations of the beam emerging from glass plate will be along OR In effect, the half wave plate turns the plane of polarisation of the incident light through an angle 2θ. If the principal plane of the Nicol N_2 is aligned parallel to OP, the plane polarized light emerging from the glass tube will pass through the glass plate of the half shade plate and that part appears brighter. On the other hand light coming out of the quartz plate is partially obstructed and the corresponding field of view appears less bright. If the principal plane of N_2 is parallel to OQ the quartz half will appear brighter than the glass half. Thus, the two halves of the plate are unequally illuminated. When the principal plane of N_2 is parallel to AB, the two halves appear equally bright and when it is parallel to CD, the two halves are equally dark.

To find the specific rotation of a solution, the analyser is first set in the position for equal darkness without solution in the tube G. The reading on the circular scale is noted. Next, the tube is filled with the optically active solution of known concentration. The field of view is now partially illuminated. The analyser is rotated till the field of view becomes equally dark. The reading on the circular scale is noted again. The difference between the two scale readings gives the angle of rotation of the plane of polarisation caused by the solution. Knowing the values of θ, l, and c, the specific rotation is obtained using

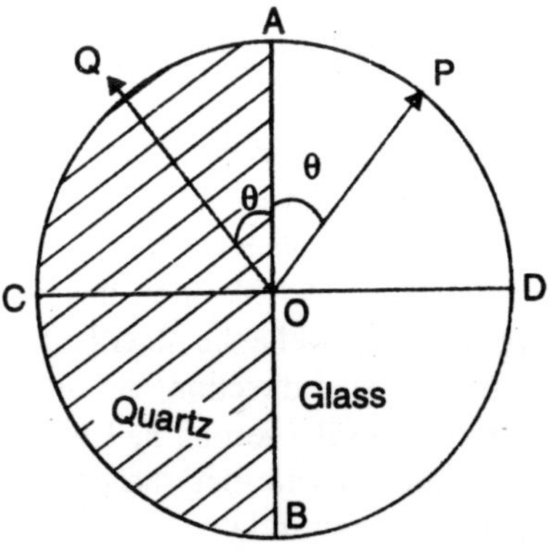

Fig. 1.58

the formula (44). Or otherwise, knowing the value of the specific rotation, the concentration of the solution can be determined with the help of the equation (44).

In the actual experiment, different concentrations of solutions are taken and the corresponding angles of rotation are determined. A graph is plotted between concentration C and the angle of rotation θ. The graph is a straight line (Fig. 1.59).

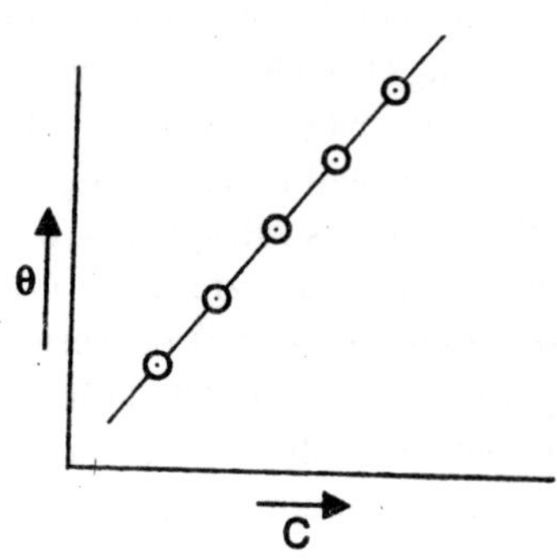

Fig. 1.59

Using the value of the slope in eqn. (44), the specific rotation of the optically active substance is calculated.

BIQUARTZ

Instead of half shade plate, a biquartz plate is also used in polarimeters. It consists of two semicircular plates of quartz each of thickness 3.75 mm. One half consists of right-handed optically active quartz, while the other is left-handed optically active quartz.

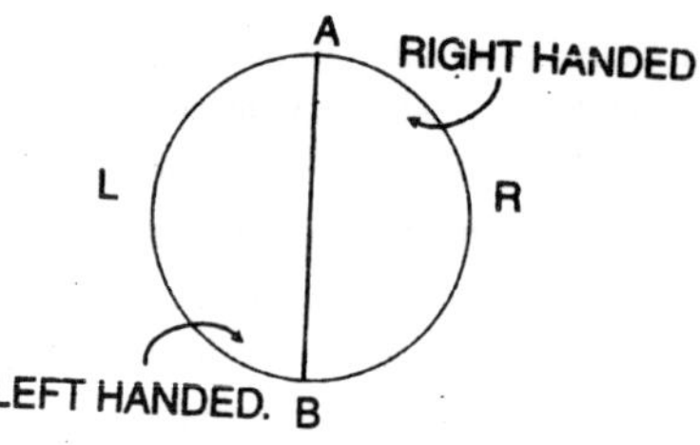

Fig. 1.60

If white light is used, yellow light is quenched by the biquartz plate and both the halves will have the tint of passage. This can be adjusted by rotating the analyser N_2 to a particular position. When the analyser is rotated to one side from this position, one half of the field of view appears blue, while the other half appears red. If the analyser is rotated in the opposite direction, the first half which was blue earlier appears now red and the second half which was earlier red appears blue now. Therefore, by adjusting the position of the analyser, the field of view appears equally bright with tint of passage.

LIPPICH POLARIMETER

Laurent's polarimeter suffers from the defect that it can be used only for light of a particular wavelength for which the half wave plate has been designed. Lippich developed a polarimeter that can be used for light of any wavelength.

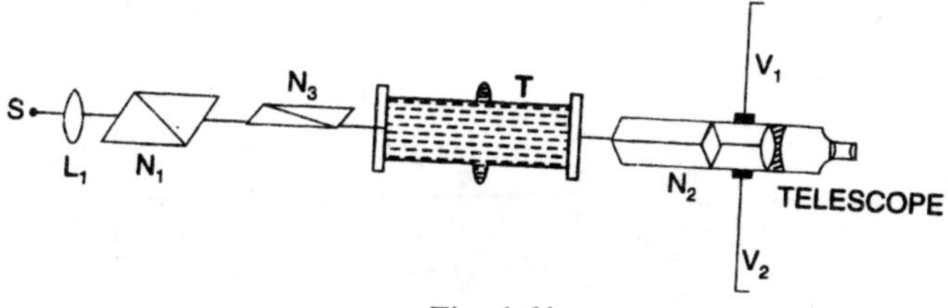

Fig. 1.61

The schematic of the polarimeter is shown in Fig. 1.61. It consists of two Nicol prisms N_1 and N_2. Behind N_1, there is a Nicol prism N_3 that covers half the field of view. The Nicols N_1 and N_3 have their planes of vibration inclined at a small angle. Suppose the plane of vibration of N_1 is along AB and that of N_3 is along CD (Fig. 1.62). The angle between the two planes is θ. When the analyser N_2 is rotated such that the plane of vibration of N_2 is along AB, the left half will be brighter than the right half. If the analyser N_2 has its plane of vibration along CD, the right half will be brighter than the left half. YY' is the bisector of the angle AOC.

Therefore, when the plane of vibration of the analyser N_2 is along YY', the field of view is equally illumined. For as light rotation of the analyser, either to the right or to the left, the field of view appears to be of unequal brightness. Therefore, by rotating N_2 the position for equal brightness of the field of view is obtained.

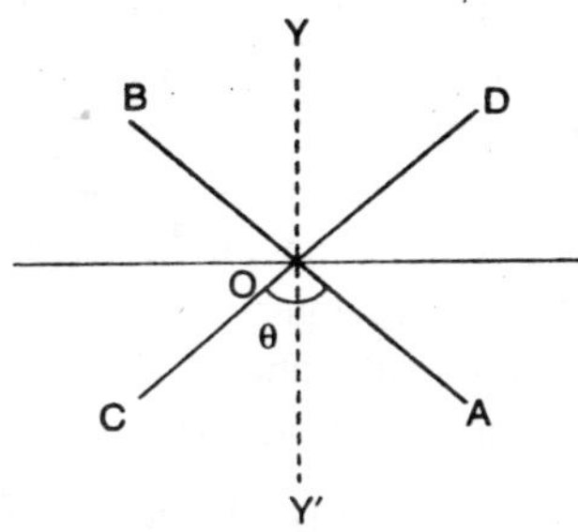

Fig. 1.62

Three-field system : In the improved form, Lippich polarimeter has a three-field system. The defect in the two field system is that if the eye is off the axis, even for the position of equal brightness of the field of view, one side appears brighter than the other. Just behind N_1 there are two Nicol prisms N_3 and N_4, Fig. 1.63.

The planes of vibration of N_3 and N_4 are parallel to each other and make a small angle with the plane of vibration of N_1. For a particular position of the analyser N_2 the field has three parts. The central portion is illuminated by light, which has passed through N_1 and N_2 while the other two portions, which are equally bright, are illuminated by light passing through N_1, N_2 and one of the prisms N_3 and N4.

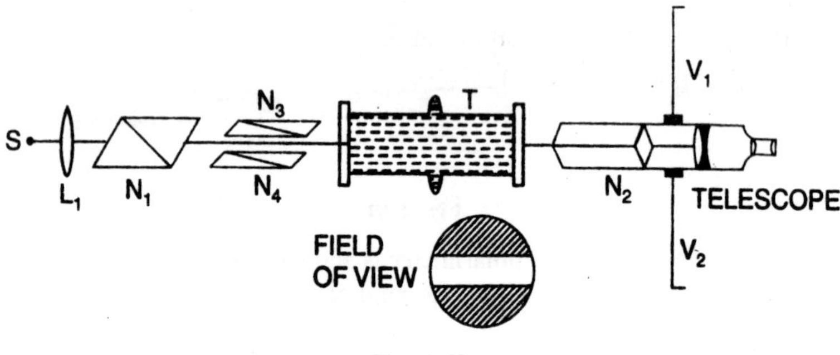

Fig. 1.63

ARTIFICIAL DOUBLE REFRACTION

Isotropic transparent materials such as glass do not exhibit double refraction under ordinary circumstances. However, they acquire the optical properties of a umaxyl crystal under the action of external forces. Consequently, they exhibit double refraction. The appearance of double refraction under the influence of an external agent is known as *artificial double refraction* or *induced birefringence.* The direction of the optical axis in such materials will be collinear with the direction of the external force. The action of the external force is to cause distortion of the molecular arrangement within the material and thereby transform the isotropic substance into an anisotropic substance. The induced birefringence disappears as soon as the external force ceases to act.

The property of double refraction can be induced in an isotropic material under the action of

(i) a mechanical strain

(ii) an electric field or

(iii) a magnetic field.

The materials which experience a change in their optical behaviour under the action of an electric field are called *electro-optic materials* and the resulting optical effects are known as *electro-optic effects*. Similarly, the materials that get influenced by a magnetic field are called magneto-optic materials and the resulting optic effects are known as magneto-optic effects. The *electro-optic materials* play a very important role in modern technology.

Anisotropy Induced by Mechanical Strain

Materials such as glass become double refracting when they are subjected to mechanical strain. Experiments show that the induced birefringence is directly proportional to the stress σ experienced at a given point of the material. Thus,

$$\Delta\mu = k\sigma \qquad ...(45)$$

where k is the proportionality constant characteristic of the material. This effect was discovered by Sir David Brewster in 1816.

When the material is held between crossed polarisers P and A, as in Fig. 1.64, the field of view appears dark as long as the external force is not applied. As soon as the force is applied, coloured contours will be seen. Dark regions indicate the absence of strain in those areas. Each coloured contour shows the areas that are identically deformed. Such contours enable us assess the distribution of the stresses in the material. The mechanically induced birefringence is used to study stresses in girders, beams etc. The model of the object under investigation is made of transparent plastic material and is then loaded. Using crossed polarizer system, the stresses produced at different positions are analysed and estimated. This method of analysis is known as *photoelastic analysis* and is widely employed in civil and mechanical engineering practices.

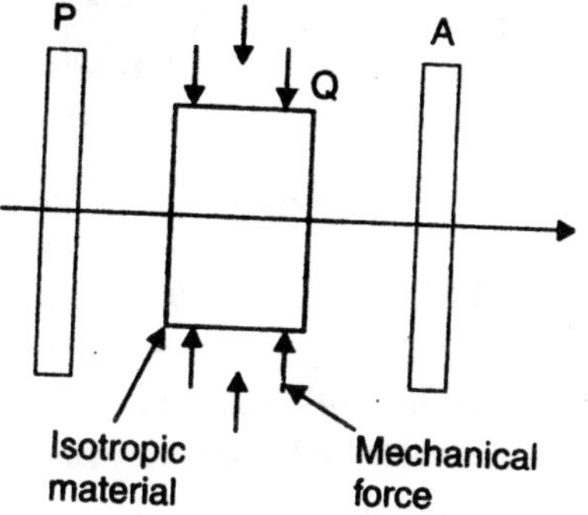

Fig. 1.64 : Birefringence is induced in an isotropic material under mechanical strain

Kerr Effect

Anisotropy induced in an isotropic liquid under the influence of an electric field is known as the Kerr effect. John Kerr discovered it in 1875.

A Kerr cell is required for studying the effect. It consists of a sealed glass cell filled with a liquid comprising of asymmetric molecules. Two plane electrodes of specific length *l* are arranged in it with their faces strictly parallel to each other. When a voltage is applied to them a uniform electric field is produced in the cell. The Kerr cell is placed between a crossed polarizer system. When -the electric field is applied, the molecules of the liquid tend to align along the field direction. As the molecules are asymmetric, the alignment causes anisotropy and the liquid becomes double refracting. The induced birefringence is proportional to the square of the applied electric field and to the wavelength of incident light. Thus,

$$\Delta\mu = J\lambda E^2 \quad ...(46)$$

where K is known as the Kerr constant.

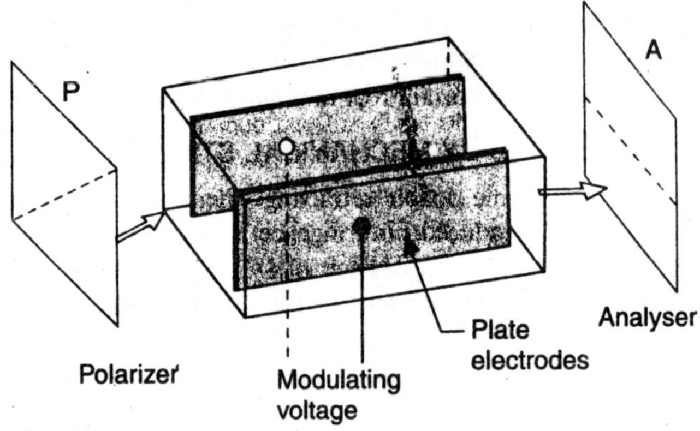

Fig. 1.65 : Kerr effect. Birefringence is induced in a liquid subjected to an electric field.

Among the liquids, nitrobenzene ($C_6H_5NO_2$) is found to have the highest value for the Kerr constant. Therefore, Kerr cells use nitrobenzene.

Kerr cell is used as an electro-optic shutter in high-speed photography, and as a light chopper in the measurement of the speed of light.

Pockels Effect

This is also an electro-optic effect. F. Pockels discovered in 1893 that the application of an electric field to piezoelectric crystals makes them birefringent. Normally, piezoelectric crystals are birefringent but in certain directions do not exhibit double refraction. When an electric field is impressed along these directions, double refraction is induced along these directions also.

A Pockels cell consists of a piezoelectric crystal, for example lithium niobate placed between crossed polarisers. Transparent electrodes (thin conducting coatings of tin oxide or indium) are deposited on opposite sides of the crystal. The crystal is oriented in such a way that its optic axis lies along the direction of the electric field applied between the electrodes. The transparent electrodes ensure free propagation of light through the crystal. A Pockels set up is shown in Fig. 1.66.

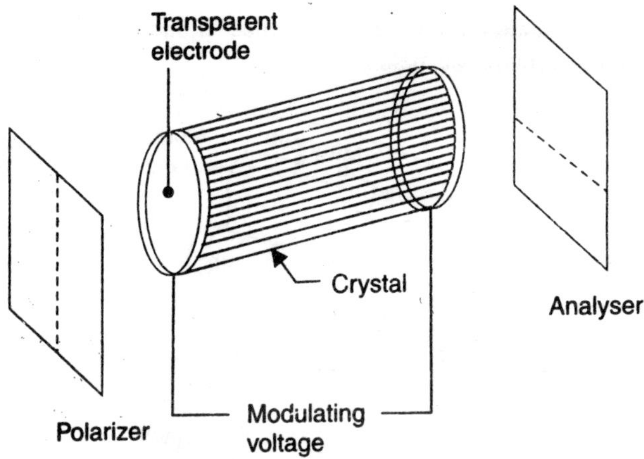

Fig. 1.66 : Pockels effect. Birefringence is induced in a piezoelectric crystal subjected to an electric field.

The birefringence induced in the crystal is proportional to the strength of the applied field. Thus,

$$\Delta\mu = kE \qquad ...(47)$$

where k is a constant characteristic of the material. Eqn. (47) shows that Pockels effect is a linear effect.

The total birefringence of the cell is initially made equal to $\lambda/1$. When the electric field is increased, the beam is transmitted or hindered depending on the phase difference between the o-ray and e-ray. The device switches on and off periodically. Pockels cells are used in fast switching applications and in fibre optics. It can be used to obtain amplitude, frequency or phase modulation.

A Pockels cell is simple in construction and requires a small voltage of the order of 1.5 kV where as a Kerr cell is complicated in construction and requires higher voltages of the order of 15 kV. The piezoelectric crystals of ammonium dihydrophosphate (ADP) and potassium dihydrophosphate (KDP) are widely used in Pockels cell.

Kerr and Pockels cells are widely used as electro-optic shutters for Q-switching of lasers.

Cotton-Mouton Effect

The Cotton-Mouton effect is a magneto-optic effect. An isotropic material acquires the optical behaviour of a uniaxial crystal under the action of an external magnetic field. The set up is shown in Fig. 1.67

Cotton-Mouton effect. Birefringence is induced in a material subjected to a magnetic field.

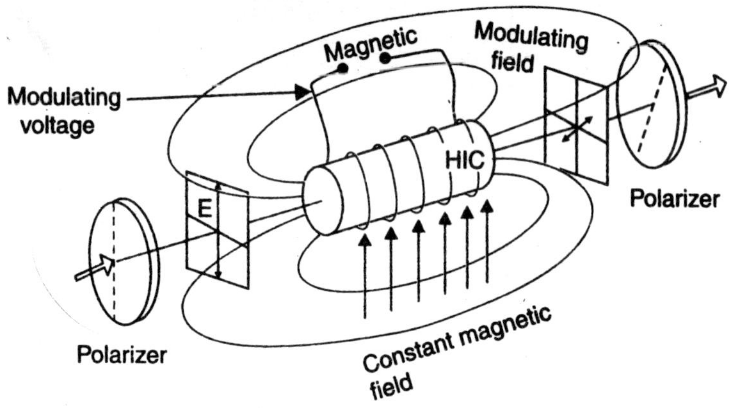

Fig. 1.67

The induced birefringence is governed by the relation

$$\Delta\mu = C\lambda B^2 \qquad \text{...(48)}$$

where C is a constant characteristic of the material. The magnitude of the induced birefringence is usually very small.

Faraday Effect

Optically inactive substances acquire the ability of rotating the plane of polarisation of light when they are subjected to a magnetic field, parallel to propagation direction. Michael Faraday discovered this effect and hence it is called Faraday effect.

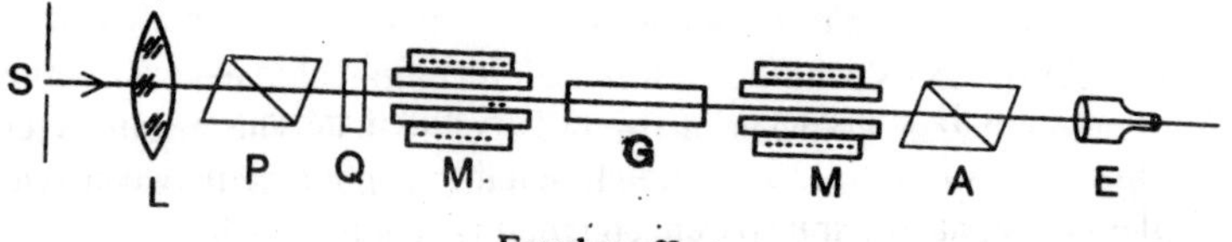

Faraday effect

Fig. 1.68

The angle of rotation θ of the plane of polarisation is proportional to the length of the path of light in the material and to the strength of the applied magnetic field. Thus,

$$\theta = VlH \qquad \text{...(49)}$$

where V is known as Verdet constant.

The angles θ of rotation are not very large. For magnetic field strengths of the order of 10^6A/m and $l = 0.1$ m, θ is about 1° to 2°.

One of the interesting problems encountered in satellite communications is the Faraday effect. As radio waves pass through the ionosphere, their plane of polarisation is rotated by the ionised particles in conjunction with the Earth's magnetic field. A horizontally polarised wave becomes vertically polarised because of the Faraday rotation in the ionosphere. The reception problem is solved by using an antenna with circular polarisation, which ensures that the waves are received satisfactorily, no matter how they have been rotated.

LCDs

Liquid crystal display (LCD) devices constitute one of the most beautiful applications of polarisation phenomenon. LCDs are widely used as alphanumerical read-outs in wristwatches, calculators, clocks, electronic instruments, video games, etc.

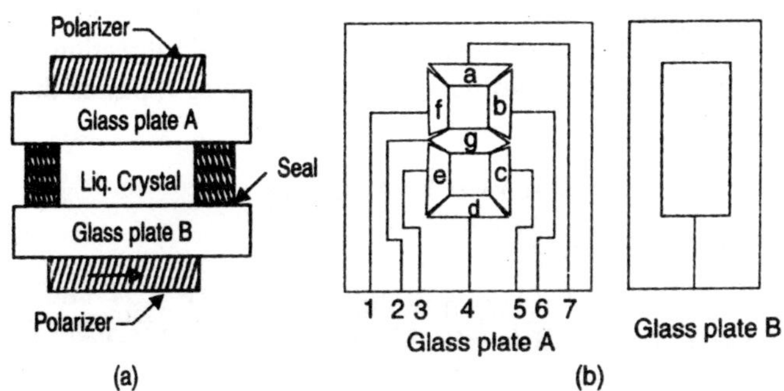

Fig. 1.69

An LCD consists of a liquid crystal material, which is double refracting, of about 10 μm thick suitably supported between two thin glass plates having transparent conducting coatings on their inner surfaces (Fig. 1.69 a). The conducting coating is etched in the form of a digit or character, as shown in (Fig. 1.69 b). The assembly of glass plates with liquid crystal material in between is sandwiched between two crossed-polarizer sheets.

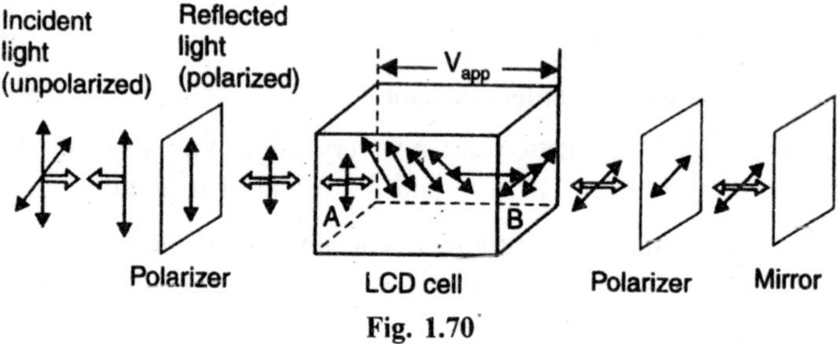

Fig. 1.70

During the fabrication of LCDs the liquid crystal molecules are aligned in such a way that their long axes undergo a 90° rotation, from plate A to plate B, as illustrated in Fig. 1.70. It is called a twisted molecular arrangement. When natural light is incident on the assembly, the front polarizer converts it into linearly polarised light. As the linearly polarised light propagates through the LCD, the optical vector is rotated through 90° by the twisted molecular arrangement. Therefore, it passes unhindered through the rear polarizer whose transmission axis is perpendicular to that of the front polarizer. A reflecting coating at the back of the rear polarizer sends back the light, which emerges unobstructed by the front polarizer. Consequently, the display appears uniformly illuminated. When a voltage is applied to the device, the molecules between the elecodes untwist and align align the field direction. As a result, the optical vector does not undergo rotation as it passes though that regin.

SOLVED EXAMPLES

Example 1:

The refractive indices of glass and water are 1.54 and 1.33 respectively. Calculate the polarizing angle for a beam incident from:

(i) Water to glass,

(ii) Glass to water,

(iii) Glass to air,

(iv) Air to glass.

Solution:

For case (i), $\mu = 1.54/1.33 = 1.155$

$i_p = \tan^{-1}(1.155) = 49°8'$

For case (ii), $\mu = 1.33/1.54 = 0.866$

$i_p = \tan^{-1}(.866) = 40°52'$

Similarly for, case (iii), $i_p = 33°$, case (iv) $i_p = 57°$.

Example 2:

At a particular incidence angle the coefficient of amplitude reflection of glass plate is 2% for | | vibrations and 8% for ⊥ vibrations. Calculate the degree of polarization of beam passing through a pile of 10 plates at this incidence angle, neglecting absorption.

Solution:

10 plates have 20 surfaces. At each surface 98% of one and 92% of the other vibration is transmitted (in amplitude). If a is the initial amplitude in each plane (| | and ⊥ to plane of incidence), the emergent beams have amplitudes

$$a_1 = a\text{x}\,(0.98)^{20} = .667a,$$

$$a_2 = a\text{x}\,(0.92)^{20} = .180a.$$

The corresponding intensities are (on squaring)

$$I_1 = I_0\text{x}\ 0.445$$

$$I_2 = I_0\text{x}\ 0.036$$

$$\text{Degree of Polarization} = \frac{I_{max} - I_{min}}{I_{max} + I_{min}} = \frac{0.445 - 0.036}{0.445 + 0.036} = 0.85.$$

Example 3:

An analyzer examines two adjacent plane polarized beams A, B whose planes of polarization ae mutually perpendicular. In one position of the analyzer, beam B shows zero intensity. From this position a rotation of 27° shows the two beams as matched. Deduce the intensity ratio IA/IB of the beams.

Solution:

In the zero position, as described, the vibrations transmitted by the analyzer are parallel to the vibrations in beam. A Hence, in the position 27° from this,

$$I_A \cos^2 27° = I_B \cos^2 (90° - 27°) = I_B \sin^2 27°$$

$$I_A/I_B = \tan^2 27° = 0.26$$

(This principle is used in the intensity comparison in many instruments).

Example 4:

Calculate the angular deviations of E and O rays passing through a 60° prism of $NaNO_3$ in minimum deviation condition, optic axis being perpendicular to the base.

Solution:

Using the equation above with Å = 60°

$$\sin \frac{1}{2}(60+\delta_0) = 1.587 \times 0.5 = 0.794$$

$$\Rightarrow \quad \delta_0 = 45°12'$$

$$\sin \frac{1}{2}(60+\delta_E) = 1.336 \times 0.5 = 0.668$$

$$\Rightarrow \quad \delta_E = 23°48'$$

[Note that the difference between μ_o and μ_E is much larger than the variations of μ with wavelength in most cases. For extra-dense flint glass difference of μ from violet to red light is less than 0.03].

Example 5:

Let a double image prism like be made of a material for which $\mu_0 = 1.662$, $\mu_E = 1.474$. Deduce the angular separation of the emergent beams.

Solution:

The angle of incidence at R is 45°. The angles of refraction are:

$$\sin^{-1}\left(\frac{1}{\sqrt{2}} \text{x} \frac{1.662}{1.474}\right) = 52°50';$$

$$\sin^{-1}\left(\frac{1}{\sqrt{2}} \text{x} \frac{1.474}{1.662}\right) \; 38° \; 50'$$

These rays fall on the opposite face at angles of incidence + 7°50' and -6°10' respectively. The angles of emergence are

$$\sin^{-1} (1.474 \sin 7°50') = 11°34'$$

$\sin^{-1}(1.662 \sin -6°10') = -10°12'$

The angle between the emergent beams therefore 21°46'.

Example 6:

Unpolarized light falls on two polarizing sheets placed one on top of the other. What must be the angle between the characteristic of the sheets if the intensity of the transmitted light is one third intensity of the incident beam?

Solution:

Intensity of the light transmitted through the first polarizer $I_1 = I_0/2$, where I_0 is the intensity of the incident unpolarized light.

Intensity of the light transmitted through the second polarizer is $I_2 = I_1 \cos^2\theta$ where θ is the angle between the characteristic direction of the polarizer sheets.

But $\quad I_2 = I_0/3$ (given)

$\therefore \quad I_2 = I_1 \cos^2\theta = I_0 \cos^2\theta/2 = I_0/3$

$\therefore \quad \cos^2\theta = 2/3$

$\therefore \quad \cos\theta = 0.8165$

$\therefore \quad \theta = 35.3°$

Example 7:

If the plane of vibration of the incident beam makes an angle of 30° with the optic axis, compare the intensities of extraordinary and ordinary light.

Solution:

Intensity of the extraordinary ray $I_e = E^2 \cos^2\theta$

Intensity of the ordinary ray $I_o = E^2 \sin^2\theta$

$$\frac{I_e}{I_o} = \frac{\cos^2\theta}{\sin^2\theta} = \frac{\cos^2 30°}{\sin^2 30°} = 3$$

Example 8:

A plane-polarized light is incident perpendicularly on a quartz plate cut with faces parallel to optic axis. Find the thickness of quartz plate, which introduces phase difference of 60° between e- and o-rays.

Solution:

The path difference between the waves is given by D = $(\mu_e - \mu_o)d$

The phase difference between the waves is given by d = $\frac{2\pi}{\lambda}\Delta$

$$D = \frac{60^\circ}{360^\circ}\cdot\lambda = \frac{\lambda}{6}$$

$$\therefore \qquad d = \frac{\lambda}{6(\mu_e - \mu_o)}$$

$$d = \frac{5400\text{Å}}{6(1.553 - 1.544)} = \frac{0.54}{0.054}\ \mu m = 10\mu m.$$

Example 9:

Calculate the thickness of double refracting plate capable of producing a path difference of λ/4 between extraordinary and ordinary waves. Given: $\lambda = 5890\text{Å}$,

$\mu_o = 1.53$, $\mu_e = 1.54$.

Solution:

$$d = \frac{\lambda}{4(\mu_e - \mu_o)} = \frac{5890\text{Å}}{4(1.54 - 1.53)}$$

$$= \frac{0.589}{0.04}\ \mu m = 14.7\mu m$$

Example 10:

Calculate the least thickness of a calcite plate which would convert plane polarized light into circularly polarized light. Given $\mu_o = 1.658$, $\mu_e = 1.486$ *and wavelength of light is 5890Å.*

Solution:

Plane polarized light gets converted into circularly polarized light by a suitably oriented quarter wave plate. Its thickness is given by

$$\therefore \qquad d = \frac{\lambda}{4(\mu_o - \mu_e)} = \frac{5890}{4(1.658 - 1.486)}$$

$$\text{or} \qquad d = 0.856\mu m.$$

Example 11:

For calcite, $\mu_o = 1.658$, $\mu_E = 1.486$ *for sodium yellow. Calculate the thickness of the thinnest QW plate of calcite for sodium yellow.*

Solution:

$$d = \frac{5893\times10^{-8}\,cm}{4(1.658-1.486)} \times \frac{5893}{4\times.172} x10^{-8} = 0.86 \times 10^{-4}cm$$

(Obviously such thin plate will have to be supported between glass plates).

Example 12:

A quarter-wave plate is meant for $\lambda_0 = 5.893 \times 10^{-5}$ cm. What phase retardation ϕ will it show for $\lambda = 4.358 \times 10^{-5}$ cm? (Neglect changes of μ_0 and μ_E with λ).

Solution:

Assuming that QW plate is of the smallest thickness,

$$|\mu_0 - \mu_E|\, d = \frac{1}{4} \times 5.893 \times 10^{-5}\ cm$$

Phase retardation for $\lambda = 4.358 \times 10^{-5}$ cm

$$\phi = \frac{2\pi}{4.358\times10^{-5}} \times \frac{5.829\times10^{-5}}{4} = 0.67\pi \text{ radian.}$$

(Thus a QW plate for yellow is *not* a QW for other colours).

Example 13:

Plane polarized light passes through a calcite plate with its optic axis parallel to the faces. Calculate the least thickness of the plate for which the emergent beam will be plane polarized. Given $\mu_o = 1.6584$, $\mu_e = 1.4864$ and wavelength of light is 5000Å.

Solution:

When a plane-polarized beam is incident on a half wave plate, the emergent beam will also be plane polarized but the plane of polarization undergoes a rotation through an angle 2θ.

$$d = \frac{\lambda}{2(\mu_o - \mu_e)} = \frac{5\times10^{-7}\,m}{2(1.6584-1.4864)} = 1.45\ \mu m$$

Example 14:

The rotation in the plane in a certain substance is 10°/cm. Calculate the difference between the refractive indices for right and left circularly polarized light in the substance. Given λ = 5893Å.

Solution:

$$d = \frac{2\pi}{\lambda}[\mu_R - \mu_L]d \quad \therefore \quad [\mu_R - \mu_L] = \frac{\delta\lambda}{2\pi d}$$

It is given that $\frac{\delta}{d} = 10^\circ = \frac{10\times 2\pi}{360^\circ} = \frac{\pi}{18}$ radian/cm

and $\lambda = 5893\text{Å} = 5893 \times 10^{-8}$ cm.

$$\mu_R - \mu_L = \frac{\pi}{18}\cdot\frac{5893\times 10^{-8}\,\text{cm}}{2\pi} = 1.6 \times 10^{-6}.$$

Example 15:

A 200 mm long tube containing 48 cm³ of sugar solution produces an optical rotation of 11° when placed in a saccharimeter. It the specific rotation of sugar solution is 66°, calculate the quantity of sugar contained in the tube in the form of a solution.

Solution:

It is given that $\theta = 11^\circ$,

$l = 200$ mm $= 20$ cm,

$S = 66^\circ$, and $V = 48$ cm^3.

$$C = \frac{10\theta}{lS} = \frac{10\times 11^\circ}{20\text{cm}\times 66^\circ} = 0.0833 \text{ g/cm}^3$$

Mass of sugar in solution

$M = CV = 0.0833$ g/cm^3 $\times$ 48 cm^3 = 4 grams.

Example 16:

20 gm of cane-sugar is dissolved in water to make 50 cc of solution. A 20 cm length of this solution causes + 53°30′ optical rotation. Calculate the α.

Solution:

$\theta = 53.5^\circ$, $c = 20/50 = 0.40$ gm/cc, $l = 2$ decimeters.

$$\therefore \alpha = \frac{\theta}{cl} = \frac{107}{2}\times\frac{1}{0.40} = 66.9\ \frac{\text{deg}}{\text{dm}}\cdot\left(\frac{\text{gm}}{\text{cc}}\right)^{-1}$$

Example 17:

If 20 cm length of a certain solution causes right-handed rotation 38°, and 30 and path of another solution causes left-handed rotation 24°, what optical rotation will be caused by 30 cm length of a mixture of the above solutions in volume ratio 1:2? (The solutions are not chemically reactive).

Solution:

The 30 cm length of the mixture may be treated as 10 cm of the first solution and 20 cm of the second (ratio 1 : 2).

$\therefore$ optical rotation by first is $38° \times 10/20 = 19°$ (+)

optical rotation by second is $24° \times 20/30 = 16°$ (–)

$\therefore$ Total rotation $= 19° - 16° = 3°$ (+)

Example 18:

Quartz has specific rotation 21.7° per mm for l = 5893Å. A plane polarized beam is passed through a dextro-rotatory quartz plate of thickness 1.92 mm. Deduce the optical rotation.

Solution:

$$\theta = \alpha l = +21.7\ \frac{\text{deg}}{\text{mm}} \times 1.92\ \text{mm}$$
$$= +41.7\ \text{deg.}$$

EXERCISES

1. A material which is optically active but is not double refracting is put in the path of :

   (i) an elliptically polarized light,

   (ii) a partially plane polarized light.

   Discuss the nature of the transmitted beam in each case.

2. An unpolarized beam of light is passed through a double image prism. Then one beam is passed through a quartz plate to rotate the plane of polarization by 90°, and this is then superposed with the other beam. Discuss if interference fringes would occur. Howwould things change if the starting beam itself were plane polarized?

3. Outline Fresnel's explanation of optical rotation. Discuss :

   (i) the dependence of rotation on λ,

   (ii) the experimental evidence in support of the assumptions.

4. In a Laurent half-shade device should the optic axis of the HW plate be parallel to the diameter? Could the portion covered by

the simple glass plate be left just vacant? Where should the telescope of the polarimeter be focused and why?

5. In measuring concentration with a polarimeter, a sample tube of length 20.0 ± 0.1 cm is found to show an optical rotation + 32.6 ± 0.1 ± with $\lambda = 5893$ A. Tne spednc rotation is known for this A, tobe + 58.2 ± 0.1° per dm per gm/cc, Deduce the range of error in the concentration as obtained from these measurements.

6. The settine of a Laurent half-shade in a polarimeter was such that the optic axis of half-shade made 10° angle with the principal plane of the polarizer. Deduce the transmitted intensity and 'contrast for $\Delta\theta = 0.1.°$' How would these change if the setting is changed to 5°?

7. If the Optic axis of a uniaxial crystal is taken as z-axis, write down the equation for the extraordinary wave-surface at unit time in terms of V_o and V_E. For extraordinary ray in xz plane at angle 30° from x-ad deduce the (i) ray velocity, and (ii) the corresponding wave velocity and direction of the wave-normal.

8. Using Huygens' construction, show that a beam incident along the normal on a plate of biaxial crystal would split into two beams with a lateral separation proportional to, thickness of the plate. What other factors determine this separation?

9. A 'polarizer' is that which would convert ordinary unpolarized light to plane polarized light. The ideal polarizer is that which would give 50% of the original light as plane polarized one. Discuss this 50% business.

10. Define 'ordinary' and 'extraordinary' refractive indices for an uniaxial anisotropic material. For the extraordinary beam the ray and wave normal generally do not coincide. Explain.

11. What do you understand by the term polarization of light?

12. Distinguish between polarised and unpolarized light.

13. Describe the process of production of plane polarized light by reflection.

14. Unpolarized light falls on two polarizing sheets so oriented that no light is transmitted through the combination. If a third polarizing sheet is placed between them, can light be transmitted? Explain.

15. Explain how a quarter wave plate functions to produce circularly and elliptically polarised light.
16. Explain the working of quarter wave plate.
17. What are the optical devices required to prodüce circularly polarized light from unpolarized light? Explain how they are used to produce circular and elliptical polarised light.
18. What is a half wave plate? Explain its action on polarized light incident on it with its electric vector E making an angle θ with the optic axis of the half wave plate.
19. Discuss the production and detection of circularly polarized light.
20. Explain the phenomenon of double refraction in calcite crystal.
21. What is double refraction?
22. Describe the construction of a Nicol prism and show how it can be used as a polarizer and analyser.
23. Describe in brief the phenomenon of birefringence. Discuss briefly Huygens' theory of double refraction.
24. Explain the terms:
    (i) Double refraction
    (ii) optic axis
    (iii) positive and negative crystals.
25. Explain the propagation of ordinary and extraordinary wave fronts in a calcite crystal for normal incidence with optic axis:
    (i) Parallel to the direction of propagation
    (ii) Normal to the direction of propagation
26. Using Huygens principle, construct refracted beams and wave fronts in calcite crystal when:
    (i) Incident ray is normal to crystal surface, optic axis parallel to the crystal surface and lies in the plane of incidence;
    (ii) Incident ray makes an angle with normal while optic axis is as above.
27. Calculate the thickness of :
    (i) a quarter wave plate, and
    (ii) a half wave plate.

Given that $\mu_e = 1.553$, $\mu_o = 1.544$
and $\lambda = 5000Å$.

28. Plane polarised light is incident on a piece of quartz cut parallel to the axis. Find the least thickness for which the o-ray and the e-ray combine to form plane polarised light.
Given that $\mu_e = 1.5533$, $\mu_o = 1.5442$
and $\lambda = 5 \times 10^{-5}$ cm.

29. Calculate the thickness of a double refracting plate capable of producing a path difference of $\lambda/4$ between e-and o-waves.

30. Determine the specific rotation of the given sample of sugar solution in the plane of polarisation is turned through 13.2°. The length of the tube containing 10% sugar solution is 20 cm.

31. Determine the path and wave fronts of ordinary and extra ordinary rays inside a double refracting crystal on the basis of Huygens' theory when the optic axis is inclined to the incident surface.

32. Show that (coherent) light waves represented by equations

$$E_1 = i\, E_x \cos(\omega t - kz) \text{ and}$$
$$E_2 = j\, E_y \cos(\omega t - kz + \theta)$$

give rise generally to an elliptically polarized wave that can become linearly and circularly polarised wave under special conditions.

33. What is a quarter wave plate? Deduce its thickness for a given λ in terms of its refractive indices.

34. Explain what is circular polarized light? How is it produced in laboratory with the help of a quarter wave plate?

35. Distinguish between polarised and unpolarized light. Discuss the process of production and detection of elliptically polarised light.

36. How can we experimentally distinguish between plane polarised, circularly polarised and elliptically polarised light?

37. What is Babinet's compensator? Explain how it can be used to analyse elliptically polarised light?

38. What is a Babinet's compensator? What are its advantages over a quarter wave plate?

39. What is meant by :
    (i) plane polarised light,
    (ii) circularly polarised light, and
    (iii) elliptically polarised light. How are they produced, and detected?
40. Discuss double image polarising prisms:
    (i) Rochon's prism and
    (ii) Wollaston's prism.
41. Give the construction and working of Laurent's half shade polarimeter.
42. Give the construction and theory of Lippich polarimeter. How will you determine the specific rotation of a given solution with its help?
43. Explain Fresnel's theory of rotation of the plane of polarisation. How would you increase the sensitiveness of a pair of crossed Nicol prisms?
44. Show that for incidence at Brewster's angle, the angle between thel dent and refracted rays is 90°.
45. Sunlight reflected from still water in a lake is observed throu) Polaroid. When the polaroid is rotated' in its own plane the inte changes. Explain.
46. Calculate the polarizing angle for an air-diamond interface when light is incident (i) from air, (ii) from diamond. (Given n of diamond = 1.8,
47. A biaxial crystal has wave-velocities V_o and V_E. The optic axis is designated as z-axis. For a wave with wave-normal atong the Jr-axis, deduce the velocities and directions ofE vectors for the two polarized waves possible.
48. State the laws of rotatory polarisation. Give Fresnel's hypothesis for rotatory polarisation and derive a formula for the rotation of quartz. Give the experimental verification of this formula.
49. A transparent plate is given. Using two Nicol prisms how would you find whether the given plate is a quarter wave plate, a half wave plate or a simple glass plate.

50. A black dot is marked on a white paper. It is then viewed through a calcite crystal from the top.
    (i) How many images are expected to be seen and why?
    (ii) Is it possible to see one image only? If the answer is yes, explain the reason for it.
51. On introducing a polarimeter tube 25 cm long and containing sugar solution of unknown strength, it is found that the plane-of polarisation is rotated through 10°. Find the strength of the sugar solution in g/cm^3. Given that the specific rotation of sugar solution is 60° per decimetre per unit concentration.
52. How will you orient the polarizer and analyser so that a beam of natural light is reduced to (i) 0.5 (ii) 0.25 (iii) 0.75 and (iv) 0.125 of its original intensity?
53. A solution of concentration 6.0 gm per 100 cm^3 used in a tube of length 33 cm causes 14° 30' rotation in the plane of polarisation of light of $\lambda = 5500$Å. Deduce the specific rotation. Also estimate the rotation it would cause for λ = 4500 Å.
54. State the reasons to conclude that optical rotation in liquids :
    (i) is not a surface effect,
    (ii) has molecular origin. Discuss the origin of optical activity in liquids and solutions.
55. 20 cm path of a solution containing 10 gm of material A per 100 cm^3 causes optical rotation + 25°, and 30 cm path of a solution containing 30 gm of material B per 100 cm^3 causes optical rotation –15.° Deduce the optical rotation due to 40 cm path of a solution containing 8 gm of A and 6 gm of B in 50 cc of the solution. State any assumption made.

2

CRYSTAL STRUCTURE AND DIFFRACTION BY CRYSTALS

INTRODUCTION

Almost all solids have regularity of atomic arrangements, *i.e.*, they are crystalline. (The few exceptions are called 'glasses', or glassy solids). In some cases this regularity extends upto only a few thousand atoms on each side, so that the regularity of arrangements is not seen in the external looks. In other cases, like canesugar or common salt, the regularity of atomic arrangement extends to $\sim 10^7$ or more atoms in each direction very frequently, so that individual crystals have dimensions of a few mm on each edge. Then the regularity of flat faces with definite angles becomes obvious.

Study of the crystal structure of different solids is itself a very important aspect of solid state chemistry and physics. However, presently our interest is to see how crystalline structure of solids is used as a diffraction

grating. The closest *ruled* gratings have a grating space e $\approx 10^{-4}$ cm = 10000 A, while X-rays have wavelengths of the order of a few angstroms. Bragg discovered that the regularity of atomic arrangements in crystals had a spacing of just the right order for determination of wavelengths of X-rays. We shall later see that even beams of electrons, neutrons, etc., show wave-like behaviour and diffraction from crystals can' be used to measure the wavelengths whiyh they show up. For a clearer understanding we shall include some details of crystal structure before we proceed to describe crystal spectrography.

SPACE LATTICES

A lattice is a framework of geometrical points in three dimensional Space described by the translation operation:

$$T = n_1 a + n_2 b + n3c \qquad ...(1)$$

where a, b, c are three *unit vectors*, and n_1, n_2, n_3 take all possible integral values. The parallellopiped formed by a,b,c as the concurrent edges is called the *unit cell* of the lattice. A unit cell repeated with translations T fills all space. Fig. 2.1 shows a two-dimensional space lattice, and Fig. 2.2 shows a three-dimensional unit cell.

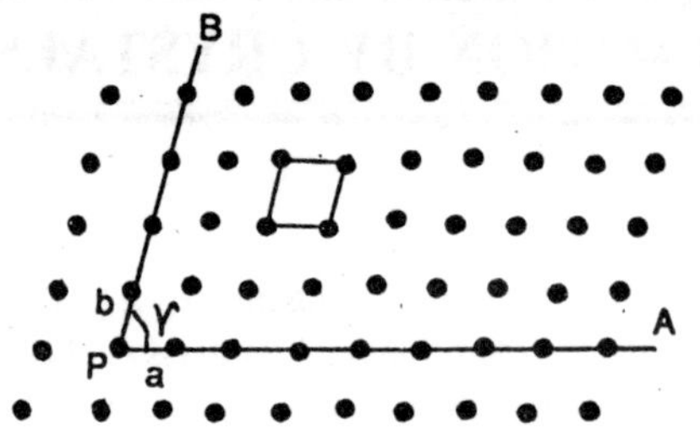

Fig. 2.1 : A two-dimensional space lattice.

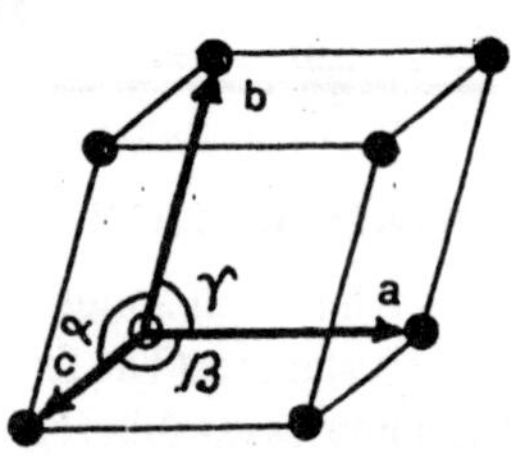

Fig. 2.2 : A unitcell three-dimensional lattice.

Seven Systems of Crystals

On the basis of the shape of the unit cell all crystals are classified into seven *systems*. The least symmetry is in triclinic crystals, and the largest symmetry is in cubic crystals. We shall limit our discussion largely to the *cubic system* alone. This is not only the simplest, but is also amongst the commonest in nature. Amongst the elements, well over half crystallise in the cubic system.

Haw is it that the geometrical variety of unit cells is limited to just seven? The answer is like this: With $a \neq b \neq c$, if two angles become equal, they have to be 90° otherwise no three dimensional repetitive arrays can be formed. Similarly with two axes equal ($a = b$), the only repetitive structures possible are when two angles are 90° or 120°. Finally with $a = b = c$ all three angles also have to be equal, otherwise 3-dimensional repetitive structure cannot be formed. (We omit proofs for these). It is for these reasons that the geometrical variety of unit cells is limited to just seven. Not all combinations a,b,c and α, β, γ are allowed.

Fourteen Bravais Space Lattices

Consider the three unit cells shown in Fig. 2.3. All three have a cubic shape. The first has lattice points at the corners only; the second has an extra lattice point at the centre of *each* face; the third has an extra point at the centre of its volume. These fall under *one system* (cubic system), but *three different space lattices.*

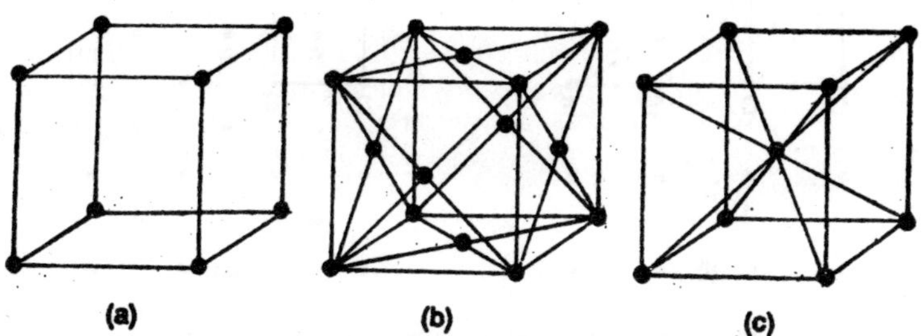

Fig. 2.3 : Unit cells in simple cubic (cubic P), face-centred cubic (cubic F) and body-centred cubic (cubic I) space lattices of Bravais.

Bravais showed that there could be in all 14 different lattices under the seven systems. When there is no extra lattice point in the unit cell besides the corners, the cell is called a *primitive cell* (P). When there are lattice points at the face-centres or at the body centre, symbols F and I respectively are used. Symmetry considerations decide which of these can be had in each system. For instance, in the cubic system" we have P, F and I lattices. In triclinic system only primitive cells (P) can occur.

Cubic P, cubic F, and cubic I lattices are also referred to as sc (simple cubic), *fcc* (face-centred cubic) and *bcc* (body-centred cubic) lattices respectively. One way of looking at the bcc lattice is to consider it as one *sc* lattice plus one more shifted by $\left(\frac{1}{2}, \frac{1}{2}, \frac{1}{2}\right)$. Similarly *fcc* lattice is equal to one *sc* lattice plus *three* more shifted by $\left(0, \frac{1}{2}, \frac{1}{2}\right)\left(\frac{1}{2}, 0, \frac{1}{2}\right)$ and $\left(\frac{1}{2}, \frac{1}{2}, 0\right)$. [All coordinates here are expressed in units of a].

Unit Cell and Primitive Cell

In Fig. 2.4 (i) we have a 2-dimensional lattice. We could choose ABCD or ABEC orABFE... as the unit cell, with appropriate translation (eq. 1) to fill all space. The choice? That which represents the *maximum symmetry* available is chosen as the unit cell. ABCD is thus the choice.

In Fig. 2.4 (ii) is another situation. The *smaller* unit is ABEF, but *smallest with maximum symmetry* is ABCD. In this case ABCD is chosen as the *unit cell.* ABEF is called *primitive cell.* A unit cell may have multiple lattice points per cell; a primitive cell has just one.

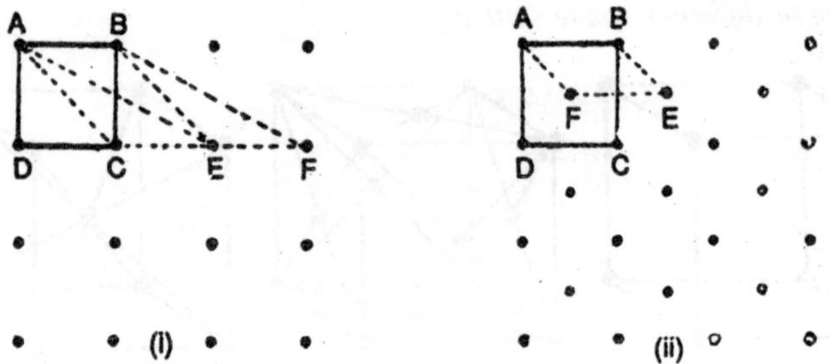

Fig. 2.4 : To explain the choice of (i) unit cell, (ii) primitive cell.

CRYSTAL STRUCTURE

The lattice is only any array of *geometrical* points or a framework in which atoms are to be arranged. In an actual crystal, each point has around it an assembly of atoms, which is called *the basis.* In Fig. 2.5 at (a) we show *the lattice*, at (b) we show *the basis* comprising two different atoms, and at (c) we show the crystal structure obtained by introducing the basis at each lattice point. Thus one may symbolically write:

LATTICE + BASIS = CRYSTAL STRUCTURE

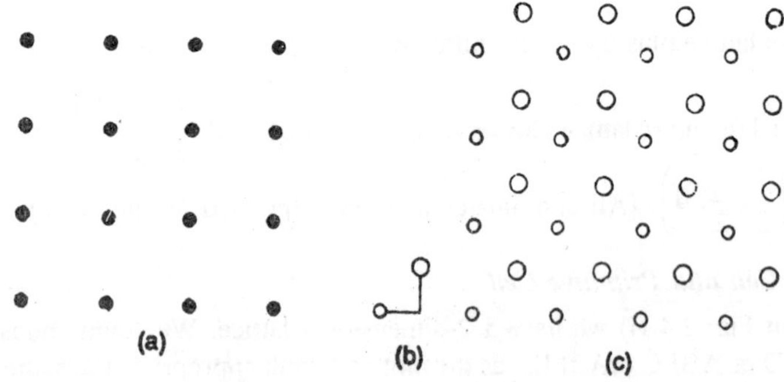

Fig. 2.5 : (a) Crystal lattice, (b) basis, (c) crystal structure.

In other words, the basis repeated according to a translation operation (eqn. 1) gives us the crystal structure. Note that for a compound containing several atoms in each molecule, (and also water of crystallization) the crystal *structure* will in general be quite complicated, but if attention is centred on equivalent atoms of one kind only, we get in each case the Bravais *lattice*. There are only 14 such *lattices*, though the number of crystal *structures* is unlimited

(i) CsCI Structure. The unit cell for this is shown in Fig. 2.6.

If the corner atoms are Cs, the body-centre has Cl, and vice-versa. Each unit cell contains 8/8=1 atom of one kind, and 1 atom of the other kind, giving in all 1 *molecule* per unit cell.

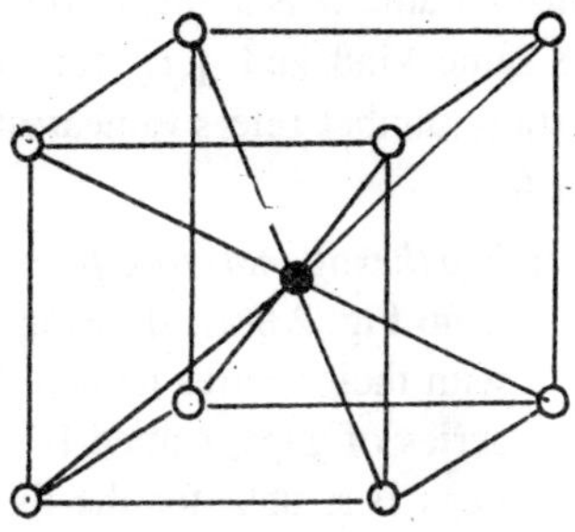

Fig. 2.6 : The CsCI structure; one unit cell is shown.

(ii) *NaCl Structure :* This has a face-centred cubic (fcc) lattice and the basis is $Na^+ - Cl^-$ with distance a/2 along (say) the x-axis. In Fig 2.7 if solid dots show Na^+ ions, the hollow circles show Cl^-, or vice-versa.

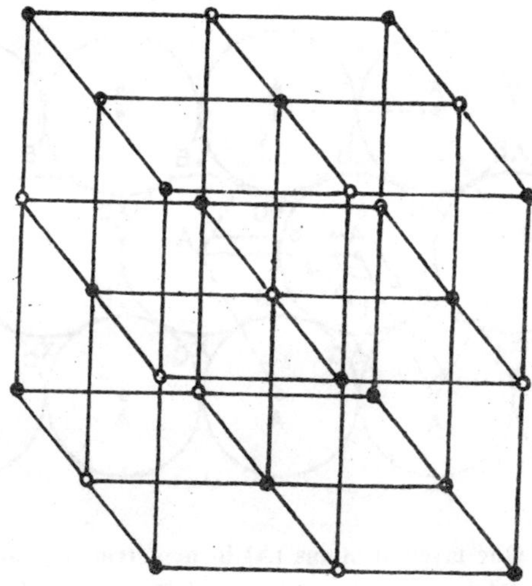

Fig. 2.7 : The Nad structure. One unit cell is shown.

It is to be emphasised that a crystal with CsCI structure looks like having a bcc lattice, but really has a simple cubic (cubic P) lattice. The Cl atoms alone are arranged in cubic P lattice. The Cl atoms alone are also arranged in cubic P lattice; and the two arrangements are relatively shifted by an amount $n_1(a/2)$ i + $n_2(a/2)$ j + $n_3(a/2)$ k.

If you ignore the Na^+ and Cl difference the Nad lattice looks like a simple cubic one. But one kind of atoms alone show the fcc lattice. If lattice parameter is a, the nearest neighbour distance is a VI for atoms of the same kind and $a\sqrt{2}$ for atoms of the opposite kind. The co-ordination number refers to nearest neighbours (here of the other kind) and is 6.

(iii) *hcp (hexagonal close-packed) Structure* : This may be understood from Fig. 2.8(a), showing one layer of spherical atoms packed, with their centres named A. For the second layer there are two series of gaps, named B's and Cs. If the second layer Gils the gaps B's, then for the third layer there will be gaps over the positions of A's and C's. Beyond that there are two choices. We may have successive layers with atom-centres arranged as ABCABCABC... or as ABABABAB... The first arrangement leads only to an fee lattice (we will not go to prove it). The arrangement ABABABAB... gives the hcp structure.

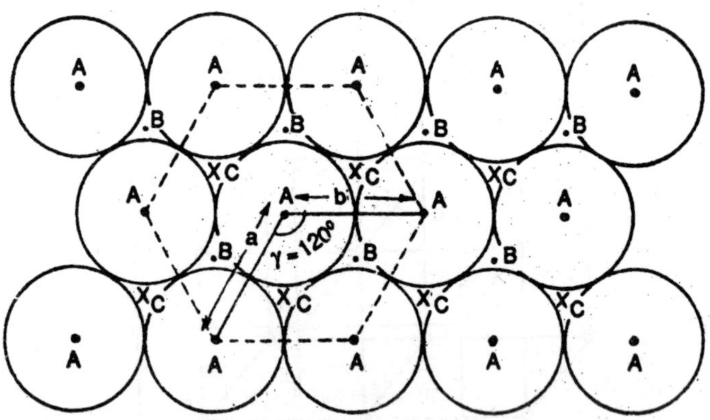

Fig 2.8(a) : One layer of atoms (A) in hcp structure, and possible locations of atoms in the next layer: B or C.

The name hexagonalin hcp comes from the hexagon base shown dotted in Fig. 2.8(a). But a parallelogram with base a = b and $\gamma = 120°$ is the smallest which describes the symmetry. With $\alpha = \beta = 90°$ the unit cell is shown in Fig. 2.8(b). The length c may be calculated from geometry and comes to $2a\sqrt{\frac{2}{3}} = 1.63d$. In an actual hcp crystal the ratio c/a may differ from 1.63 a due to nature of the bonds.

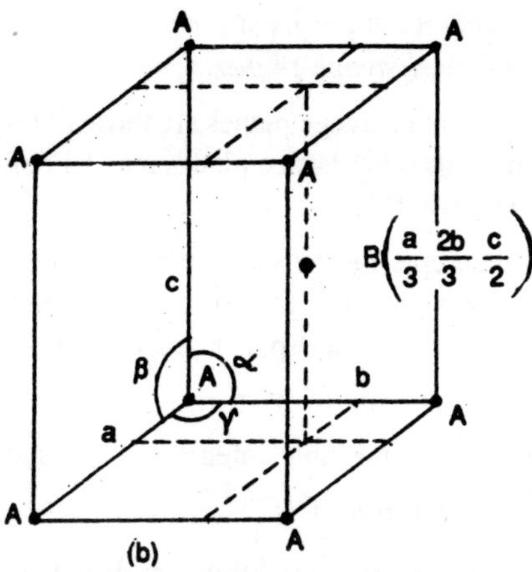

Fig. 2.8(b) : Unit cells of hcp structure a = b, c = 1.63 a, $\alpha = \beta = 90°$, $\gamma = 120°$... The dotted lines are to help locating B.

(iv) *ZnS Structure :* This is the same as hcp structure but with atoms of two different kinds (like Zn and S) in alternate layers. The unit cell thus has IZn + IS atom = 1 molecule per unit cell. Usually $c/a \neq 1.63$.

(v) *Diamond Structure :* This has fcc lattice with basis (000); $\left(\frac{1}{4}, \frac{1}{4}, \frac{1}{4}\right)$. One eighth part of/cc unit cell (shown in Fig. 2.7) is reproduced in Fig. 2.9, with an extra C atom at $\left(\frac{1}{4}, \frac{1}{4}, \frac{1}{4}\right)$. This is one-eighth of the unit cell of diamond structure. Note the tetragonal C-C bonds with bond length $a\sqrt{\frac{3}{4}}$ and bond angles 109°28'.

LATTIÇE PLANES AND MILLER INDICES

A crystal lattice may be considered as an aggregate of a set of parallel equidistant planes passing through the lattice point. These are called lattice planes. For a given lattice these sets of planes can be chosen in an umber of different ways; the spacing between the successive planes accordingly vanes, as also the density of lattice points per unit area in each plane.

A crystal can be easily split or *cleaved* along these lattice planes, particularly the planes of high density of lattice points. Hence lattice planes are also called *cleavage planes.*

The most important cleavage planes are those defined by (a, b) (b, c) and (c, a) But numerous other lattice planes may be defined. The method of specifying them is a follows:

(i) Let the intercepts by the given plane on the three crystal axes be in the ratio pa : qb : rc where a, b, c are the corresponding unit vector lengths, and p,q, r are integers.

(ii) Take the reciprocals of p, q, r.

(iii) Get the smallest possible integers h, k, *l* such that

$$h : k : l = p^{-1} : r^{-1} \qquad \text{...(2)}$$

Now these numbers h, k, *l* are known as *Miller indices* of the given see of planes, and we specify the plane as (h k *l*). In terms of co-ordinate geometry the family of (h k *l*) planes is described by:

$$\frac{hx}{a} + \frac{ky}{b} + \frac{lz}{c} = n \quad (n = \text{integer}) \qquad \text{...(3)}$$

For planes parallel to the xy plane we have p = q ∞ and r finite. Therefore, the Miller indices are (0 0 1). Similarly, the indices (0 1 0) and (1 0 0) represent lattice planes parallel to zx and yz planes respectively.

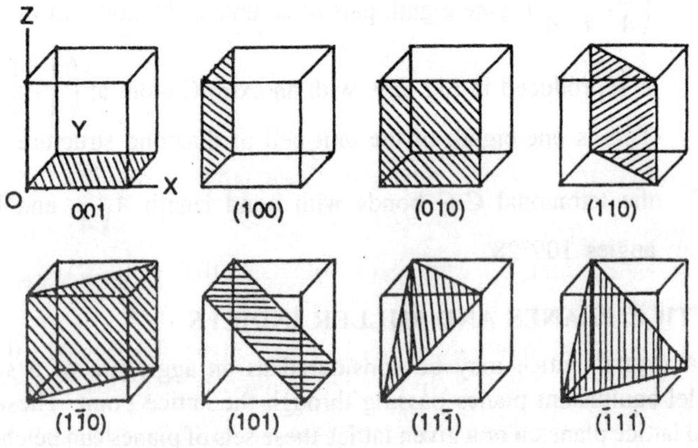

Fig. 2.9 : Some important lattice planes in cubic crystals.

Consider (1 1 0) plane. It is parallel to z-axis and intercept along x and y axes are equal in terms of the respective primitives. Similarly, for (21) plane the intercept along x-axis is 1/2 times that along they-axis. For (3 2 4) plane the x–, y–, z– intercepts are in the proportion $\frac{1}{3}, \frac{1}{2}, \frac{1}{4}$, which means 4:6:3.

If an intercept is negative the corresponding Miller index is written with after at the top, like $\bar{3}$. Thus a plane $(1\bar{3}0)$ is not the same as (130) But $(\bar{1}\bar{3}0)$ is the same as (1 3 0), because if all *Miller indices are reversed in sign we get only a-parallel plane*, as may be seen from Eqn. (3).

Lattice Planes in bcc and fee Lattices

In a simple cubic lattice the largest spacings d are given by the smallest values of $h^2 + k^2 + l^2$. Thus largest three spacings are as follows:

planes	spacing	ratio
{100}	a	1.00
{110}	$\frac{a}{\sqrt{2}}$	0.71
{111}	$\frac{a}{\sqrt{3}}$	0.58

In the fcc lattice additional lattice ; oints appear at the locations {a/2, all, a}. Now the {100} and {110} planes do not cover the face-centred points and we have to draw additional planes half-way through in each case.

However, the (111} planes take care of the additional points without introducing extra planes.

A look at Fig. 2.9, will make this dear. Mathematically the decision follows from eq. (3), which for cubic lattice becomes

$$hx + ky + lz = na \qquad ...(4)$$

The {a/2, a/2, 0} points do not satisfy this for the {100} and {110} planes; they do for the {111} planes.

Some authors specify the half-way planes by using Miller indices {200} and {220}, which is a modification of step (iii) to define h, k, *l*.

. This conveniently gives the spacings of Miller planes in fcc lattice.. The lowest three $h^2 + k^2 + l^2$ now correspond to the planes as listed below:

planes {111}	spacing $\frac{a}{\sqrt{3}}$	ratio 1.00
{200}	$\frac{a}{\sqrt{2}}$	0.87
{220}	$\frac{a}{2\sqrt{2}}$	0.61

For the bcc lattice, one may see similarly that additional planes arise half-way in the cases of {100} and {111} orientations, but not in the case of {110}. Hence the first two planes become {200} and {222} and hence planes {110} have the largest spacing. The three largest spacings are as follows:

planes {110}	spacing $\frac{a}{\sqrt{2}}$	ratio 1.00
{200}	$\frac{a}{2}$	0.71
{222}	$\frac{a}{2\sqrt{3}}$	0.41

DIFFRACTION OF X-RAYS BY CRYSTALS

When X-rays fall on a crystal, each electron of each atom scatters a part of the beam, and we have to consider the sum-total of the scattered radiation. This may be seen in three steps:

(i) The amplitude of the scattered wave resulting from Z different electrons of a *given atom* is not just Z-fold the amplitude of scattered wave per electron; it is AZ-fold, whereas is called the atomic structure factor. It depends on the distribution of electrons in the atom, A of the X-rays used and the direction of observation relative to the incident light.

(ii) The amplitude of the scattered wave resulting from the different atoms in a unit cell of the crystal is not just $\Sigma A_i Z_i$ summed over the atoms in the unit cell, but is $f\Sigma A_i Z_i$, where f is called the *crystal structure factor*. It depends on the relative placements of the-atoms in a unit cell, on λ of the X-rays used, and on the direction of observation.

(iii) The amplitude of the scattered wave from a crystal as a whole is then the resultant arising from the *lattice arrangement* of different unit cells. For some chosen directions the contributions from all unit cells addup in phase. These form the maxima, and their positions depend on the lattice parameters, the λ value, and direction of observation.

Laue followed the procedure of treating the unit cell as the scatterer and summing up over the crystal as a whole. In chosen directions alone are the scattered beams from the assembly of unit cell in *phase agreement*, and in these directions, the maxima occur as *Laue spots*. Their *positions* and *intensities* lead us to (he lattice arrangement and the structure details within each unit cell. In fact, even the charge density distribution within atoms is obtained.

But we will use a simpler (limited) procedure used by Bragg, who centered attention on *atomic planes*.

Bragg's Law : In Fig. 2.10 we show a set of parallel atomic planes of a crystal with spacing a. The wave-normal of the incident beam is OP at *glancing angle* θ (that is the practice in X-rays), and the direction of observation is A_1Q_1 at glancing angle θ' on the other side. The atoms (in the planes) are not shown, because the result is in dependent of their locations within a given plane.

Now, if $\theta' = 0$ corresponding to specular reflection (*i.e.* A_1Q_1 also lies in the plane of incidence) then it is simple to see that the scattered waves from all the atoms in a given plane reach the observer in the same phase, and therefore reinforce.

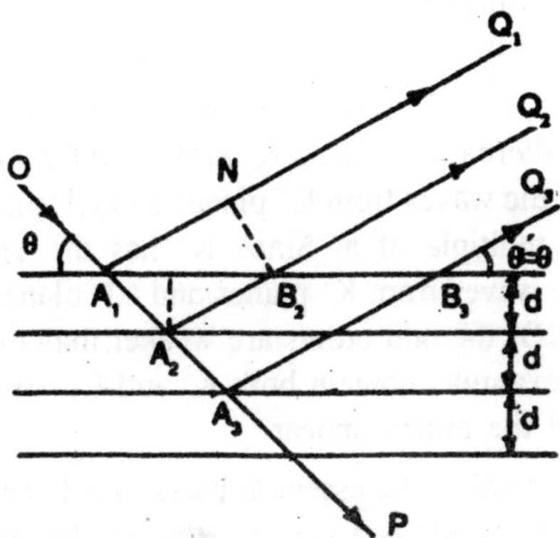

Fig. 2.10 : Brass's Law.

The total scattered waves from successive planes have a path difference

$$p = A_1A_2B_2 - A_1N$$

where N is the foot of perpendicular from B_2 on A_1Q_1. We have

$$p = \frac{2d}{\sin\theta} - 2d \cot\theta \cos\theta = \frac{2d}{\sin\theta}(1-\cos^2\theta) = 2d \sin\theta$$

For reinforcement this must be an integral multiple of λ. Hence

$$2d \sin\theta = n\lambda \qquad ...(5)$$

Thus specular reflection condition *plus* Bragg's law means that waves *from all unit cells in all the lattice planes* reach the observer in agreement of phase.

An ordinary mirror behaves with visible light exactly the same way—each atom of the mirror being the centre of scattering. But for visible light the penetration is not large and $\lambda >> d$. Hence Bragg's condition plays no role and light of all wavelengths shows a maximum at $\lambda' = \lambda$.

If the crystal belongs to a compound, then the planes in Fig. 2.10 may be treated as belonging to one kind of atoms. Eqn. (5) gives the maxima so far as contribution from one kind of atoms are concerned. The atomic planes for other kinds of atoms have a relative shift. We shall consider only a diatomic crystal. Just two kinds of situation arise: (i) for some (h k *l*) values the planes may cover both the kinds of atoms, and (ii) for other (h k *l*) values the planes containing the other kind of atoms lie half-way between those containing the first kind of atoms.

As one example, in KC*l* (whose structure is of NaCI type, Fig. 2.7) the {1 1 1} planes alternately contain K^+ alone and Cr^- alone. Therefore for odd values of n the waves from K^+ planes and Cl^- planes have phase difference an odd multiple of π. Since K^+ has the same number of electrons as CF the waves from K^+ planes and Cl^- planes totally cancel out for orders. In KBr the odd orders are weaker than even orders. But {1 0 0} plane for example, contain both K^+ and Cl^- (or Br^-) ions, and for these planes all the orders appear.

Bragg Spectrometer : The essential parts of a Bragg spectrometer are shown in Fig. 2.11. AB is a narrow straight channel between two brass blocks. This serves as a narrow 'slit' ($\perp$ to plane of paper) as well as 'collimator' for the incident X-rays. C is a single crystal, usually of NaCI with (100) face. I is an ionization chamber.

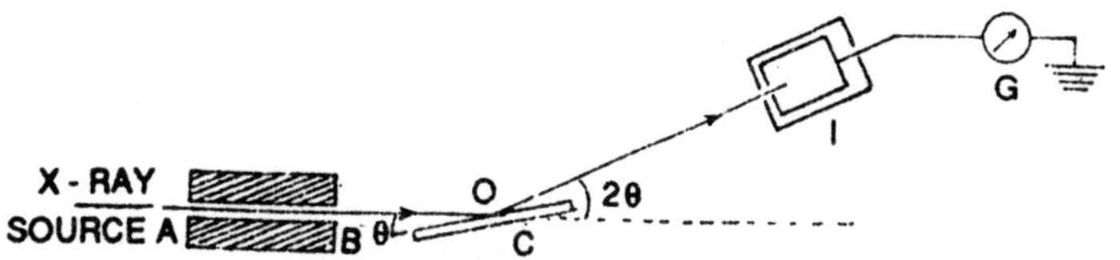

Fig. 2.11 : Schematic diagram of a Brag; Spectrometer.

The ionisation chamber and the crystal can be rotated about an axis at O, perpendicular to the plane of paper, in such a way that the angular relation θ and 2θ as shown in the diagram is always maintained.

As the angle 2θ is changed, the current through the galvanometer G shows some peaks, as represented schematically in Fig. 2.12. This figure shows two close lines in first and second order. From the observed 2θ, and known d_{100}, along with suitable n, we can calculate λ values for the two lines. The lines are actually superposed over a continuous spectrum of X-rays. From this the intensity distribution in the continuous spectrum can also be deduced.

It will thus be seen that crystals provide natural (three-dimensional) gratings for the analysis of an X-ray spectrum. The maximum wavelength for which a crystal can be thus used is obviously limited at Id.

Monochromator for X-rays : The characteristic X-rays from an X- ray tube have a fixed set of A values characteristic of the target in the tube. A "white" spectrum accompanies it.

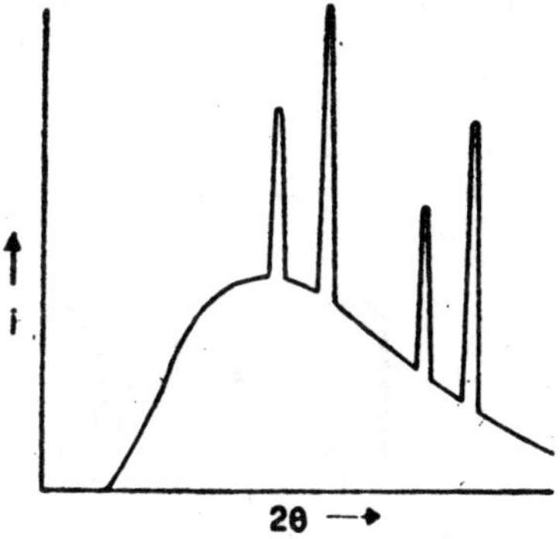

Fig. 2.12 : A typical Bragg spectrum.

Nowadays, there are also X-ray sources with only the white spectrum. How to get monochromatic X-rays from them?

Bragg's law provides the answer. Reflect the beam from a chosen crystal (h k *l*) plane at an adjustable grazing angle. Only X-rays of wavelengths given by

$$A = (2d \sin \theta)/n, (n = 1, 2, 3...)$$

are reflected, all others are scattered aside. Using another crystal for reflection, at a chosen angle, eliminates all but the one for which

$$A = (2d \sin \theta)/n = (2d' \sin \theta')/n' \quad ...(6)$$

where n' is also an integer.

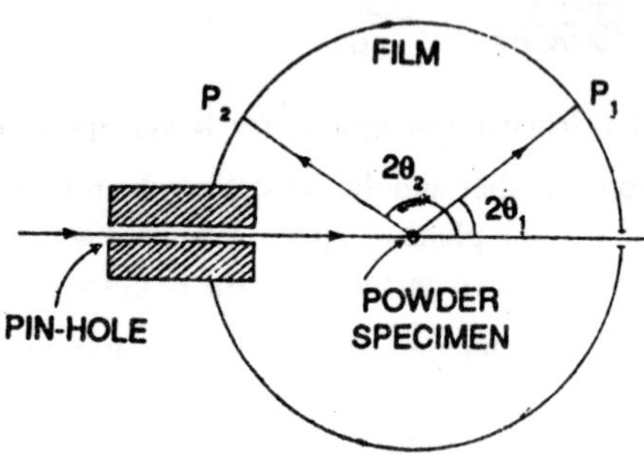

Fig. 2.13 : Powder camera for analysis of crystal structure.

The Powder Method. In this method (Fig. 2.13) a monochromatic X–ray beam is passed through a pin hole (not slit) in a block of brass, and allowed to fall on a *powder specimen.* A photographic film-strip is put along a circle concentric with the specimen and the pattern recorded is like the one shown typically in Fig. 2.14. (The central spot is masked off). Alternatively, one could put a plane film on the front side alone, which gives a pattern of concentric circles.

Fig. 2.14 : A powder pattern from X-rays on a cylindrical film strip.

A 'powder' is an assembly of a large number of microcrystals oriented in all possible directions. Let d_1, d_2... be the prominent lattice spacings tor the given specimen. Then for the single wavelength of incident X-rays chosen for the experiment there will be a few angles θ_1, θ_2... etc., for which Bragg equation (5) is satisfied with n = 1 or 2. From the aggregate of micro-crystals only those crystals which have these proper orientation give Bragg reflections, which occur at the appropriate 2θ angles from the incident beam. Thus there are cones of Bragg reflection with semi-angles $2\theta_1$, $2\theta_2$, $2\theta_3$, $2\theta_4$... The photographic film records the intercepts of these cones.

From the measured 2θ values, known λ value, and properly sorted n values, we deduce the various lattice spacings (*i.e.*, d *values*) of the given crystal. Thus *powder method is used for determining the various d-values for crystal of a given material, and hence in crystal structure analysis.* In contrast the Bragg method is used for determining X from known d (saydioo). Hence, it is used in X-ray spectroscopy.

The large angle beam (*i.e.*, P_2 in Fig. 2.13) is often called "back reflection." Its one advantage is in the accurate determinations of small changes in d, with (say) temperature.

The Laue Method : It uses a pencil of *"white"* X-rays, falling on a *single crystal* (Fig 2.15), one of whose symmetry axes is kept parallel to the pencil. On a photographic film kept normal to the incident beam at some distance we get several spots. These are called *Laue spots.* Let us see what they mean.

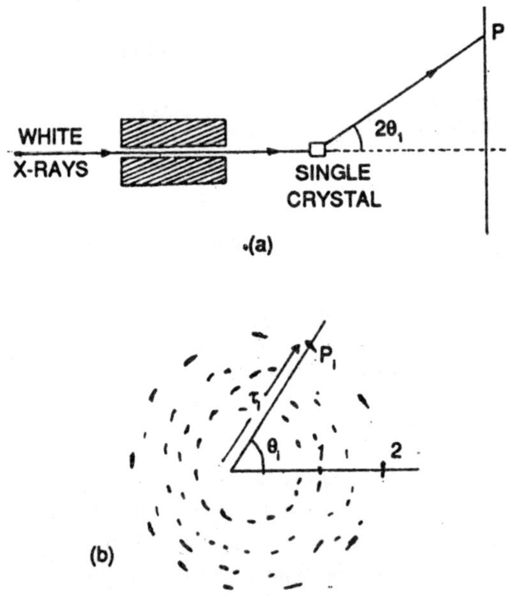

Fig. 2.15 : (a) Laue setting (b) Laue spots.

The crystal has its sets of lattice planes of spacings d_1, d_2, d_3... d_i... in fixed orientations with respect to the incident beam. None of these planes may have its normal in the plane of Fig. 2.15; hence no X-ray reflection may occur in this plane. But if we consider a plane rotated by angle ϕ about the beam axis from the figure plane, then at a discrete set of orientations ϕ the desired condition for X-ray reflection will be satisfied and we gel a spot. An example is P_i in Fig. 2.15(b) occurring at angle ϕ_i. Its radial distance r_i leads to deduction of θ_i, but not to d_i since the wavelength is not known.

However, the complete set ϕ_i and θ_i values for the system of Laue spots enables us to deduce the-set of d_i values of the crystal, and hence the structure. The relative intensities of the spots lead to further details. But we will not go to describe them here.

DIFFRACTION OF MATERIAL PARTICLES

In the year 1925 Davisson and Germer reported the results of an experiment on the scattering of a beam of electrons impinging on a single crystal

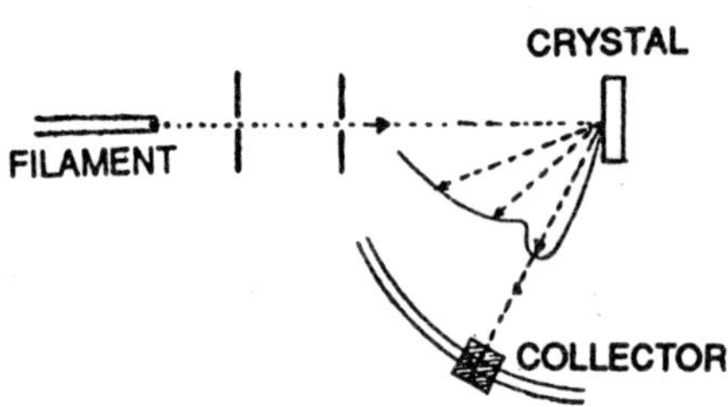

Fig. 2.16 : Davisson-Germer experiment. The electron currents scattered in various directions are shown by lengths of broken lines. The curved solid line represents the distribution.

of nickel (Fig. 22.18). In high vacuum, a beam of electrons is accelerated by a potential of ~50 volts. The beam, narrowed by two diaphragms, falls on the single crystal of nickel normal to the {1 1 1} face. Many of the electrons flow into the crystal, while others bounce back. The intensity of electrons bouncing in different directions is measured by a small collector, whose angular position can be varied. In the collector a retarding potential just short of the accelerating voltage is applied so that only those electrons which bounce with full initial energy are recorded.

The unexpected result of the experiment is that the scattered electrons show a pattern with a well-defined maximum and minimum, resembling a diffraction pattern. When the accelerating potential is varied, the maximum appears most distinctly at 54 volts and is at angle 50° from the incident beam of electrons (Fig. 2.17).

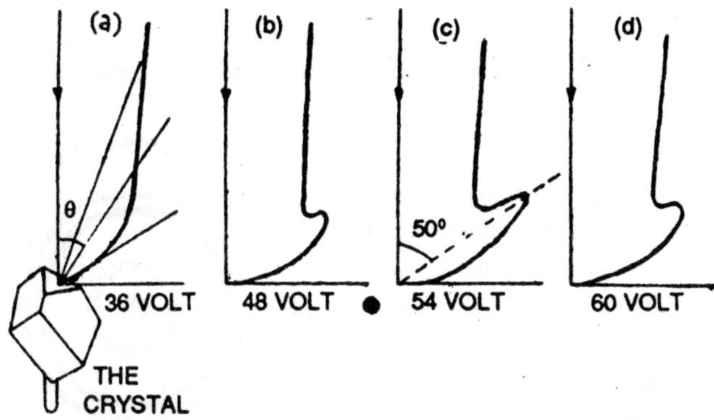

Fig. 2.17 : Electron-diffraction pattern for different accelerating voltages. (Radial distance is proportional to intensity in the concerned, direction).

To explain this, let us tentatively attribute a wavelength λ to the electron beam. As discussed for Bragg reflection, the crystal planes responsible for the maximum must be equally inclined with the directions of incidence and observation. This gives the grazing angle as $\theta = 90° -1/2(50°) = 65°$. The nickel crystal had a prominent lattice plane actually at this angle from the beam. Further, the spacing was d = 0.91 A, and using Bragg's equation, the wavelength would be (assuming first order).

$$A = 2 \times 0.91 \times 0.9063 = 1.65.4$$

In 1923 Louis de Broglie had stipulated theoretically that material particles of linear momentum? should show a wave-like behaviour with A given by

$$\lambda = \frac{h}{P} \quad \text{...(7)}$$

where h is Planck's constant. One finds that computation for the 54-volt electron gives λ = 1.67 A, in close agreement with the value attributed by Davisson and Germer's experiment.

G.P. Thomson in 1927 observed electron diffraction by *transmission* of a fast electron beam through a thin foil of gold. A narrow pencil of *electrons* accelerated through several thousand volts was sent through a gold film so thin that most of the electrons passed out. Now the 'film' is really composed of numerous very small crystals of gold, so that it acts like a 'powder' and with X-rays it produces a ring-shaped diffraction pattern. The remarkable thing is that it produced a very much similar pattern with the electron beam also (Fig. 2.18). The relative dimensions of the two patterns enable us to compute the experimental value of λ to be attributed to the electron beam; it is found to agree with the theoretical value.

Diffraction of Other 'Particles' : Diffraction experiment's with beams of protons, deuterons and other charged particles have been performed, and all of them establish the validity of the de Broglie relation (7). Due to larger masses in these cases one needs a smaller accelerating voltage to get a wavelength of the proper order.

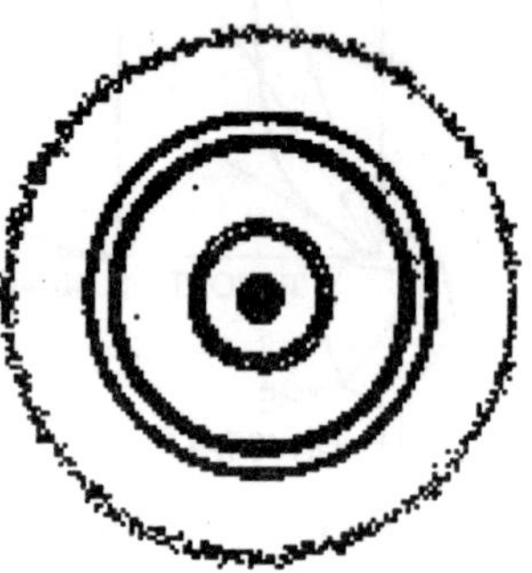

Fig. 2.18 : Transmission pattern of 48000-volt electrons through a silver foil 500 A thick.

Neutral particles like *neutrons* also show diffraction in the same manner. A velocity-selector sorts out neutrons of a narrow range of momentum P, and special detectors have to be used to observe them.

APPLICATIONS OF ELECTRON AND NEUTRON DIFFRACTION

The theoretical significance of wave-nature exhibited by material particles will be treated in the next chapter. About the use of electron beams and ion beams in microscopy we have spoken 15.8. They are now widely used to see the details of the order of a few angstroms—close to molecular sizes.

Here we will discuss the use of electron diffraction and neutron diffraction for studying the crystalline structure of materials. The procedure is the same as with X-rays, but fortunately these methods have complementary character. Let us see this.

X-ray interaction occurs through electromagnetic nature of light atoms on the Z electrons of the atom. This interaction is weak; therefore the X-rays penetrate deep, and X-ray diffraction gives information about the entire interior of a solid. The information is more about the higher Z atoms (like Pb, V, etc.) than about the lighter atoms (like H, C, O, etc.).

In contrast, interaction of electrons occurs through electrostatic force This being a much stronger interaction the beam does not peneyate deep This is not a disadvantage always, because if you want to study the structure of thin films, or surface films on a solid, or even the structure of liquid near the surface (which can be different from that in the interior), then electron diffraction is the best. Electrons also have a magnetic moment, and therefore its behaviour also depends on the magnetic character of the surface layers. That is a great advantage.

Neutrons have no charge and their penetration in the solid is very deep-even more than X-rays. But the interaction of neutrons does not depend on Z; it depends on the fact that it has a magnetic moment μ. So if there are isotopes with different μ, or neighbouring Z atoms with different μ then neutron scattering experiments will be able to distinguish between them. Another fact is that, for a given interaction strength, the momentum sharing is more in collisions with particles of comparable mass (like H, D, Be, He etc.), than with heavier or much lighter ones. For this reason neutron diffraction gives far more information about the nuclei of the lighter atoms in the crystal lattice than what X-ray. diffraction does.

Thus, for material in bulk X-ray diffraction gives best information for high Z atoms and ignores the effect of μ of the target atoms/ions. Neutron diffraction gives information dominantly depending on μ of the nuclei of target ions/atoms, and more so for the lightest atoms. Electron diffraction gives no information about the bulk material, but about thin films and surface layers, it is the only method. The complimentary features make a combination so useful.

SOLVED EXAMPLES

Example 1:

Alpha iron has a bcc lattice with lattice constant 2.86A°. Deduce the (i) co-ordination number, (ii) nearest neighbour distance, and (iii) number of atoms per unit cell.

Solution:

With any atom treated as reference atom, the nearest neighbours in a bcc lattice have positions

$$\left(\pm\frac{1}{2},\pm\frac{1}{2},\pm\frac{1}{2}\right)\text{a}$$

Thus the number of nearest neighbours is 2 × 2 × 2 = 8. This is the *co-ordination number*. The nearest neighbour distance is given by

$$d^2 = (a/2)^2 + (a/2)^2 + (a/2)^2 = a\sqrt{3/3}$$

In a unit cell in *bcc* lattice there are 8 atoms at the corners and 1 at the body centre. Since each corner atom is shared by 8 cells, there are (8/8) + 1 = 2 atoms per unit cell.

Example 2:

The density of sodium chloride is 2.18 g/cc. Given that the crystal has fcc space lattice deduce the lattice constant a. (Avogadro number = 6.02 × 10^{23} per g-mole).

Solution:

Volume of unit cell is a^3 cc and mass in each unit cell is therefore $a^3 \times 2.18$ g.

Now molecular weight of NaCl is 58.5, and in *fcc* lattice there are 4 molecules per unit cell. Hence mass in each unit cells is

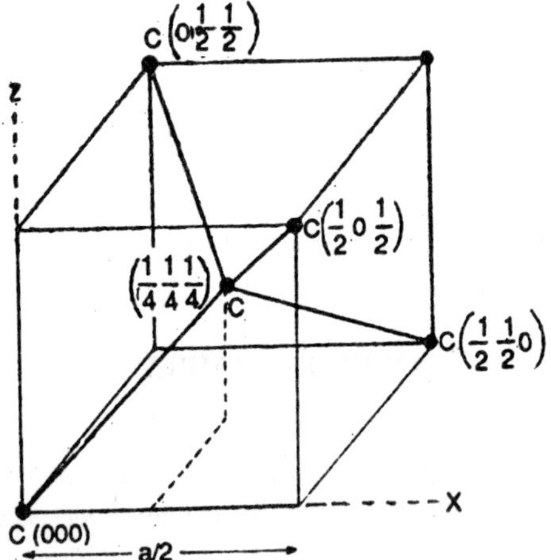

Fig. 2.19 : Diamond structure. One-eighth of a unit cell is shown.

$$(4 \times 58.5), (6.02 \times 10^{23})\text{g}$$

Equating the two expressions.

$$a^3 \times 2.18 = \frac{4 \times 58.5}{6.02 \times 10^{23}} \Rightarrow a = 5.63 \times 10^{-8} \text{ cm.}$$

Example 3:

Packing fraction in crystals is defined as the ratio of volume of the atoms in a unit cell and the volume of unit cell. Compute it for (i) sc, (ii) fcc, and (iii) hcp structure of elemental solids, treating the atoms as spherical.

Solution:

If atomic diameter is d, the volumes occupied by the atoms in a unit cell in the three cases are:

(i) $[\pi d^3/6] \times 1$ (ii) $[\pi d^3/6] \times 4$ (iii) $[\pi d^3/6] \times 2$

Unit cell parameters for the cases are:

(i) $a = d$,

(ii) $a = d\sqrt{2}$,

(iii) $a = b = d$, $c = 2d\sqrt{2/3}$, $\lambda = 120°$

Hence the unit cell volumes are:

(i) d^3, (ii) $2\sqrt{2}\,d^3$, (iii) $\sqrt{2}\,d^3$

The required packing fractions therefore are

(i) $\pi/6$, (ii) $\pi/3\sqrt{2}$, (iii) $\pi/3\sqrt{2}$

[Note that fcc has the same packing fraction has hcp. Only the arrangements are ABCABC ... in fcc and ABABAB ... in hcp as stated before].

Example 4:

Calculate the glancing angle on the cube face {100} of a rocksalt crystal (a = 2.814A) corresponding to second order reflection for X-rays of λ = 0.710A.

Solution:

For {100} planes, the spacing d is equal to a.

$$2 \times 2.8114 \times 10^{-8} \sin\theta = 2 \times 0.710 \times 10^{-8}$$

$$\theta = \sin^{-1}\left[\frac{0.710}{2.814}\right] = 14°22'.$$

Example 5:

In a Bragg spectrometer a photographic film is set perpendicular to the incident beam and at a distance 25.0 cm from the axis of the NaCI crystal. A line is observed at a distance 8.472 cm from the direct beam. Treating this as a second order reflections from {110} face, deduce the wavelength. [a = 5.63A].

Solution:

$$\tan 2\theta = \frac{8.472}{25.0},$$

$$\therefore \theta = 9°22' \text{ (check it)}$$

Now, for {110} reflections, $d = a\sqrt{2}$

$$\therefore \lambda = \frac{2d\sin\theta}{n} = \frac{2\times 5.63\times \sin 9°22'}{\sqrt{2}\times 2} = 0.651\text{A}.$$

Example 6:

In a triclinic crystal[3] *a cleavage plane makes intercepts 2.93, 4.47 and 2.35 units along the three crystallographic axes, the corresponding primitives being 3.05, 6.99 and 4.90 A. Deduce the Miller indices of the cleavage plane.*

Solution:

$$p:q:r = \frac{2.93}{3.05}:\frac{4.47}{6.99}:\frac{2.35}{4.90}$$

$$= .961:\ .640:\ .480 = 6:4:3$$

$$\therefore p^{-1} : q^{-1} : r^{-1} = \frac{1}{6}:\frac{1}{4}:\frac{1}{3} = 2:3:4$$

Hence the Miller indices of the plane are (234).

Example 7:

In a tetragonal lattice a = b = 2.42A, c = 1.74A. Deduce the lattice spacings between (1 0 1) planes.

Solution:

We have

$$d_{101} = \left(\frac{1^2}{(2.42)} + \frac{0^2}{(2.42)^2} + \frac{1^2}{(1.74)^2}\right)^{-1/2}$$

= 1.41A (check it).

Example 8:

Show that the (1 1 0) planes do not cover all the face centre points in a fcc lattice.

Solution:

Face centered points are represented by {ra/2, sa/2, ta} where r and s are odd integers and t is any integer. The curly bracket here means

$$\left(\frac{ra}{2}, \frac{sa}{2}, ta\right), \left(ta, \frac{ra}{2}, \frac{sa}{2}\right)$$

and $$\left(\frac{sa}{2}, ta, \frac{ra}{2}\right)$$

(1 1 0) planes refer to h = 1, k = 1, l = 0. Substitution for the three sets of face-centered points,

$$\frac{ra}{2}+\frac{sa}{2} = na, \; ta + \frac{ra}{2} = na,$$

and $$\frac{sa}{2} + ta = na$$

Only the first of these holds true (since r and s ae odd integers), which mans that the other sets of the face-centered points are *not* covered by the (1 1 0) planes.

Example 9:

X-rays of λ = 0.741A are incident on a powder specimen. A plane film placed beyond the powder at distance 20.0 cm shows the smallest concentric rings of radii 7.22, 10.74, 17.06 and 32.50 cm. Interpret the observations regarding crystal structure.

Solution:

We deduce in succession tan 2θ, θ and sin θ for the four cases

$$\sin\theta = 0.1723,\ 0.2439,\ 0.3458,\ 0.4879$$

We note: that the last two value fo sin θ are very nearly two times the first two values. Hence it is *reasonable* to take the first two rings as the first order (n = 1) images and the last two as the second – order (n = 2) images.

From Bragg equation the two d values are related by

$$\frac{d_1}{d_2} = \frac{\sin\theta_2}{\sin\theta_1} = \frac{.2439}{.1723} = 1.416$$

This is close to $\sqrt{2}$ and we therefore conclude that $d_1 = d_2\sqrt{2}$.

In the cubic system the crystal could have sc or bcc; arrangement, but not fcc.

From the given λ, we get

$$d_1 = \frac{0.741}{2\times.1723} = 2.15A;$$

$$d_2 = \frac{7.31}{2\times.2439} = 1.52A.$$

Example 10:

The smallest angle for strong reflection of a beam of neutrons with a family of crystallographic planes of spacing 3.84A is 30°. Calculate the wavelength of the neutron beam and also the speed of neutrons.

Solution:

From the Bragg equation $2d \sin\theta = n\lambda$.

$$2 \times 3.84 \times 10^{-8} \times 0.500 = 1\lambda$$

We put n = 1 since the angle is the smallest,

$$\therefore \quad \lambda = 3.84 \times 10^{-8} \text{ cm}$$

Now, by de Broglie relation

$$\lambda = \frac{h}{p} = \frac{h}{mv}$$

$$\therefore \quad v = \frac{h}{\lambda m} = \frac{6.63\times10^{-17}}{3.84\times10^{-8}\times1.67\times10^{-24}}$$

$$= 1.03 \times 10^5 \text{ cm/sec.}$$

Example 11:

Deduce the primitive vectors and primitive cell volume for a bcc lattice.

Solution:

In Fig. 2.20 the dotted lines show that cubic unit cell, with the body centre at O. The vectors of the primitive cell are lines from O to coroners

A, B, C, (chose *any* one corner of the cube and go over across face-diagonals to get the other two corners; all choices are equivalent).

The co-ordinates required are

$$O\left(\frac{1}{2}, \frac{1}{2}, \frac{1}{2}\right) \; A\ (1, 0, 0) \; B\ (1, 1, 1) \; C\ (0, 0, 1)$$

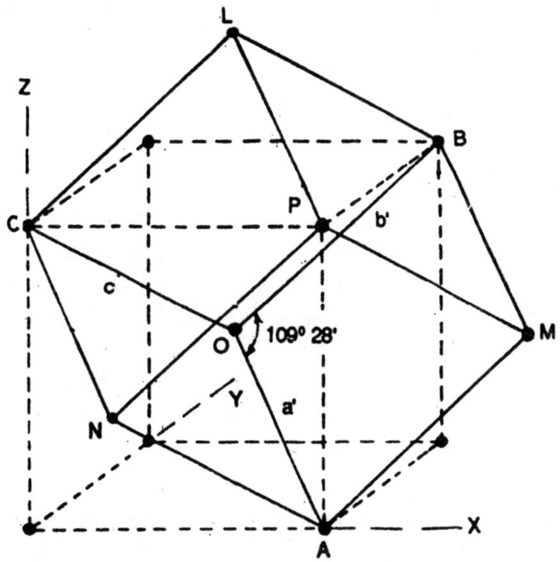

Fig. 2.20 : Primitive cell the bcc lattice.

So the vectors are (in units of a)

$$OA = a' = \left(\frac{1}{2}i - \frac{1}{2}j - \frac{1}{2}k\right),$$

$$OB = b' = \left(\frac{1}{2}i + \frac{1}{2}j + \frac{1}{2}k\right)$$

$$OC = c' = \left(-\frac{1}{2}i - \frac{1}{2}j + \frac{1}{2}k\right)$$

These lead to $a' = b' = c' = \frac{\sqrt{3}}{2} a.$

$$\cos\lambda = \frac{a', b'}{a'b'} = \frac{\frac{1}{4} - \frac{1}{4} + \frac{1}{4}}{3/4} = -\frac{1}{3} \Rightarrow l = 109°28'$$

$$\text{volume} = c'.a' \times b' = \frac{1}{2}a^3 \qquad \text{(check it).}$$

EXERCISES

1. CsCI has density 3.98 g/cc and molecular weight 90.5. It has a cubic P lattice with Cs^+ at $\left[\frac{1}{2}\frac{1}{2}\frac{1}{2}\right]$ positions relative to Cl^-. Compute :

   (i) the lattice constant a,

   (ii) the co-ordination number,

   (iii) the nearest neighbour distance.

2. If a,b,c arc the primitives along orthogonal axes, show that the spacing between consecutive lattice planes (h k l) is given by

$$d_{hkl} = \left[\frac{h^2}{a^2} + \frac{k^2}{b^2} + \frac{l^2}{c^2}\right]^{-1/2}$$

3. NaCl has fee lattice with a = 5.63Å. Deduce the spacings of {100} and {110} planes. Also compute the dimensions and shape of the primitive cell.

4. KCl has the same crystal structure as NaCl and both are ionic. But for reflections by {111} planes the odd order spectra are absent in KCl while they are only weak in NaCl. Discuss this. The atomic numbers of K, Na and Cl are 19,11 and 17 respectively.

5. In reflection of ordinary light the only condition required is $\theta' = \theta$, whereas in Bragg reflection an additional condition required is $2d \sin \theta = n\lambda$. Discuss this and also other differences between reflection of light from a surface and of X-rays from a lattice.

6. In a Bragg spectrometer maxima are observed at angles of deviation 11°56', 24°02', and 36°24' from the direct beam. The incident X-rays are monochromatic and of λ = 0.612A. Deduce the conclusions about these maxima.

7. The powder of a material which has cubic P lattice is examined with monochromatic X-rays, using a cylindrical film to record spectra all around the specimen. If λ of X-rays is 0.812-4 and a = 1.920A, compute the angular positions of the spectra of all the orders due to the lattice planes {100},{111}, and {211}.

8. Define atomic structure factor. Explain why its consideration is significant when A is of the order of atomic dimensions (or smaller), but insignificant for large A.
9. Define crystal structure factor. Show that for CsCl the odd order maxima from (100) planes will be weak on this account.
10. Which of the three beams –X-rays, neutrons, electrons—is best suited to study the :
    (i) structure of surface layers,
    (ii) positions of the lighter atoms,
    (iii) positions of heavier atoms, and
    (iv) positions of magnetic nuclei. Explain the reason in each case.
11. A crystal involving equal number of atoms A and B has CsCI structure in the ordered system but in the disordered system A and B occupy the sites randomly. A and B have nearly equal atomic numbers, but have different magnetic moments. By what experiment will you decide the extent of ordering? Explain.
12. Gold has atomic weight 197 and density 193 gm/cc. Deduce an order-of-magnitude value of spacing between the atoms in solid gold. (Given N = 6.0×10^{23} per gm mole).
13. Distinguish between cubic I and fee lattices. Compute the number of atoms of one kind in a unit cell in each case.
14. In a CsCI structure the positions of Cs atoms are main + $n_2$oj + $n_3$nk. Write down the positions of the Cl atoms, and the translation vector from any Cs atom to any Cl atom, or vice versa.
15. Show that for the primitive cell of fee lattice the edges are $a\sqrt{2}$ each with angles 60° between the concurrent edges. Also deduce the volume per primitive cell?
16. Define packing fraction' and show that its value is $\pi\sqrt{\frac{2}{16}}$ for the diamond structure and $\pi\sqrt{\frac{3}{8}}$ for an elemental solid in bcc lattice.
17. The X-ray powder pattern of a simple solid of the cubic system is obtained on a plane film 20.0 cm away from the specimen.

The radii of the first few rings are found to be 6.95, 10.28, 16.20, 22.2, and 29.7 cm. Deduce the Bragg angles and find whether the lattice is sc, bcc or fee.

18. Show that the de Broglie wavelength of On electron accelerated to V volt is given by

$$\lambda = \frac{12.27}{V^{1/2}} \text{ A}^{\circ}$$

if Vis not too large. By what factor would the number 12.27 change if the particle were a proton?

19. Deduce the de Broglie wavelength corresponding to :

(i) 54-volt electron,

(ii) 5400-volt electron,

(iii) 0.05 eV neutron.

21. Show that neutrons of speed 1.635×10^2 cm/sec have de Broglie wavelength 0.600×10^8 cm. Deduce the kinetic energy of these neutrons in eV unit.

3

MECHANISM OF LIGHT EMISSION

QUANTUM OPTICS

Einstein postulated that electromagnetic radiation is made up of photons indicating thereby that the electromagnetic field itself is quantized. Photons are the basic and discrete units of energy. Each photon has an energy E = hv and momentum $p = \frac{hv}{c} = \left(\frac{h}{2\pi}\right)k$. Photons are stable, electrically neutral and massless elementary particles. They always travel with the speed of light 'c' and exist only at that speed.

A photon has the mass $m = \frac{hv}{c^2}$ which is the mass of the electromagnetic field and is not associated with the rest mass because photons at rest do not exist. Photons always travel with the speed of light whether in vacuum or in a medium. Note that the velocity of photons in a medium is different from the velocity of the propagation of the wave front of light in the medium. Photons are emitted when atoms make a transit from a higher energy state to a lower energy state. They are also emitted upon the acceleration or retardation of charged particles, in the decay of small particles, and in the annihilation process of electron-positron pairs.

Photons have spin angular momentum of 1η and obey Bose-Einstein statistics. Therefore, they are called *bosons*. Bosons do not obey Rauli exclusion principle and therefore any number of photons can occupy the same state. When a very large number of photons occupy the same state, the inherent discreteness of the light beam disappears and the beam appears to be a bunch of continuous electromagnetic waves. A monochromatic plane wave may be regarded as a stream of photons having high population density, all photons occupying the same state.

There is an intimate relationship between the wave and corpuscular nature of photons. Photons exhibit interference and diffraction phenomenon which are the result of their redistribution in space. The intensity of light reaching any point of space is a measure of the number of photons striking this point. The quantum properties of light are due to the fact that the energy, momentum and mass of electromagnetic radiation is concentrated in photons.

Quantum optics deals with the discrete nature of the emission, propagation and interaction of light with a substance. It is chiefly concerned with the generation of coherent light and detection of light. Lasers, light emitting diodes (LEDs) and various types of photo-detectors are quantum optic devices.

INTRODUCTION

Radiation comes in a broad range of frequencies grouped into radiowaves, microwaves, infrared rays, visible rays, ultraviolet rays, x-rays and γ-rays. They are all essentially electromagnetic waves and therefore, the mechanism of their emission should be the same. In all cases electromagnetic waves originate in the oscillations of electric charge. The simplest device, which produces electromagnetic waves, is an *oscillating dipole*. A dipole consists of two opposite charges. When these opposite charges vibrate to and fro along a straight line, they constitute an oscillating dipole. In terms of the simple Bohr's theory light emission from atoms is explained as due to electron's transition from an excited state to the ground state. According to quantum-mechanical considerations, the electric dipole moment of the atom is the source of light and one can calculate the probability of electron making a transition. We may visualize that during the act of light emission the orbital electron makes its downward transition through a gradually damped oscillatory motion occurring at the specific resonance frequency. In this chapter, we briefly describe some of the important developments that led to the understanding of atomic structure and mechanism of light emission.

OSCILLATING ELECTRIC DIPOLE

Let us consider an oscillating dipole which consists of equal positive and negative charges +q and –q located at A and B respectively (Fig. 3.la). The charges vibrate along the line AB with simple harmonic motion, the charge of one sign being out of phase with that of the other by 180°. Let us further assume that the oscillations occur at constant amplitude and there is no damping.

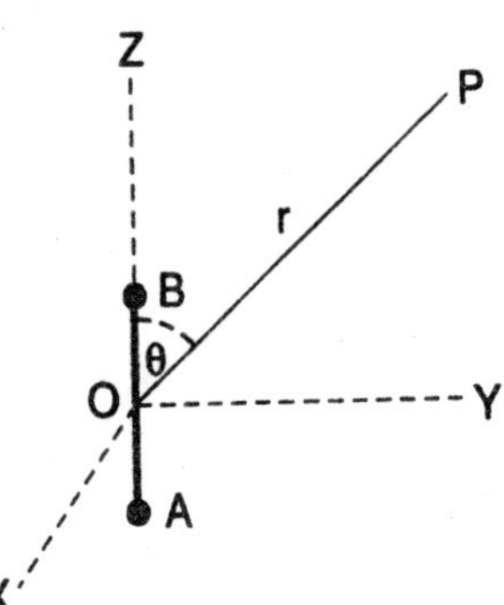

Fig. 3.1(a) : Oscillating dipole

If A_1 is the amplitude of the vibration of the charge + q, the displacement of the charge at any instant is

$$z_1 = A_1 \cos \omega t$$

The displacement of the negative charge vibrating with the same frequency but out of phase by 180° with the first is given by

$$z_2 = A_2 \cos (\omega t + \pi) = -A_2 \cos \omega t$$

where A_2 is the amplitude of vibration of the second charge. At any instant the distance between the charges is

$$z_2 - z_2 = (A_1 + A_2) \cos \omega t$$

The electric moment of the dipole is

$$q (z_1 - z_2) = p_o \cos \omega t \text{ where } p_o = q (A_1 + A_2)$$

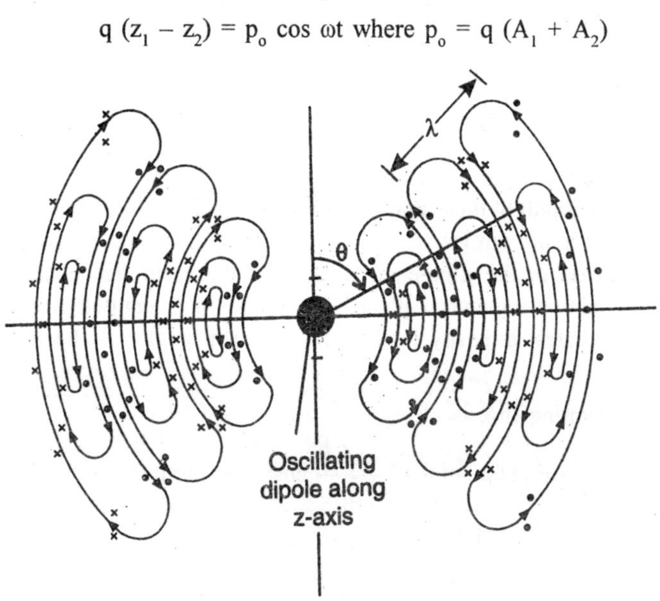

Fig. 3.1(b) : Representation of the electric field and the magnetic field in a plane containing an oscillating electric dipole.

According to Maxwell's equations, the electric intensity at any point around the dipole is made up of two components. Close to the dipole the field at any instant may be treated as if static. At a distance r from the centre of the dipole large compared with AB

$$E_r = \frac{2p_o \cos\theta}{r^3}\cos\omega t$$

and

$$E_\theta = \frac{p_o \sin\theta}{r^3}\cos\omega t.$$

The magnetic field near the dipole is

$$H = \frac{\omega p_o \sin\theta}{cr^2}\sin\omega t$$

at right angles to the rz-plane. The lines of magnetic force are therefore circles about the z-axis.

At a great distance from the dipole E_θ. has a negligible magnitude compared with –g, and the electric field intensity is given by

$$E_\theta = \frac{4\pi^2 p_o \sin\theta}{\lambda^2 r}\cos(\omega t - kr) \qquad ...(1)$$

and the magnetic intensity is given by

$$H_\Phi = \frac{4\pi^2 p_o \sin\theta}{\lambda^2 r}\cos(\omega t - kr) \qquad ...(2)$$

in a direction at right angles to the rz-plane, *i.e.*, in the direction of increasing azimuth Φ measured about the z-axis. Therefore, the lines of magnetic force are circles around the z-axis. Both E_θ and H_Φ are zero along the axis of the dipole on account of the factor sin θ and are maximum in the equatorial plane. The radiation is emitted from the dipole in greatest intensity in directions at right angles to the line of oscillation.

The Pointing flux is

$$S = \frac{c}{4\pi} E_\theta H_\Phi \frac{4\pi^2 cp_o^2 \sin^2\theta}{\lambda^2 r^2}\cos^2(\omega t - kr) \qquad ...(3)$$

in the direction of radius vector.

The energy radiated per unit time is given by

$$I = \int_0^\pi S.2\pi r^2 \sin\theta \, d\theta = \frac{32\pi r^4 cp_o^2}{3\lambda^2}\cos^2(\omega t - kr)$$

The mean rate of radiation is

$$I_{ave} = \frac{16\pi^4 c\, p_o^2}{3\lambda^4} = \frac{16\pi^4\, v^4\, p_o^2}{3c^3} \quad ...(4)$$

THERMAL RADIATION

In majority of the cases, light is produced as a consequence of their temperature. Therefore, they are said to be thermal sources and the radiation emitted by the source by virtue of its temperature is called *thermal radiation.* Thermal radiation is electromagnetic in nature and its energy is smoothly distributed over all wavelengths. Therefore, a thermal source produces *continuous spectrum*. The intensity and the predominant wavelength of radiation vary with the temperature of the body. At low temperatures the radiation mainly lies in the infrared region. As the temperature of the body is increased, the component of maximum intensity shifts to a higher and higher frequency. For example, the filament of an incandescent bulb appears dark at room temperature and as current is passed, it gets heated up. As the current increases through the filament, it appears initially red, orange gradually and then yellow and finally it emits white light. At temperatures above 1000°C, a heated body is capable of giving out energy in the form of waves of all possible wavelengths. Normally the amount of radiation emitted by a hot body depends on factors such as the properties of its surface.

Kirchhoff's Law

Bodies radiate energy either in the invisible form or in the visible form. Similarly, they absorb light to varying degrees. A body that is, a good light-absorber appears black. In 1859, Gustav Kirchhoff formulated the following law of radiation.

If a body is in thermal equilibrium, the energy it absorbs equals the energy it gives away in the form of radiation.

The law is expressed mathematically as

$$\frac{E\,(v,T)}{A(v,T)} = \varepsilon(v,T) \quad ...(5)$$

where E (v, T) is the emissive power of the body, A (v, T) is its absorptive power and ε (v, T) is the emissive power of a perfect blackbody.

A *perfect blackbody* is a body that absorbs all the radiation that is incident on it, no matter what the radiation frequency is. Conversely, when a perfect blackbody is heated it emits radiation at all frequencies. Thus, it is a good radiator as well as a good absorber. Transparent and reflective materials are poor radiators and poor absorbers. *An object that is a good radiator at a given wavelength is also a good absorber at the same wavelength.* This is another form of Kirchhoff's law.

The emissive power represents the ability of a body to radiate efficiently and is also associated with its ability to absorb radiation. Only for a true blackbody, $\varepsilon = 1$. For other bodies, the emissivity usually lies between 0.2 and 0.9, and for highly polished metals e may be as low as 0.01.

Laws of Blackbody Radiation

(i) Stefan-Boltzmann Law

The Stefan-Boltzmann law is an empirical relationship obtained by Stefan and later derived theoretically by Boltzmann. It states that the total radiation emitted from a blackbody at temperature T is proportional to the fourth power of the absolute temperature of the body T^4.

$$I = \sigma T^4 \qquad ...(6)$$

where σ is called Stefan's constant having a numerical value of 5.67 $\times$ 10^{-8} W/m^2–K4.

(ii) Wien's Law

If we plot the distribution of radiant energy as a function of wavelength at different temperatures, we obtain a set of curves as shown in Fig. 21.2. All curves have a peak and the peak is displaced towards the shorter wavelengths as the temperature rises. Wien's displacement law states that the peak wavelength, λ_{max} at which the maximum emission occurs for any given temperature is inversely proportional to the absolute temperature of the body. Thus,

$$\lambda_{max} = \frac{2.8978 \times 10^{-3}}{T} \text{ mK} \qquad ...(7)$$

The shift to shorter wavelengths agrees with common experience.

THE ULTRAVIOLET CATASTROPHE

In practice, a perfect black body is made by taking a hollow sphere (cavity) and drilling a small hole in it (Fig. 3.3). The hole acts a perfect absorber. Light entering the cavity undergoes multiple reflections at the walls and gets trapped inside the cavity. Consequently, the hole appears perfectly dark. Converselv. when the cavity is heated, the radiation produced in the cavity comes out through the aperture and contains all the wavelengths. Therefore, the hole acts as a perfect emitter. The experimental results showed that at a given temperature the radiation energy density initially increases with frequency, then peaks at around a particular frequency and after that decreases finally to zero at very high frequencies, as shown in Fig. 3.2. The spectral distribution of that radiation is a function of temperature alone and the material as such plays no role.

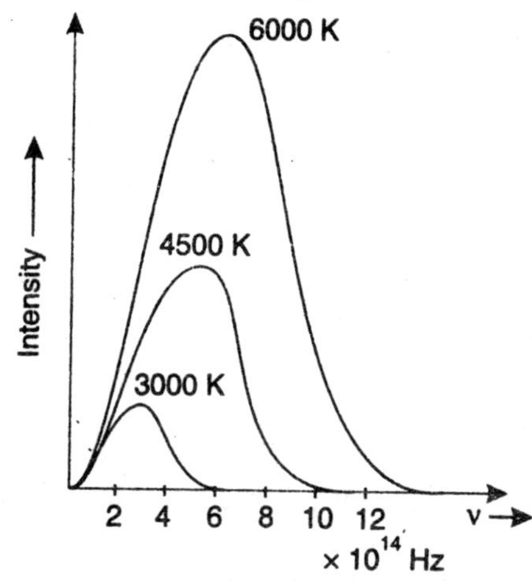

Fig. 3.2 : Spectral energy distribution of black body radiation. Area under a curve represents the total radiation.

Various efforts were made to calculate theoretically the frequency distribution of thermal radiation. At that period of time electron was discovered but the inner structure of the atom was unknown. The generally accepted model was that a blackbody was made up of a huge number of atoms and the atoms are regarded as small harmonic oscillators. The random thermal motion of atoms within the walls generates electromagnetic waves, which is the *thermal radiation* emitted from the walls of the cavity. The radiation emitted by the atoms is reflected back and forth by the cavity walls to form a system of standing waves for each frequency present. There would be many modes of vibration present in the cavity space. Finally, when thermal equilibrium is attained, the average rate of emission of radiant energy by atomic oscillators in the walls equals the rate of absorption of radiant energy by the interior walls. According to Boltzmann's *principle of equipartition of energy*, each simple harmonic oscillator has an average thermal energy of 'kT' at thermal equilibrium. Rayleigh assumed that each of the standing waves ought to have energy 'kT and derived an expression for the energy density of radiation distribution within the cavity.

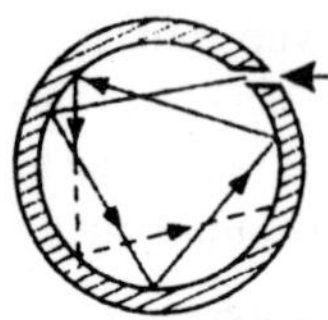

Fig. 3.3 : ***A black body*** **: A black body absorbs all light and reflects none of the light incident on it. A spherical cavity blackened inside and completely closed except a narrow aperture serves as an ideal black body. Light entering the cavity is trapped inside by multiple reflections from the walls. When heated, the black body would emit more from a unit area than any other body at a given temperature.**

$$R(v) = \frac{8\pi v^2 kT}{c^3} \qquad ...(8)$$

This result is known as Rayleigh-Jeans law. The energy density calculated with the above formula agrees well with the experimental results at long wavelength end of the spectrum but goes toward infinity at the short wavelength end. It predicted that the intensity of thermal radiation should increase with square of frequency. It implied that the radiation emitted by a hot body should have a large portion of UV rays. This is contrary to our experience and violates the law of conservation of energy. This contradiction came to be known as ultraviolet catastrophe. The failure of Rayleigh-Jeans formula presented a crisis and gave the first indication of inadequacy of classical physics.

THE PLANCK'S RADIATION LAW

After long years of struggle, Planck succeeded in 1900 to obtain the correct mathematical law for distribution of energy in the blackbody radiation. He recognized that the reason for the ultraviolet catastrophe was that Rayleigh assumed that the standing waves in the cavity of the body consist of a fundamental and also an infinite number of harmonic modes of vibration. Each of these modes was assumed to have an energy kT, leading to the absurd result that the total energy in the cavity would be infinite. Therefore, Planck put forward a revolutionary postulate that an oscillating atom can absorb or reemit energy only in quantities that are integer multiples of 'hv'. All other values of energy are forbidden. The indivisible discrete unit of energy hv is called an energy quantum.

In general, the possible values of the energy of an oscillator of frequency v are

$$E = nhv \quad (n = 1,2,3,...) \qquad ...(9)$$

where n is called the *quantum number* of the oscillator, h is a constant now known as Planck's constant.

Incorporating the discreteness in the emission of energy, and using Maxwell-Boltzmann distribution law Planck obtained the energy density in blackbody radiation as

$$E(v) = \frac{8\pi hv^2}{c^3}\left[\frac{1}{e^{hv/kT} - 1}\right] \qquad ...(10)$$

This is *Planck's radiation law*. It is seen from the formula that the exponential term in the denominator prevents E (v) from rising to infinity at small wavelength values. Planck's formula gives complete fit with the experimental observations at all wavelengths and for all temperatures.

The existence of discrete energy values (Fig. 3.4) represented a departure from the classical physics and our everyday experience. If we take a mass-spring harmonic oscillator, it can receive any amount of energy form zero to some maximum value (Fig. 3.5). Thus, in the realm of classical physics energy always appears to occur with continuous values and energy exchange between bodies involves any arbitrary amounts of energy.

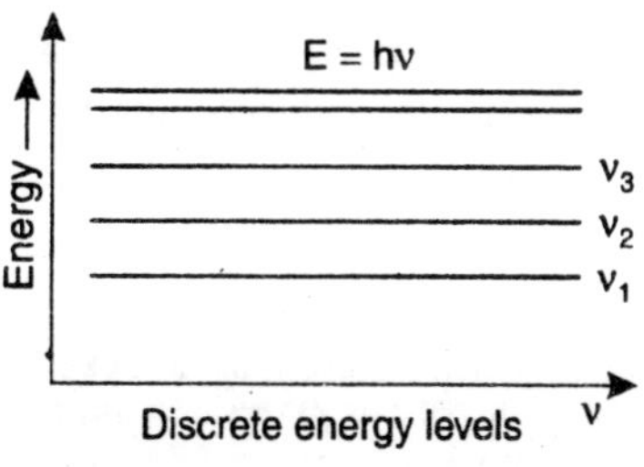

Fig. 3.4 : An atomic oscillator could have only discrete amount of energy, according to Planck's hypothesis.

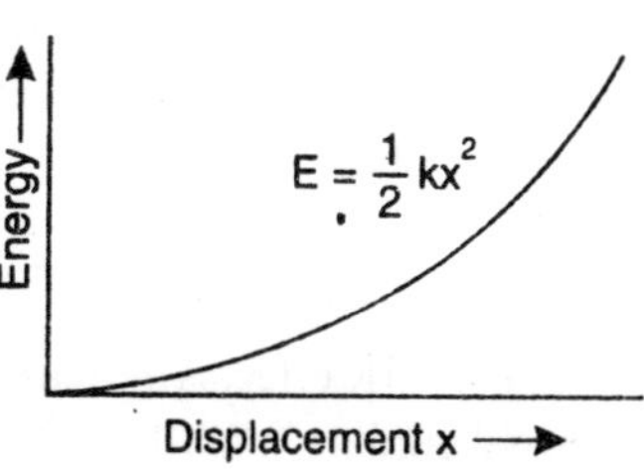

Fig. 3.5 : A classical oscillator such as a spring can have any amount of energy from zero to some maximum value; and the energy distribution is continuous.

Planck's hypothesis of discrete quanta of energy was the origin of the quantum physics. Thus, the investigations on the continuous spectrum emitted by a blackbody gave birth to quantum physics.

THE PHOTON

Max Planck introduced the concept of discontinuous emission and absorption of radiation by bodies but he treated the propagation of radiation through space as occurring in the form of continuous waves as demanded by electromagnetic theory. Einstein refined the Planck's hypothesis and invested the quantum with a clear and distinct identity. The energy quanta are named *photons*.

1. According to Einstein the quantization of energy, which is present in the emission and absorption processes, is retained as the energy propagates through space. A light beam is regarded as a stream of photons travelling with a velocity 'c'.
2. An electromagnetic wave having a frequency contains identical photons, each having an energy hv. The higher the frequency of the wave, the higher is the energy content of each photon. Thus, y-ray and .x-ray photons are more energetic compared to optical photons while the photons of r.f. frequencies are the feeblest.
3. The intensity of a monochromatic light beam / is related to the concentration of photons present in the beam.

$$I = Nh\nu \qquad ...(11)$$

4. When photons encounter matter, they impart all their energy to the particles of matter and vanish.

Two experiments, namely the photoelectric effect and Compton effect, have provided the convincing experimental evidence for the existence of photon.

PHOTOELECTRIC EFFECT

Photoelectric effect is the phenomenon in which electrons are knocked off a metal surface when light is incident on it (Fig. 3.6). The metal is said to be a photosensitive material and the liberated electrons are called *photoelectrons*.

Electronic emission increases with the intensity of the radiation falling on the metal surface, since more energy is available to release electrons. But a characteristic frequency dependence is also observed. For each substance there is a minimum frequency v_0 of light and for light of frequency less than v_0, photoelectrons are not produced, however intense the incident light may be. The minimum frequency v_0 is called *threshold frequency*. The photoelectrons that are emitted in the process are found to have a range of energies. The maximum kinetic energy of the photoelectrons varies with the frequency of the incident light (Fig. 3.7) and is independent of the intensity of light.

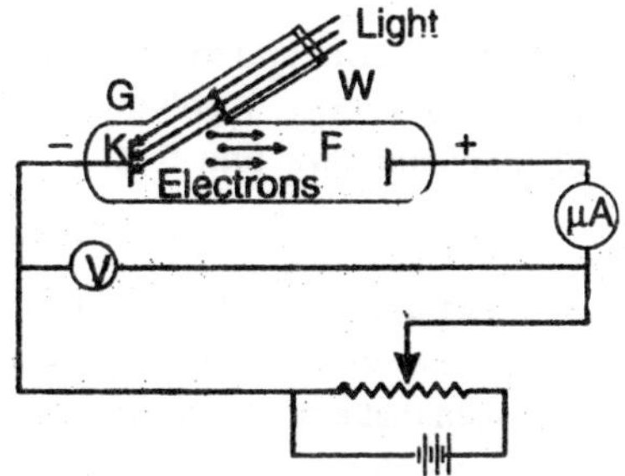

Fig. 3.6 : Schematic of the arrangement to study photoelectric effect.

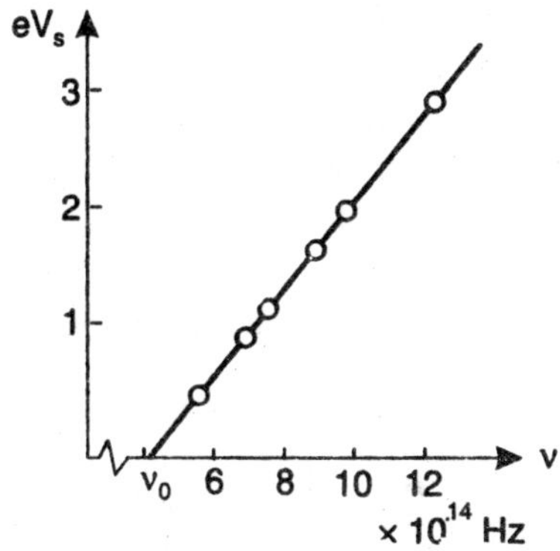

Fig. 3.7 : A plot of K.E.$_{max}$ of photoelectrons with frequency. The kinetic energy is equal to zero at threshold frequency v_0. Above threshold, the K.E. of photoelectrons is directly proportional to the frequency of the incident light.

These features could not be explained on the basis of electromagnetic wave theory of light. Einstein used Planck's quantum concept to explain the photoelectric effect. Einstein regarded light as a stream of photons. In a beam of monochromic light of frequency v, all photons have the same energy hv. When the photons encounter electrons in a metal, they give up their energy to the electrons. However, the electrons are held within the metal by a *potential barrier* at the surface. In order to escape from the metal, an electron should have enough energy to overcome the potential barrier at the surface. It requires an input energy that is equal to the work function, W_0. This work function is specific for the metal. When an electron absorbs a photon, the energy gained by it is equal to hv. If $hv \geq W_0$, the electron will escape from the metal. Thus, out of the total energy hv, a small portion, W_0, is spent on surmounting the potential barrier and the balance energy ($hv-W_0$) is given to the electron as kinetic energy. Applying the law of conservation of energy to this phenomenon, we find that

$$hv = W_0 + 1/2\ n\upsilon^2_{max} \qquad ...(12)$$

The above equation is known as *Einstein's photoelectric equation.* It explains all the features of the photoelectric effect. From this equation it is seen that a photoelectron is emitted from the metal only if the energy of the incident photon hv ≥ Wy. If $hv < W_0$, electrons cannot surmount the barrier at the metal surface and emerge from the metal, however intense the incident light may be. The minimum energy that a photon should have to dislodge an electron from the metal, without imparting it any kinetic energy (K.E. = 0), is given by $hv_0 = W_0$. Hence, photoelectric effect can occur only when the frequency of the incident light is equal to or greater than the minimum value v_0, which is given by $v_0 = W_0/h$.

COMPTON EFFECT

In 1922 Arthur H. Compton studied scattering of x-rays. From classical standpoint, X-rays are just electromagnetic waves and their frequency after the scattering should be the same as before. However, when x-rays passed through certain substances, in addition to the incident x-rays, x-rays of different frequency are observed (Fig. 3.8). The new radiation is interpreted as the x-ray radiation scattered by free electrons in the substance. The wavelength of the scattered x-rays is longer than that of the incident x-rays. Further, the wavelength of the scattered x-rays is different for each direction of scattering (Fig. 3.9).

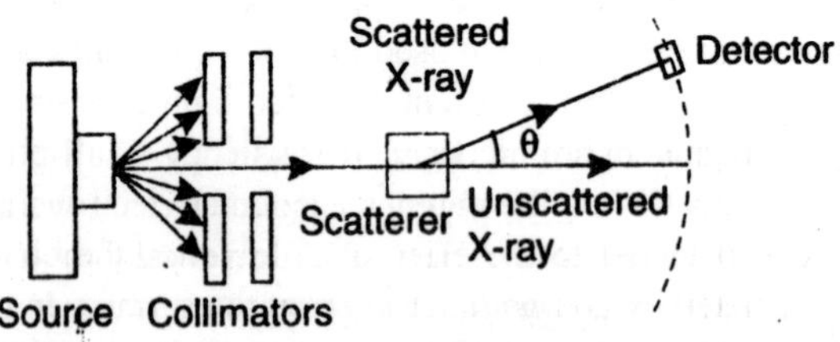

Fig. 3.8 : Schematic of the experimental arrangement for the study of Compton effect.

Compton visualized the scattering of x-rays by an electron as a collision between a photon and the electron and during the collision process there occurred an exchange of energy and momentum. By using the laws of conservation of energy and momentum, we can calculate the change in the frequency of photon as a function of the scattering angle. We obtain the following relation in terms of the wavelength.

$$\Delta\lambda = \lambda_f - \lambda_i = \frac{h}{m_0 c}(1 - \cos\theta) \quad ...(13)$$

where θ is the scattering angle. This simple formula was found to be in a perfect agreement with experimental results (Fig. 3.9).

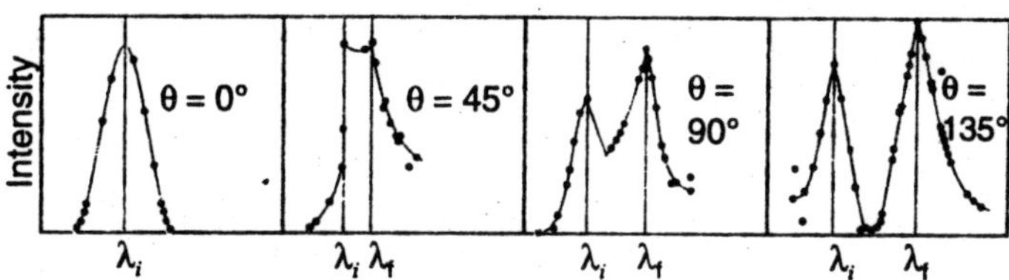

Fig. 3.9 : The variation of intensity X-ray as a function of wavelength, at different angles λ_i corresponds to the unmodified (primary) component and λ_f to modified component. Note that the Compton shift $\Delta\lambda = \lambda_i - \lambda_i$ increases with increasing scattering angle θ.

Photoelectric effect and Compton effect established beyond doubt the discrete nature of light and the existence of photons. While the photoelectric effect showed that the photons transport energy, the Compton effect indicated that photons carry momentum. Henceforth photons were granted the full-pledged status of particles.

SPECTRUM AND SPECTRAL LINES

The next question we are concerned is about how the photons are emitted. In 1666 Newton discovered that different coloured rays of light were refracted at different angles when sunlight coming through a slit was passed through a prism. When a lens was kept in the path of the band of colours, a series of coloured images of the slit were found on a screen. Newton called these series of coloured slits a *spectrum*. In general, solid materials and gases at high pressure produce *continuous spectra*, whose investigations led to quantum hypothesis by Planck.

In contrast, vapours and gases at low pressure, where the atoms or molecules are far apart and do not significantly interact, emit *line spectra*. Each chemical element has unique spectral lines that are characteristic of the element (Fig. 3.10). The set of all characteristic frequencies is designated as the *emission spectrum* of the substance. Some elements, such as hydrogen and neon, have relatively few lines; others, such as iron have many thousand lines. In the spectrum, the lines follow each other in an *orderly* fashion. A group of such lines is called a *series*. The study of line spectra led to the understanding of the structure of the atom and the mechanism of emission of photons by atoms.

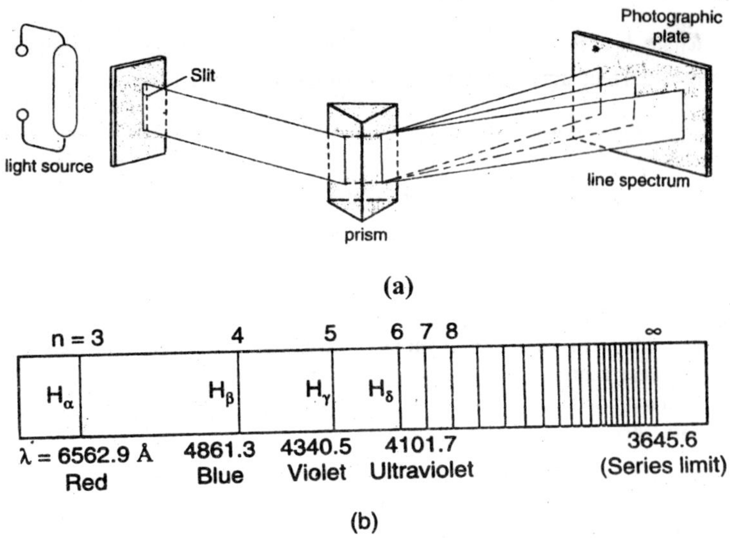

Fig. 3.10 : (a) Light emitted by an element in gaseous form passed through a prism or grating and recorded on a photographic film. (b) A series of s emitted by hydrogen atom. The lines marked as H_α, H_β, H_γ and H_δ belong to B and lie in the visible region.

In 1859, Gustav Kirchhoff summed up his observations on line spectra as follows:

1. Light from a hot body on passing through a prism gives a continuous spectrum with no lines.
2. The same light when passed through a cool gas yields continuous spectrum but with certain wavelengths of light removed.
3. If the light emitted by a hot gas is viewed through a prism, one observes a series of bright lines at specific wavelengths against an otherwise dark background.

These are known as Kirchhoff's three laws of spectroscopy. If the same gas is used in 2 and 3 the absorption and emission lines appear at the same wavelengths in both the cases.

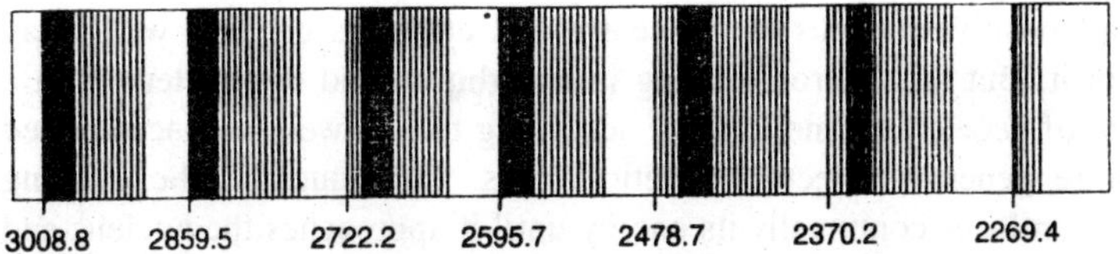

Fig. 3.11 : Simplified molecular spectrum of nitrous oxide (NO) from 3100Å to 2100Å.

The emission spectra of atoms, molecules, and solids differ considerably from each other. Atomic spectra lie mostly in the visible and ultraviolet regions and the lines are sufficiently spaced apart and hence atomic spectra are line spectra. Molecular spectra extend from the far infrared up to the ultraviolet and are composed of closely spaced lines which appear as bright bands in a spectroscope having low resolving power. Hence molecular spectra are called *band spectra.* It is already learnt that solids give continuous spectra.

The studies on atomic spectra led to the formulation of atomic models. In 1885 Johann Balmer found that for the hydrogen atom the wavelengths of the regular array of spectral lines are well described by a simple empirical formula,

$$\lambda = 3646\left[\frac{n^2}{n^2-4}\right]\text{Å} \quad n = 3,4,5.... \qquad ...(14)$$

which was later rearranged by J. Rydberg in the following form.

$$\lambda = R\left[\frac{1}{2^2}-\frac{1}{n^2}\right] \qquad ...(15)$$

where R is a constant and known as Rydberg constant and has the value

$$R = 1.0973732 \times 10^7 \text{ m}^{-1}. \qquad (16)$$

The reciprocal of wavelength I/A, is called the wave number.

ATOMIC STRUCTURE

J.J. Thomson discovered electron in 1897. He put forward the first ever model of the atom. Thomson represented it as a positively charged sphere with electrons immersed in it here and there. When unperturbed, both electrons and the positive charge were believed to be at rest. When in motion the electrons lose their energy in the form of radiation. Basing on the results of α-scattering experiments conducted by him, Rutherford proposed 'planetary model' of the atom. The atom consists of a relatively small positive nucleus at the centre and electrons orbiting well away from it. But an electron moving in an orbit around the nucleus is in a state of accelerated motion and according to Maxwell, any accelerated charge generates electromagnetic waves. Consequently, the orbiting electron loses continually its energy until it approaches the nucleus and falls into it. This leads to the prediction that an atom exists for only about 10^{-8}s, but in fact atoms have remarkable stability. Secondly, if an electron is rotating around the nucleus, it radiates light whose frequency is the same as that of the rotation. As the electron spirals and falls onto the nucleus the light frequency continually increases and therefore the radiation given out by Rutherford atom must be continuous. However, the atomic spectra were in fact a set of discrete lines. In 1913, Niels Bohr removed the deficiencies of Rutherford model by incorporating some adhoc quantum hypotheses into it.

Bohr explained the origin of line spectra in general terms on the basis of two central ideas. One is the concept of photon and the other is the concept of energy levels of atoms. Bohr combined these two concepts nicely to explain how light is emitted and why the spectral lines are arranged in an orderly fashion.

Bohr's Model of Atom

Bohr fused the quantum idea with the purely mechanical model of Rutherford and introduced the following three revolutionary ad hoc postulates.

1. Electrons revolve about a nucleus only in certain special orbits called stationary orbits, though an infinite number of orbits are mechanically allowed. While moving in the permitted orbits, electrons do not emit or absorb electromagnetic radiation though they are in accelerated motion. Hence the atom is stable.
2. The allowed electron orbits are those for which the angular momentum is an integral multiple of $\hbar$, where h is Planck's constant. The angular momentum of electron is $I\omega = (mr^2)(\upsilon/r) = m\upsilon r$. Accordingly

   $$m\omega r = n\hbar \quad (n = 1, 2, 3, ...) \qquad ...(17)$$

   where n is called the *principal quantum number*.
3. The atom radiates energy only when the electron jumps from one of the upper allowed orbits, with energy say Ey to a lower allowed orbit, with energy say E_2. The change in energy during

the transition is given by

$$E_2 - E_1 = h\nu \qquad ...(18)$$

Note that the frequency of the emitted light is not determined by the motion of the electron but is governed by changes in its motion.

Bohr obtained values of the energies of various states of hydrogen atom by assuming that an electron with velocity 'υ' revolves in a circular orbit of radius 'r' around the nucleus (Fig. 3.12). For a dynamically stable orbit, the centripetal force experienced by the electron equals the force of electrical attraction between the nucleus and the electron. Thus,

$$\frac{m\upsilon^2}{r} = \frac{e^2}{4\pi\varepsilon_0 r^2} \qquad ...(19)$$

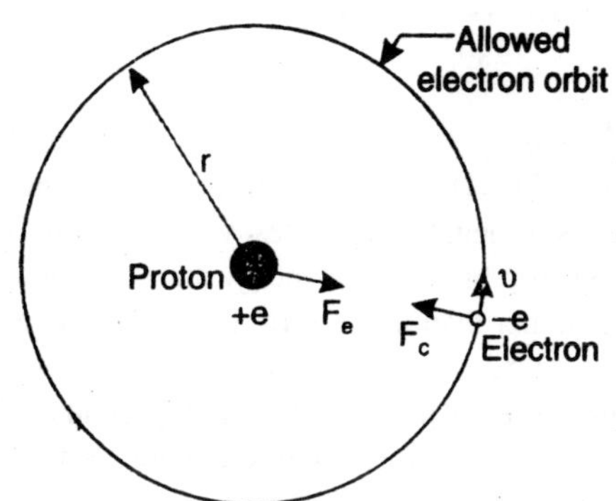

Fig. 3.12 : Bohr's model of the hydrogen atom. The radius of the orbit is quantized in this model.

Using (17) and (19) to eliminate υ, we obtain for the radius of the orbits as

$$\therefore \qquad r = \frac{\varepsilon_0 h^2}{\pi m e^2} n^2 = r_0 n^2 \qquad ...(20)$$

where r is the radius of the first Bohr orbit (n = 1) given by

$$r_0 = \frac{\varepsilon_0 h^2}{\pi m e^2} = 0.53 \text{ Å} \qquad (21)$$

The total energy for a specific electron orbit is obtained by adding the kinetic energy and the potential energy of the electron with respect to the nucleus. Thus,

$$E = \frac{1}{2} m\upsilon^2 - \frac{e^2}{4\pi\varepsilon_0 r} \qquad ...(22)$$

where the first term represents the kinetic energy of electron due to its angular motion and the second term is the potential energy due to electrical attraction. Substituting the value of r from (20) and eliminating υ using (17), we get

$$E = -\frac{e^2}{8\pi\varepsilon_o r} \qquad ...(23)$$

Using the value of r into the above equation (23), we find that electron can have only certain discrete negative values of energy E_n given by

$$E_n = -\frac{me^2}{8\varepsilon_o^2 h^2}\cdot\frac{1}{n^2} = -\frac{E_o}{n^2} \qquad ...(24)$$

where $$E_o = \frac{me^2}{8\varepsilon_o^2 h^2} = 13.6 \text{ eV} \qquad ...(25)$$

Energy Levels

We find from (21.24) that the electron in a hydrogen atom can have any one of a series of negative energies, which are referred to as *energy states*. It is easier to represent the energy states of an atom in the form of *energy levels*, rather than in the form of orbits (Fig. 3.13). The words energy state and energy level are used interchangeably. (To be more specific, the hydrogen atom has two states in its ground level corresponding to –13.60 eV, eight states in its –3.40 eV level, that is the first excited level and so on).

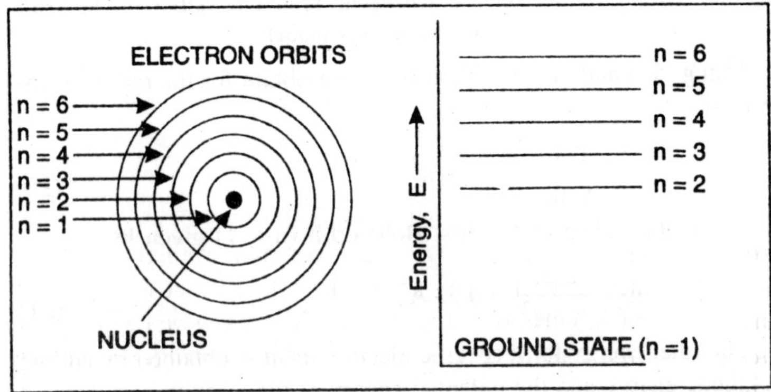

Fig. 3.13 : Electron orbits and the corresponding energy levels of the hydrogen atom

An energy-level diagram of hydrogen is shown in Fig. 3.14.

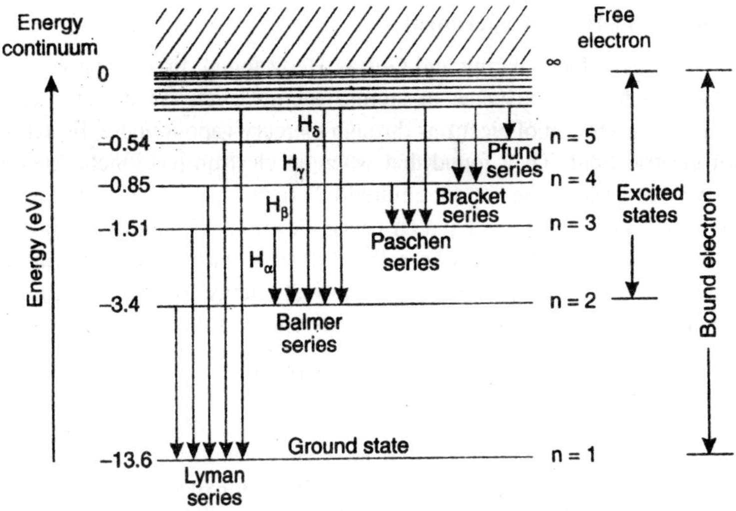

Fig. 3.14 : Energy level diagram of the hydrogen atom.

Each horizontal line represents an allowed energy state. The vertical arrows between levels represent various discrete transitions. The lowest energy level is the level in which the electron revolves in the innermost orbit (n = 1); that level is called the *ground state.* For hydrogen, the ground sate is at –13.6 eV. When the hydrogen gas is heated or subjected to an electric discharge, the atoms are raised to higher, *excited states.* For example, the first excited level is at –3.4 eV (n = 2) and so on. The actual value of the discrete negative energy indicates the amount of energy required to remove the electron completely away from its orbit around the nucleus, and it is referred to as *ionization* energy.

The above picture can be extended to atoms of all elements. Each atom has a set of possible energy levels. An atom can have an amount of internal energy equal to any one of these levels, but it cannot have energy intermediate between two levels. *All isolated atoms of a given element have the same set of energy levels, but atoms of different elements have different sets.*

Every atom has a *ground level* that represents the minimum internal energy state that the atom can have. An atom shut off from outside influences will always lie in the ground level. All the levels higher to it are *excited levels.* If the atom is disturbed by exposing it to some radiation, the total energy of the electron is increased and the atom is raised to an excited level.

The Frank-Hertz Experiment

In 1914, James Frank and Gustav Hertz found direct experimental evidence for the existence of atomic energy levels. Frank and Hertz studied the motion of electrons through mercury vapour under the action of electric field. They found that when an electron has kinetic energy 4.9 eV or greater, the vapour emitted UV light of wavelength 0.25 mm. Suppose mercury atoms have an energy level 4.9 eV above the lowest energy level. An atom can be raised to this level by collision with an electron; it later drops back to the lowest energy level by emitting a photon. According to eqn. (18) the wavelength of the photon should be

$$\lambda = \frac{hc}{E} = \frac{(4.136 \times 10^{-15}\ \text{eV.s})\ (3 \times 10^{8}\ \text{m/s})}{4.9\,\text{eV}} = 0.25\ \mu\text{m}$$

This is equal to the measured wavelength, confirming the existence of this energy level of the mercury atom. Similar experiments with other atoms yielded the same kind of evidence for atomic energy levels.

Atomic Transitions

The transition of atom from one energy state to another is accomplished by a transfer of energy. If energy is supplied to the system consisting of atoms, the atom is raised from a lower energy state E_1 to a higher excited state, E_2. Such a transition, $E_1 \rightarrow E_2$ is called absorption. The transition from a lower state to an excited state can occur only if the difference in energy is exactly equal to the photon energy hv. The atom does not stay in an excited state indefinitely. It usually returns on its own to the lower state after about 10 ns by emitting a photon. The downward transition $E_2 \rightarrow E_1$ is called *emission.* During the emission process, as the atom returns from a higher energy state E_2 to a lower energy state E_1, it emits a quantum of energy, hv. If the photon energy is $hv = \frac{hc}{\lambda}$, then conservation of energy gives

$$hv = \frac{hc}{\lambda} = E_2 - E_1 \qquad ...(26)$$

Vertical lines in Fig. 3.14 show the jumps that electrons can make from one level to another. The location of these transitions, or *emission lines*, can be shown on a wavelength graph. The ensuing pattern corresponds exactly to that recorded by the spectrographic plates. Thus, *position* is one of the important properties of spectral lines.

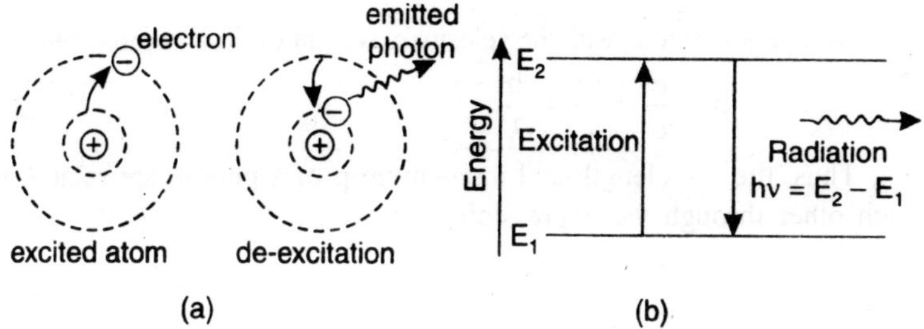

Fig. 3.15 : (a) Absorbing energy, electron jumps from an inner orbit to an outer orbit. When the electron returns to the inner orbit, it emits the same energy in the form of a photon, (b) energy level representation of excitation and de-excitation of the atom.

The average time spent by the atom in an excited level is called the *lifetime* of the level. The lifetime is usually of the order of 10^{-8}s. Besides the short-lived excited states, there are states with average lifetimes greater than 10 milliseconds or even as long as several seconds. They are called *metastable states*.

Limitations of Bohr Theory

Bohr's theory employed a semi-classical model where quantum postulates are introduced into a mechanical model. But it explained with remarkable success the origin of line spectra in case of hydrogen atom, which is a one-electron atom. However the theory could not give correct predictions for two electron atoms. The Bohr hypothesis established the relation of wavelengths to energy levels, but it did not provide any general principles for predicting the energy levels of a particular atom. It also did not explain why angular momentum is quantized and why and when atoms make transitions. Further, it cannot determine what the intensity of the spectral line will be. Inspite of such limitations, Bohr's theory contributed immensely to the understanding of atomic structure. A more general understanding of atomic structure and energy levels is provided by the quantum mechanics.

DE BROGUE HYPOTHESIS

We learnt earlier that a light beam consists of electromagnetic waves and in this chapter, we have learnt that it consists of a stream of photons.

Thus, we find that light has dual nature and behaves as waves sometimes and as particles at other times.

As a photon travels with the velocity c, we can express its momentum as

$$p = \frac{E}{c} = \frac{h\nu}{c} = \frac{h}{\lambda} \qquad ...(27)$$

Thus, the wavelength and momentum p of a photon are related to each other through the expression

$$\lambda = \frac{h}{p} \qquad ...(28)$$

The quantities ν and λ are wave properties and the quantities E and p are particle properties. The relations (27) and (28) demonstrate that the wave and particle natures of a photon are intimately tied up to each other.

In 1924, de Broglie suggested that the wave-particle dualism need not be a special feature of light. The relation (28) between the momentum and the wavelength of a photon must be a universal relation applying to photons and material particles alike. A particle of mass 'm' moving with a velocity υ carries a momentum

$$p = m\upsilon$$

and it must be associated with a wave of wavelength

$$\lambda = \frac{h}{p} = \frac{h}{m\upsilon} \qquad ...(29)$$

The waves associated with moving particles are called *matter waves or de Broglie waves*. The relation $\lambda = h/m\upsilon$ is known as de *Broglie equation* and the wavelength is called the *de Broglie wavelength.*

If the de Broglie hypothesis is valid, then the waves associated with matter should suffer diffraction. A beam of electrons diffracted by a crystal should show interference phenomena. Davisson and Germer obtained, in 1927, the first experimental evidence that gave convincing proof of the wave nature of matter.

De Broglie's Justification of Bohr's Postulate

One of the postulates that Bohr used in formulating a model of atom is that the angular momentum L of the electron revolving in a stationary orbit is quantized. Thus,

$$L = n\hbar \quad \text{—Bohr's, postulate} \qquad ...(30)$$

The above postulate of Bohr follows directly from the concept of matter waves. If a stretched string is fastened at both ends and is made to vibrate, standing waves are formed provided the length of the string is an integral number of half-wavelengths of the disturbance. If the string is formed into a circular loop, the condition for standing waves is that the circumference of the loop should be an integral number of whole wavelengths of the disturbance. Thus, if r is the radius of the circular loop,

$$2\pi r = n\lambda \qquad n = (1, 2, 3, ---) \qquad ...(31)$$

We may regard the stationary electron orbits in an atom to be analogous to the circular loop of string. We conclude that stationary electron wave pattern can form in the orbit if only an integral number of electron wavelengths fit into the orbit, as shown in Fig. 3.16.

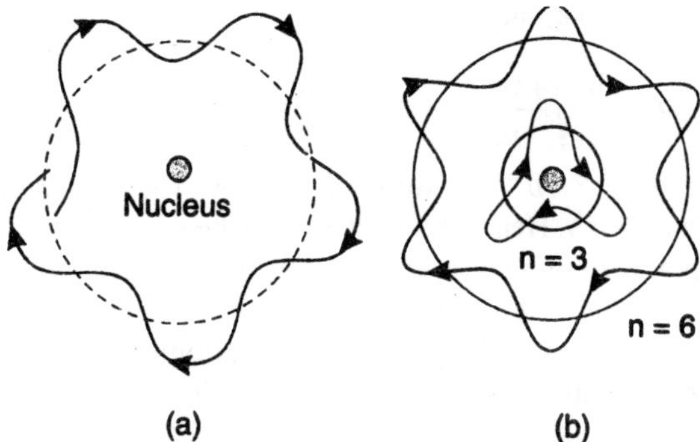

Fig. 3.16 : (a) When the circumference of the circular orbit it not equal to an integral multiple of wavelengths, the electron wave does not form standing waves and the orbit is not allowed. (b) The allowed orbits accommodate an integral multiple of electron waves.

The above equation (31) can be applied for electron waves, taking λ as de Broglie wavelength of electron waves.

The de Broglie wavelength of electron wave is given by

$$\lambda = \frac{h}{m\upsilon}$$

where υ is the speed of the electron in the orbit. Using de Broglie wavelength into eqn. (31), we obtain

$$2\pi r = \frac{nh}{m\upsilon}$$

$$\therefore \qquad m\upsilon r = \frac{nh}{2\pi}$$

But the quantity 'mυr' is the angular momentum, L, of electron in the orbit of radius r. Thus,

$$L = m\upsilon r$$

It, therefore, follows that

$$L = n\,\hbar \qquad ...(32)$$

which is precisely Bohr's postulate. De Broglie thus demonstrated that the quantization of angular momentum is a direct consequence of wave nature of electron.

Let us calculate the wavelength of the electron in the first orbit of hydrogen atom. The electron speed υ in the orbit is given by

$$\upsilon = \frac{e}{4\pi\varepsilon_0 mr}$$

$$\therefore \qquad \lambda = \frac{h}{e}\sqrt{\frac{4\pi\varepsilon_0 r}{m}} \qquad ...(33)$$

Taking $r = 5.3 \times 10^{-11}$ m, we get $\lambda = 3.3$Å. The circumference of the orbit is $2\pi r = 3.3$Å. Hence, the first orbit of the electron in a hydrogen atom corresponds to one complete electron wave joined on itself. The de Broglie hypothesis thus offered a new meaning to the principal quantum number "n". n is the number of de Broglie wavelengths that fit into the circumference of Bohr allowed orbits.

HEISENBERG UNCERTAINTY PRINCIPLE

The wave nature of moving particles leads to some inevitable consequences. Classically, the state of a particle is defined by its position and momentum. At each instant, the position and momentum of a classical particle can be measured to a very high accuracy. In case of a quantum particle, there are uncertainties associated with its location and momentum.

At about the same time that Davisson and Germer conducted the experiments of wavelike properties of electrons, Heisenberg put forth his *uncertainty principle*. He suggested that the product of uncertainty in the location of the quantum particle and the uncertainty in its momentum. $\Delta x\ \Delta p$, would always be of the order of Planck's constant h. Thus,

$$\Delta x \,.\, \Delta p + h \qquad ..(34)$$

This is known as Heisenberg uncertainty principle for position and momentum, which may be stated as follows:

"It is not possible to know simultaneously and with exactness both the position and the momentum of a microparticle".

This uncertainty principle expresses a fundamental limitation in nature that also limits the precision of our measurements.

The uncertainty principle asserts that it is physically impossible to know simultaneously the exact position ($\Delta x = 0$) and exact momentum ($\Delta p_x = 0$) of a microparticle. According to it, the more precisely we know the position of the particle, the less precise is our information about its momentum. The momentum of a particle cannot be precisely specified without our loss of knowledge of the position of the particle at that time. Similarly, a particle cannot be precisely localized in a particular direction without our loss of knowledge of momentum in that particular direction. We can at best specify that certain momentum of the particle is more probable than other or that the particle is more likely to be here than there. Thus, the uncertainly principle implies that we can never define the path of an atomic particle with the absolute precision indicated in classical mechanics. To describe the quantum particle the concept of *energy* becomes important since it is related to the *state* of the system rather than to its path.

In addition to the uncertainty relation between co-ordinates and momentum of a moving particle, there is an uncertainty relation between energy and time. Suppose that we want to determine the energy of a particle and the time at which the particle has such energy. If ΔE and Δt are the uncertainties in the values of these quantities, then

$$\Delta E \,.\, \Delta t \approx \hbar \qquad ...(35)$$

Thus, it is impossible to know simultaneously and with exactness the energy of a particle and the time at which it has that energy.

In view of the wavelike properties of particles and uncertainties in determining the physical parameters, a theory taking into consideration of the *probability distribution* of parameters such as position, momentum, energy in place of *precise values* is required. Efforts in this direction were done independently by Schrodinger and Heisenberg who used different approaches namely wave mechanics and matrix mechanics but which were found to be equivalent.

WAVE FUNCTIONS

In quantum mechanics, various atomic parameters are described with the help of wave functions ψ. In general, these wave functions ψ are mathematical functions that depend on the variables necessary to specify the particular features of the particle. A wave function is a complex quantity and is not an *observable* quantity; hence it has no direct physical significance. For example, let the location of an electron be described by the wave function ψ (x, y, z, t). This wave function by itself cannot indicate us the location of the electron. It is known that the intensity of a wave motion is proportional to the square of the amplitude of the wave. Therefore, the square of the absolute value $|\psi|^2$ of the wave function gives where the intensity of the field $|\psi|^2$ is large and thus the regions of space where the particle is more likely to be found at time ψ. The function y is, therefore, called the *probability amplitude*. Thus,

Probability, P, of finding the particle in an infinitesimal volume dV(= dx dy dz) is proportional to $|\psi(x, y, z)|^2$ dx dy dz at time t.

$$\text{or } P \propto |\psi(x, y, z)|^2 dV \qquad ...(36)$$

The probability is a real quantity.

The probability of finding the particle between the limits x = a and x = b, y = g and y = h and z = p and z = q, is then

$$P \propto \int_a^b \int_g^g \int_p^q \psi^* (x, y, z)\, \psi (x, y, z)\, dx\, dy\, dz \qquad ...(37)$$

The probability that the electron is located somewhere must be unity.

$$\therefore \qquad \iiint \psi^* \psi \, dx\, dy\, dz = 1 \qquad ...(38)$$

SCHRODINGER WAVE EQUATION

In 1926 Erwin Schrodinger developed a wave equation that describes the behaviour of atomic particles. Let us assume that a particle of mass 'm' is in motion along the x-direction. Let the wave function ψ be the dependent variable of the de Broglie wave which is a function of the coordinates x and t. Analogous to the classical wave, we may expect that, ψ will be a function of (x – ωt). As $\upsilon = \psi/k$, the wave function may be written as a function of (kx – ψt).

Using the relation $p = \hbar k$ and $E = \hbar\omega$, we can write

$$\psi = f\left(\frac{px - Et}{\hbar}\right) \quad ...(39)$$

The more general wave would be a sum of a sine and cosine waves. Taking help of Euler's identity, we write the above equation in an exponential form as follows:

$$\psi = A \exp\left[\frac{i}{\hbar}(px - Et)\right] \quad ...(40)$$

We assume that the energy and momentum of the particle are constant. Differentiating the above equation with respect to x, we get

$$\frac{\partial\psi}{\partial x} = A\frac{ip}{\hbar}\exp\left[\frac{i}{\hbar}(px - Et)\right]$$

$$\therefore \qquad \frac{\partial\psi}{\partial x} = \frac{ip}{\hbar}\psi$$

Reàrranging the terms in the above equation, we get

$$p\psi = \frac{\hbar}{i}\frac{\partial\psi}{\partial x} = -i\hbar\frac{\partial\psi}{dx} \quad ...(41)$$

Differentiating the eq. (40) with respect to t gives

$$\frac{\partial\psi}{\partial t} = -\frac{iE}{\hbar}A\exp\left[\frac{i}{\hbar}(px - Et)\right] = -\frac{iE}{\hbar}\psi$$

Rearranging the terms in the above equation, we get

$$E\psi\ \frac{\hbar}{i}\frac{\partial\psi}{\partial t} = i\hbar\frac{\partial\psi}{\partial t} \quad ...(42)$$

The partial derivatives with respect to x and t are connected by means of the relation between the energy and momentum. The classical expression for the kinetic energy in terms of the momentum is

$$E_k\ \frac{m\upsilon^2}{2} = \frac{(m\upsilon^2)}{2m} = \frac{p^2}{2m} \quad ...(43)$$

The total energy and momentum are related by the expression

$$\frac{p^2}{2m} + V = E \quad ...(44)$$

where Vis the potential energy of the particle. Multiplying the eqn. (44) with w, we obtain

$$\frac{p^2}{2m}\psi + V\psi = E\psi$$

Using the relations (41) and (42) into the above equation, we get

$$-\frac{\hbar^2}{2m}\frac{\partial^2\psi}{\partial x^2} + V\psi = i\hbar\frac{\partial\psi}{\partial t} \quad ...(45)$$

The above equation is known as the *time-dependent Schrodinger wave equation.*

Knowing the form of V, eqn. (45) can be solved for the wave function ψ. In a number of cases the potential energy V of a particle does not depend on time; it varies with the position of the particle only and the field is said to be *stationary.* In the stationary problems Schrodinger equation can be simplified by separating out time and position- dependent parts. Accordingly, we can write the wave function as a product of x, ψ(x) and a function of t, φ (t).

We, therefore, write that

$$\psi(x, t) = \psi(x)\,\varphi(t) \quad ...(46)$$

Equation (45) may be written as

$$-\frac{\hbar^2}{2m}\varphi\frac{\partial^2\psi}{\partial x^2} + V\psi\varphi = i\hbar\psi\frac{\partial\psi}{\partial t}$$

Dividing the above equation with ψ_{φ}, we get

$$-\frac{\hbar^2}{2m}\frac{1}{\psi}\frac{d^2\psi}{\partial x^2} + V = i\hbar\frac{1}{\varphi}\frac{d\varphi}{\partial t} \quad ...(47)$$

If we assume that the potential energy Vis a function of x only, the entire left hand side of eqn. (47) is a function of x only while the right hand side is a function of t only. Since x and t are independent variables, both the function of x and t must be equal to a constant. The constant that each side must equal is called the *separation constant* E. Thus,

$$-\frac{\hbar^2}{2m}\frac{1}{\psi}\frac{d^2\psi}{\partial x^2} + V = E \quad ...(48)$$

and $$i\hbar\frac{1}{\varphi}\frac{d\varphi}{\partial t} = E$$

Eq. (48) may be rewritten as

$$-\frac{\hbar^2}{2m}\frac{2^2\psi}{dx^2} + V\psi = E\psi$$

or $$\frac{2^2\psi}{dx^2} + \frac{2m}{\hbar^2}(V - E)\psi = 0$$

$$\therefore \qquad \frac{2^2 \psi}{dx^2} + \frac{8\pi^2 m}{h^2}(V - E)\,\psi = 0 \qquad ...(49)$$

The above equation is called the time-independent Schrodinger equation.

Allowed Wave Functions and Energies

The time-independent wave equation is the pertinent equation for studying properties of atomic systems in stationary conditions. Wave mechanical methods of solving the problem of particle motion are essentially based on ψ functions. Appropriate wave equation is formulated by incorporating the particle mass 'w' and potential energy function V for the region in which the particle is located. The next step consists of solving the differential equation for solutions, namely for the ψ functions which will satisfy the differential equation. By solving the Schrodinger equation, we obtain the possible set of ψ functions. In case of bound particles the acceptable solutions for the differential equation are possible only for certain specified values of energy. These energy values will be the only possible results of precise measurements of the total energy of the particle. These discrete values of energy E_1, E_2, ... E_n are called eigen values or allowed values of the energy of the particle. The solutions ψ_1, ψ_2, ... ψ_n corresponding to the eigen energy values E_n are called the eigen functions. The quantization of energy thus appears as a natural element of the wave equation.

Thus, using the Schrodinger equation, it is possible to determine first the electron energies and then the wave functions. These wave functions can then be used to determine the probability distribution function $\psi^*\psi$ of the electron, for various discrete energies as it revolves around the nucleus.

PROPERTIES OF SPECTRAL LINES

We record me spectral lines with the help of spectrometers, which have narrow slits. The spectral lines are therefore linear and possess three important properties; namely *position* measured in terms of frequency, finite *width*, which is a *range* of frequencies and *intensity*, the brightness of the line.

Width of Spectral Lines

Real atoms or molecules do not emit, radiation at precise frequencies; each emission is more or less broadened by various processes, and so each line is really a small package of slightly different frequencies. Therefore, even if we make the slit of the spectrometer infinitely narrow, there is nonetheless a minimum line width. The natural shape of an emission line appears under ideal conditions when the emitting atom is at rest and is not subjected to the external forces during the process of emission. In practice, several factors contribute to the line broadening. They are broadly classified into two categories:

(i) Homogeneous and

(ii) Inhomogenenous broadening. If the broadening mechanism affects each individual atom in the sample to the same extent, then the broadening is said to be *homogeneous*. In such a case all of the atoms in the sample will have the centre frequency v_0 and the same line shape. Natural broadening and collision broadening belong to this category. On the other hand, if different atoms in the sample have slightly different frequencies for the same transition, the overall response of the sample broadens out and the broadening is said to be *inhomogeneous*. Doppler broadening and broadening due to crystal defects belong to this category.

1. *Collision broadening :* Atoms or molecules in liquid and gaseous phases are in continual motion and collide frequently with each other. When a radiation-emitting atom undergoes a collision, the emission conditions change and it is equivalent to an interruption. In the process, the phase of the emitted wave suffers random variations. Thus, each collision act leads to a random phase change, which means shortening of wave trains, as shown in Fig. 3.17. It results in the broadening of the spectral line. This broadening is called *collision broadening* since it is a consequence of the collisions of atoms.

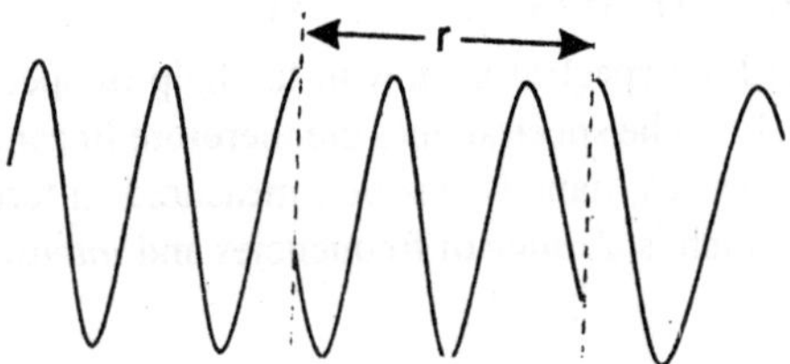

Fig. 3.17 : Random collision with other atoms cause abrupt changes in phase of the ave emitted by an atom.

2. *Doppler broadening* : Doppler effect occurs when a source and an observer are in relative motion. The frequency as measured by the observer increases if the source and observer approach each other and decreases when they recede. In a gas due to the chaotic nature of thermal motion, all the directions of the molecules' velocities relative to a spectrograph are equally probable. Therefore, the radiation received by the spectrograph contains all the frequencies in the interval from $v_0\ (1 - \upsilon/c)$ to $v_0\ (1 + \upsilon/c)$, where v_0 is the frequency emitted by the molecules and υ is the velocity of thermal motion. Hence the spectral line is broadened. The broadening of the line caused by the Doppler effect is called *Doppler broadening.* In general, for liquids collision broadening is the most important factor, whereas for gases where collision broadening is less pronounced, the Doppler effect often determines the natural line width.

3. *Natural broadening* : Even in an isolated, stationary atom the energy levels are not indefinitely sharp. According to Heisenberg uncertainty principle, if a system exists in an energy state for a limited time Δt seconds, then the energy of that state will be uncertain to an extent ΔE where

$$\Delta E \times \Delta t \approx h/2\pi$$

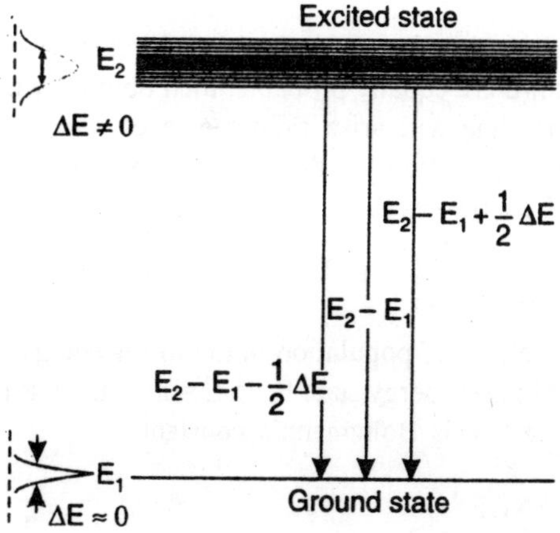

Fig. 3.18 : Natural width of energy levels.

Thus, we see that the ground state of a system is sharply defined since, left to itself, the system will remain in that state for an infinite time. Thus $\Delta t = \infty$ and $\Delta E = 0$. In contract, the lifetime of an excited state is about 10^{-8}s, which gives a value for ΔE of about 10^{-34}J.s ÷ 10^{-8} s = 10^{-26} J. A transition between an excited state and the ground state will thus have an energy uncertainty of ΔE, and a corresponding uncertainty in the associated radiation frequency of $\Delta E/h$, which we can write as

$$\Delta v = \frac{\Delta E}{h} \approx \frac{1}{2\pi \Delta t} \qquad ...50)$$

If an excited electronic state lifetime is 10^{-8}s, then $\Delta v = 10^8$Hz. Thus, the natural broadening is relatively small in magnitude and it is often masked by other mechanisms.

The Intensity of Spectral Lines

The spectral line intensities are dependent on two factors: the *transition probability*, the likelihood of a system in one state changing to another state; and the *population of state*, the number of atoms or molecules initially in the state from which the transition occurs.

1. *Transition probability* : The detailed calculation of absolute transition probabilities involves a knowledge of the precise quantum mechanical wave functions of the two states between which the transition occurs. It is often possible to decide whether a particular transition is allowed or forbidden basing on selection rules.
2. *Population of states* : If we have two levels from which transitions to a third are equally probable, then obviously the most intense spectral line will arise from the level which initially has the greater population. At thermal equilibrium the population of a set of energy levels is governed by the Boltzmann law

$$\frac{N_2}{N_1} = e^{-\Delta E/kT} \qquad ...(51)$$

where N_1 is the population in the lower energy state E_1 and N_2 in the higher energy state E_2, $\Delta E = E_2 - E_1$, T is the temperature in K, and k is Boltzmann's constant.

LUMINESCENCE

A body that emits light on account of high temperature is called *incandescent.* There are bodies that emit light due to causes other than high temperature and they are said to be *luminescent*. The emission of light from a substance, when it is stimulated by the input of energy from a source of suitable radiation, is known as luminescence. Materials exhibiting *luminescence* are broadly called as *phosphors*. There are various ways of exciting substances and so there are the following kinds of luminescence.

(i) *Photoluminescence* is the glow of bodies as a result of their irradiation by visible or uv light, x-rays or γ-rays. Tube lights use this phenomenon to produce visible light. The internal surface of the tube is coated with phosphors and these substances emit visible light of lower frequency under the action of uv light of higher frequency.

(ii) *Electroluminescence* is caused due to the passage of an electric current through a substance or the action of an electric field. The glow of a gas discharge in the tubes of advertisement is an example of this kind of luminescence.

(iii) *Cathodoluminescence* is the glow of materials due to the bombardment of the material by electrons or other charged particles.

(iv) *Chemiluminescence* is caused at the expense of energy produced in a chemical reaction. The glow of many living organisms such as bacteria and insects is due to chemiluminescene.

Many substances emit light when a beam of light is made to fall on them. When the light is emitted only as long as the exciting radiation is maintained, it is called *fluorescence*. For fluorescence, the time interval between the acts of excitation and emission is 1 to 10 ns. For example, kerosene emits a faint bluish fluorescence when illuminated by daylight.

Substances, which continue to emit light for some time after the exciting radiation is removed, are called *phosphorescent* substances. Phosphorescence is displayed for example by zinc sulphide. A coat of zinc sulphide applied to a sheet of glass makes a fluorescent screen, which will emit a faint glow for minutes after the exciting radiation is switched off.

Fluorescence occurs when the atoms of the fluorescent material absorb a portion of the incident radiation of shorter wavelength and re-emit it in the form of radiation of longer wavelength.

Luminescence in solids is closely related to impurities and lattice defects. Crystal luminescence is more important and practically useful. Whatever may the form of energy given to the luminescent material, the final stage in the process is an electronic transition between two energy levels, E_1 and E_2 with the emission of light of frequency v, where

$$v = \frac{E_2 - E_1}{h}$$

Invariably E_1 and E_2 are part of two groups of energy levels, so that, instead of a single emission frequency, a band of frequencies is usually observed.

Fig. 3.19 illustrates some processes that take place in luminescent solids. When an electron is excited from the valence band into the conduction band, a hole is left in the valence band. In a perfectly regular lattice, the electron usually returns to the valence band, although it may take some time to do so. However, if the lattice has some impurity that introduces energy levels in the forbidden region, an electron in a low lying impurity level may fill the hole in the valence band while the electron in the conduction band may fall into one of the vacant high impurity energy levels (Fig. 3.19b). These transitions generally involve photons of small energy that do not fall into the visible region. Finally, the electron may fall from higher energy impurity level to the lower energy impurity level, emitting radiation of lower energy or longer wavelength than the incident radiation; this constitutes the luminescence (Fig. 3.19c).

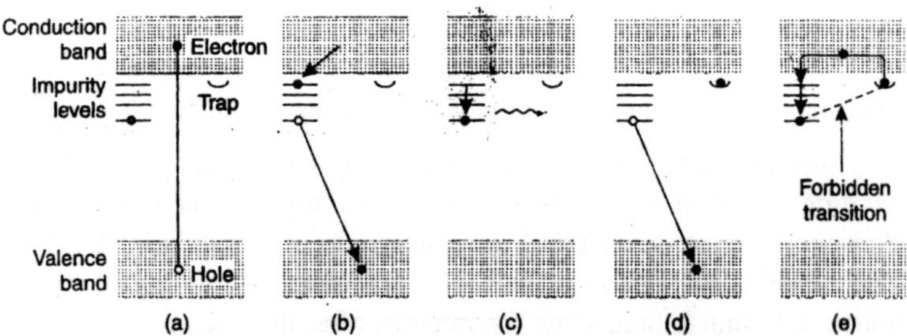

Fig. 3.19 : Mechanism of luminescence

In some instances, instead of going through the process depicted in Fig. 3.19(b), the electron in the conduction band may fall into an energy level called a *trap*. From there electron cannot jump to the ground state impurity energy level. The transition is forbidden. In such a case the electron is in a state similar to an atom in a metastable state. The trapped electron waits till it is returned to the conduction band through some other mechanism. Then it follows the steps (b) and (c). Due to the time delay involved, the process is *phosphorescence*. These substances are called *phosphors*. One such substance is zinc sulphide. Phosphors are widely used in coating the screen of CRTs used in CROs, TVs and video terminals of computers.

SCATTERING

When the electrical polarizability of molecule changes during its motion, light scattering takes place. This phenomenon also gives rise to light emission but does not involve electron transitions from a higher energy level to a lower energy level.

Raman Effect

In 1928, Sir C.V. Raman discovered that when a beam of monochromatic light was passed through an organic liquid such as benzene, the scattered light was found to contain a strong line of frequency v_0 equal to the frequency of the incident light and a few weak lines on either side of the incident line $v' = v_0 \pm v_M$. This is referred to as *Raman Scattering*. The scattering of light with change of frequency is known as *Raman effect*. The spectrum formed due to Raman effect is called *Raman spectrum* and the spectral lines obtained are called *Raman lines*. In order to observe Raman effect, the incident light should be monochromatic and very intense. Raman scattering is always accompanied by Rayleigh scattering. Raman lines at frequencies less than that of the incident frequency $(v_0 - v_M)$ are known as Stokes lines and those with frequencies greater than that of the incident frequency $(v_0 + v_M)$ are known as *anti-Stokes* lines. Experiments show that Stokes lines are far more intense than anti-Stokes lines. Overall, however, the total radiation scattered at any but the incident frequency is extremely small, and sensitive apparatus is needed for its detection and study.

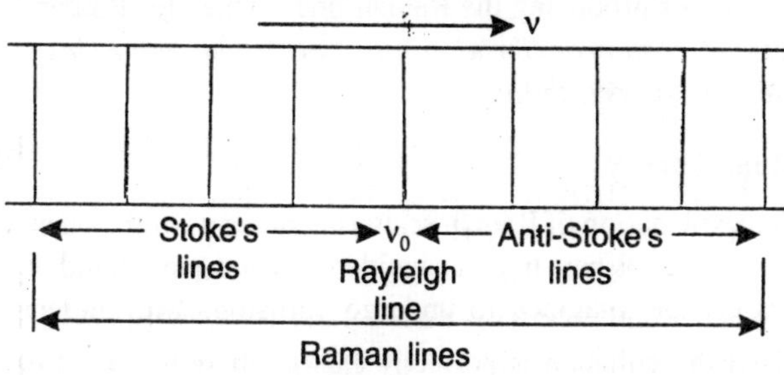

Fig. 3.20

If the frequency of the incident light is varied, the weak lines are once again observed on either side of the Rayleigh line with the same difference in frequency. It is evident that the frequency difference v' between the incident and the scattered light in Raman effect is determined by the nature of the scattering molecules and is independent of the frequency of incident light.

Classical Explanation

When a sample of molecules is subjected to a beam of light of frequency v, the induced dipole undergoes oscillations of frequency v;

$$\mu = \alpha E = \alpha E_o \sin 2\pi vt \qquad ...(52)$$

If in addition, the molecule undergoes some internal motion, such as vibration or rotation, which changes the *polaribility* periodically, then the oscillating dipole will have superimposed upon it the vibrational or rotational oscillation. As an example, let us assume a vibration of frequency v_{vib}, which changes the polarizability. Then

$$\alpha = \alpha_o + \beta \sin 2\pi v_{vib} t \qquad ...(53)$$

where α_0 is the equilibrium polarizability and p represents the rate of change of polarizability with the vibration. Using (53) into (52), we get

$$\mu = \alpha E = [\alpha_o + \beta \sin 2\pi v_{vib} t] E_o \sin 2\pi vt \qquad ...(54)$$

Expanding the above equation and using the trigonometric relation, we get

$$\mu = \alpha_o E_o \sin 2\pi vt + 1/2 \beta E_o [\cos 2\pi (v - v_{vib}) t - \cos 2\pi (v + v_{vig}) t] \qquad ...(55)$$

It means that the oscillating dipole has frequency components v ± v_{vib}, asrwell the exciting frequency v. These additional frequencies are responsible for producing the Raman lines. Thus, for Raman scattering, *a molecular rotation or vibration must cause some change in a component of molecular polarizability.*

Quantum Theory

The explanation of Raman scattering in terms of the quantum theory is very simple. When light is incident on a solid, liquid, or gas, the photons can be imagined to undergo collisions with molecules.

(i) If the collision is perfectly *elastic*, there will be no transfer of energy from the photon to the molecule or from the molecule to the photon. The photon is scattered without any change of energy. Therefore, the frequency of the scattered photon is the same as that of the incident photon. A detector placed to collect energy at right angles to an incident beam will thus receive photons of energy hv, that is the radiation of frequency V. This explains the Rayleigh line in the Raman spectrum.

(ii) However, it may happen that energy is exchanged between photon and molecule during the collision; such collisions are called *inelastic collisions*. The molecule can gain or lose amounts of energy only with the accordance with the quantum laws. That is its energy changes must be the difference in energy between two of its allowed states. That is to say, ΔE must represent a change in the vibrational and/or rotational energy of the molecule.

If the molecule gains energy ΔE (= $E_2 - E_1$) from the photon, it goes from a lower energy level E_1 to a higher energy level E_2, while the photon will be scattered with energy (hv – ΔE). Then, the scattered radiation will have a frequency (v – ΔE/h) which is less than that of the incident radiation. The resulting lines are Stokes lines located on the lower frequency side of the Raman spectrum.

Conversely, the molecule may be initially in an excited state. It may lose energy ΔE to the photon and go to a lower state after collision, while the photon is scattered with energy (hv + ΔE). Hence, the scattered radiation will have a frequency (v + ΔE/h). The resulting lines are anti-Stokes lines, which are situated on the higher frequency side of the Raman spectrum.

At ordinary temperatures, there are more molecules in the lower energy state. Therefore, transitions are more likely from lower energy state to upper energy state. Hence, energy is absorbed by the molecules from the photons. The reverse process of energy being given to photons by molecules is less likely. Therefore, Stokes lines are more intense than the anti-Stokes lines. The intensity of anti-Stokes lines increases with temperature, as a rise in temperature increases the relative population of molecules in the higher energy state.

SOLVED EXAMPLES

Example 1:

For a given metal surface the limiting wavelength for photoelectric emission if 5800Å. Deduce the work function, maximum kinetic energy of electrons ejected by light of λ = *4500A.*

Solution:

$$\text{Photon energy } h\nu = \frac{6.6\times10^{-34}\times3\times10^{10}}{4500\times10^{-8}}$$

$$= 4.40 \times 10^{-19}\text{J}$$

$$\text{Work function} = h\nu_o = \frac{6.6\times10^{-34}\times3\times10^{10}}{5800\times10^{-8}}$$

$$= 3.41 \times 10^{-19}\text{J}$$

The difference gives Ek = 0.99×10^{-19}J

Example 2:

Light of wavelength 5461A and intensity 0.20×10^{-5} watt/cm² falls on a photoelectric cell of cathode area 2.5 cm². The efficiency is 15%. Compute the saturation photocurrent.

Solution:

Photons received *per second*

$$n = \frac{0.20\times10^{-5}\times2.5}{6.6\times10^{-34}\times3\times10^{8}\times1/(5461\times10^{-10})} = 1.37 \times 10^{13}s^{-1}$$

Electrons ejected per second are 15% of this; and each electron carries a charge 1.6×10^{-19} coulomb. Hence the saturation current is

$$i = 1.37 \times 10^{13} \times 0.15 \times 1.6 \times 10^{-19} \text{ amp}$$

$$= 0.33 \text{ microamp.}$$

Example 3:

In an experiment on Compton scattering the wavelength of incident radiation is 1.872 A. Calculate the wavelength of scattered radiation at $\theta = 30°$. Also calculate the velocity of the corresponding recoil electron.

Solution:

By simple substitution for $\theta = 30°$

$$\Delta\lambda = .0242\ (1\text{-}8660) = .00324\text{A}°;$$

$$\lambda' = 1.872 + .003 = 1.875.$$

The energy difference of the incident and scattered photons gives the KE of the ejected electron. Hence

$$\frac{1}{2}mv^2 = h(v - v') = \frac{hc}{\lambda^2}(\Delta\lambda)$$

$$= \frac{6.6\times10^{-27}\times3\times10^{10}}{(1.872\times10^{-8})^2} \times (.00324 \times 10^{-8}) \text{ erg}$$

Substituting for m and solving we get

$$v = 2 \times 10^8 \text{ cm/sec.}$$

Now, from conservation of momentum, remembering that $\sqrt{}/v = 1$.

$$mv \cos\phi = \frac{hv}{c} - \frac{hv'}{c} \times .866$$

$$= \frac{hv'}{c} \times 0.134$$

$$mv \sin\phi = \frac{hv'}{c} x\ 0.500$$

$$\therefore \quad \tan\phi = \frac{0.5}{0.134} = 3.62 \Rightarrow \phi = 75°.$$

Example 4:

One sectorial series in an element is given by

$$v_n = \frac{109756}{(2-0.083)^2} - \frac{109756}{(n-0.643)^2},\ n = 3, 4, 5,...$$

Write down two more spectral series using Ritz principle, and evaluate v for the fist member of each series.

Solution:

We predict one series by changing the integer in the first term denominator to 3. Thus

$$v_n' = \frac{109756}{(3-0.083)^2} - \frac{109756}{(n-0.0643)^2},\ n = 4, 5, 6,...$$

We predict one more series by using the first running term of the given series as the fixed term, and changing the fixed term to running terms. Then

$$v_n'' = \frac{109756}{(3-0.643)^2} - \frac{109756}{(n-0.083)^2}\ n = 3, 4, 5,...$$

For the new series thus predicted the v values of the first term may be now calculated

$v' = 3.17 \times 10^3 \text{ cm}^{-1};$

$v'' = 6.85 \times 10^3 \text{ cm}^{-1}.$

Example 5:

About 0.1% of electrical energy supplied to a laboratory mercury vapour lamp of 80 watt is converted into UV light of wavelength 2500Å. Calculate the number of UV photons emitted per second by the lamp.

Solution:

Number of UV photons emitted per second

$$= \frac{\text{Energy conveted int o UV light}}{\text{Energy carried by one UV photon}}$$

$$= \frac{80\text{J / s} \times (0.1/100) \times (6.24 \times 10^{18} \text{eV / J})}{12400\text{eV} / 2500}$$

$= 10^{17}$ photons/s.

Example 6:

Deduce the average energy of Planck's oscillator corresponding to v values such that hv/kT = 0.1, 1.0, 3.0, 10.0.

Solution:

We have

$$\bar{\varepsilon} = kT \frac{x}{e^x - 1}, \text{ where } x = \frac{hv}{kT}$$

For the four given vales of x, tables show

$\varepsilon^{x-1} = 0.105,\ 1.718,\ 19.1,$ and 2.2×10^4

Hence the value of ε is given by

$$\bar{\varepsilon} = kT \left(\frac{.1}{.105}, \frac{1}{1.718}, \frac{3}{19.1}, \frac{10}{2.2 \times 10^4} \right)$$

$= kT\ (1, 0.58,\ 0.16,\ 0.00045)$

Thus for hv, < < kT we have ε = kT while for hv > > kT, ε tends to zero.

Example 7:

A surface is placed 20 cm from a 50W source of λ = 6x 10^{-5} cm. Estimate (i) the average time interval between the arrival of two photons on the same atom (sectional area – 10^{-16}cm^2, and (ii) the mean time lag *between irradiation and ejection of the first electron from the surface (area 2 cm^2, efficiency 40%).*

Solution:

Energy flux = $50/4\pi\,(0.20)^2 = 100\ \text{Jm}^{-2}\text{s}^{-1}$

Photon energy $= hc/\lambda = 6.6 \times 10^{-34} \times 3 \times 10^{+8}/6 \times 10^{-7}$

$= 3 \times 10^{-19}$ J/photon

$\therefore$ Photon flux at 20 cm $= 100/(3 \times 10^{-19})$

$= 3 \times 10^{20}$ photon $\text{m}^{-2}\text{s}^{-1}$

Now, (i) photon flux on a given atom = 3 photon s^{-1}

λ mean $\Delta t = 0.3\text{s}$

(ii) ejection rate from 2 cm^2 area $= 3 \times 10^{20} \times 2 \times 10^{-4} \times .4$

$= 2.4 \times 10^{16}$ electrons^{-1}

$\therefore$ mean $\Delta t = 4 \times 10^{-15}\text{s}$

(*Note*: One conclusion is that the chances of two photons acting together in ejecting an electron are negligibly small. Another is that the time lag between irradiation and start of emission is too small to be observable.)

Example 8:

X-rays of 0.5Å are scattered by free electrons in block of carbon through 90°. Find the velocity of recoil electrons.

Solution:

K.E. of the recoil electrons $= h\nu_i - h\nu_f$

$$= 12400\left(\frac{1}{\lambda_i} - \frac{1}{\lambda_f}\right)\text{eV}\ (1.602 \times 10^{-19}\ \text{J/e})$$

$$\lambda_f = \lambda_i + \frac{h}{m_o c}(1-\cos\theta)$$

$$= (0.5 + 0.02426)\text{Å} = 0.5243\text{Å}$$

$$\therefore \text{K.E.} = 12400\left(\frac{1}{0.5} - \frac{1}{0.5243}\right)(1.602 \times 10^{-19})\ \text{J}$$

$$\therefore u = \left[\frac{2(\text{K.E.})}{m}\right]^{1/2} = \left[\frac{2(1.84\times10^{-16}\,\text{J})}{9.11\times10^{-31}\,\text{Kg.}}\right]^{1/2}$$

$$= \left\{4.04\times10^{14}\,\frac{kg.m^2/s^2}{kg}\right\}^{1/2}$$

$\therefore u = 2 \times 10^7$ m/s.

Example 9:

Calculate de Broglie wavelength of an electron moving with velocity 10^7 m/s.

Solution:

$$\lambda = \frac{h}{mu} = \frac{6.626\times10^{-34}\,J.s}{9.11\times10^{-31}\,kg\times10^7\,m/s}$$

$$= \frac{6.626\times10^{-34}}{9.11\times10^{-24}}\,\frac{kgm^2/s^2.(s^2)}{kg.m} = 7.28 \times 10^{-11}\text{ m}$$

$\therefore \lambda = 0.72$Å

Example 10:

Compute the minimum uncertainty in the location of a mass of 2.0gm moving with a speed of 1.5 m/s and the minimum uncertainty in the location of an electron moving with a speed of 0.5×10^8 m/s. Given that the uncertainty in the momentum p for both $\Delta p = 10^{-3}p$.

Solution:

$p = mu = 2 \times 10^{-3}$ kg $\times$ 1.5 m/s $= 3 \times 10^{-3}$ kg m/s.

$\therefore \Delta p = 10^{-3}\,p = 3 \times 10^{-6}$ kg m/s

$$\therefore \Delta x = \frac{h}{2\pi\Delta p} = \frac{6.63\times10^{-34}\,Js}{2\pi\times3\times10^{-6}\,kg.m/s}$$

$$= 3.5 \times 10^{-19}\text{Å}$$

For an electron, $p = mu$

$= (9.11 \times 10^{-31}$ kg$)(0.5 \times 10^8$ m/s$)$

$= 4.55 \times 10^{-25}$ kg. m/s

$\therefore \Delta p = 10^{-3}\,p = 4.55 \times 10^{-26}$ kg. m/s

$$\therefore \Delta x = \frac{h}{2\pi\Delta p} = \frac{6.63\times10^{-34}\,Js}{2\pi\times4.55\times10^{-26}\,kg.m/s} = 23\text{Å}.$$

Example 11:

With Franck-Hertz type of experiment on sodium vapour the first spectral line to appear is the D-line, λ = 5.89x 10^{-5} cm. Deduce the first excitation potential of sodium.

Solution:

If excitation potential is V volt, excitation energy is eV joule where = 1.6×10^{-19} coul. Equating this with hv,

$$1.6 \times 10^{-19} V = hv = \frac{hc}{\lambda} = \frac{6.62 \times 10^{-34} \times 3 \times 10^{8}}{5.89 \times 10^{-7}}$$

$$V = \frac{6.62 \times 3}{5.89 \times 1.6} = 2.1 \text{ volt.}$$

Example 12:

Sodium has first two excited states at 2.1 and 3.7 eV. In a Franck-Hertz experiment electrons of energy 4.7 eV are fired in sodium gas. Deduce the possible energy values of the electrons received at the collection.

Solution:

In collisions the electrons may lose energy 0, or 2.1 eV, or 3.7 eV or 2 × 2.1 eV or (2.1 + 3.7) eV, or (2 × 3.7) eV... Out of these the values below 4.7 eV are 0, 2.1, 3.7, and 4.2 eV. Hence the possible energies of the received electrons are 4.7 eV *minus* these values. Thus

$$E = 4.7, 2.6, 1.0, 0.5 \text{ eV.}$$

EXERCISES

1. Explain the term probability density.
2. What is luminescence? Explain the difference between fluorescence and phosphorescence.
3. What is Raman effect? What are Stokes and anti-Stokes lines in Raman spectrum?
4. How is Raman effect explained on the basis of quantum theory?
5. Why are the anti-stokes lines fainter than Stokes lines?
6. The threshold frequency for photoelectric emission in copper is 1.1×10^{15} Hz. Find the maximum energy in eV when light of frequency 1.2×10^{15} Hz is incident on the copper surface.

7. Find the energy of neutron having de Broglie wavelength of 10^{-14}m. Given the rest mass of neutron is 1.6×10^{-27}kg.
8. Calculate de Broglie wavelength of an electron whose kinetic energy is 500 eV.
9. State the basic postulates of Planck's quantum theory. How did Einstein modify this theory in the light of experimental results?
10. What is photoelectric effect? How does Einstein's equation explain the features of photoelectric effect?
11. Explain de Broglie's concept of matter waves.
12. Using the concept of matter waves, obtain the Bohr's condition for quantization of angular momentum.
13. State uncertainty principle. Write its mathematical form for the following pairs of variables:

    (i) Position and momentum

    (ii) Energy and time

    (iii) Angular position and angular momentum.
14. What is Heisenberg Uncertainty principle? Explain how it is the out come of the wave description of a particle.
15. Write down Schrodinger's time dependent and time independent wave equations for matter waves. Explain, why:

    (i) The wave function ψ must be single valued and continuous function of position.

    (ii) The integral of $|\psi|^2$ over all space must equal unity.
16. What is the physical significance of wave function 'ψ'?
17. Explain the physical significance of quantum numbers n, l and m_l.
18. Discuss the space quantization of angular momentum.
19. A particle of mass 10^{-6} gm has a speed of 1m/s. If the speed is uncertain by 0.01% what is the minimum uncertainty in the position of the particle?
20. The first member of Balmer series of hydrogen has a wavelength of 6536Å. Calculate the wavelength of its second member.
21. What is meant by work function of a material?
22. What is Compton effect? Show that Compton shift is independent of wavelength of incident x-rays.

4

LASERS

INTRODUCTION

Laser is one of the outstanding inventions of the 20th century. Laser is a *photonic device*, which is actually responsible for the resurgence of interest in optical technology and for the birth of a new field, namely *photonics*. The word 'LASER' is the acronym for Light Amplification through Stimulated Emission of Radiation. However, laser is *not* a simple amplifier of light but is actually a generator of light. In fact, the device should have been called a LOSER signifying Light Oscillation through Stimulated Emission of Radiation; this name was avoided because of its bad connotation and the name laser has been retained. Though it is a source of light, laser differs vastly from the traditional light sources. It is not used for illumination purposes as we use other light sources. Laser is more akin to radio and microwave transmitters and produces a highly directional coherent monochromatic light beam. Laser is the most sought-after tool in metal-working, entertainment electronics, optical communications, bloodless surgery, weapon guidance in wars and in a wide variety of other fields.

ATTENUATION OF LIGHT IN AN OPTICAL MEDIUM

When light travels through a medium, a gradual reduction in its intensity occurs mainly because of the processes of absorption and scattering of light in the medium.

(i) Light *absorption* occurs because part of the incident light is transformed into the energy of motion of the atoms in the medium; and

(ii) light is scattered when it encounters obstacles of sizes smaller than a wavelength. The reduction in intensity with distance in a medium is called *attenuation of light.* The following relation governs the attenuation of 14 a transparent medium.

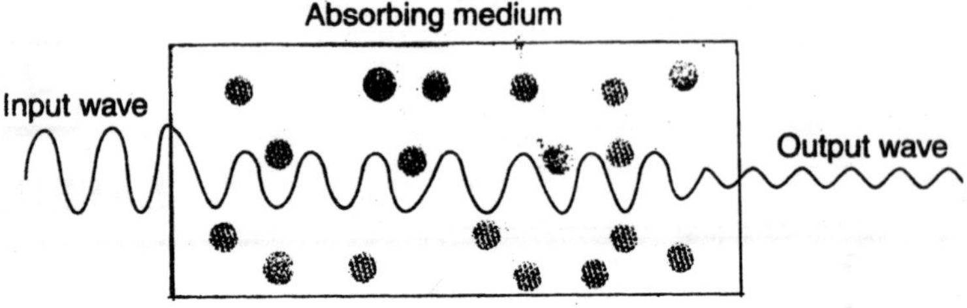

Fig. 4.1 : Attenuation of a light wave in an absorbing medium.

$$I = I_0 e^{-\alpha x} \quad ...(1)$$

where x is the distance in the medium, I_0 is the value of intensity at x = 0, and a is the *coefficient of attenuation or absorption coefficient* of the material at frequency v.

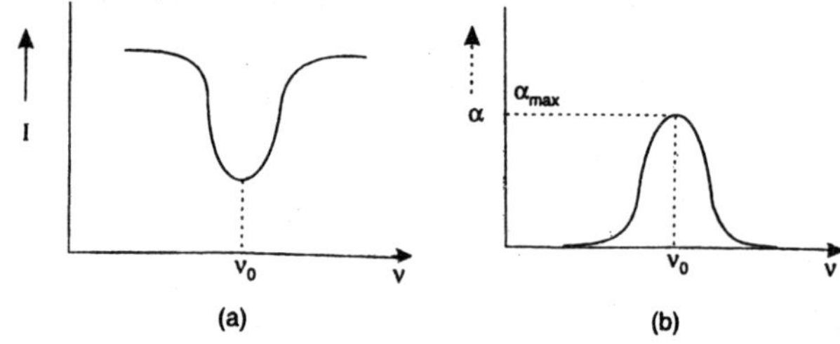

Fig. 4.2

Fig. 4.2 (a) shows the variation of intensity I with frequency V at a fixed distance x in the medium. Fig. 4.2 (b) shows the variation of the attenuation coefficient a with frequency v. In general a is a positive quantity and we say that the material has & positive *coefficient of absorption.*

INTERACTION OF LIGHT WITH MATTER

The process of the transfer of energy from atom to light is not possible from classical point of view. However, the possibility arises if the interaction of light with medium is considered from the point of view of quantum mechanics. A *laser* is a light source that utilizes the quantum processes for its operation. It is therefore, necessary to appreciate the quantum processes involved in the development of a laser.

The radiation incident on a material is viewed as a stream of photons, where each photon carries an energy E = hv. We assume that the two energy levels of the atoms in the material have an energy difference $(E_2 - E_1) = hv$. When photons travel through the medium, three different processes are likely to occur. They are absorption, spontaneous emission and stimulated emission. We study these in detail.

Absorption

Suppose an atom is in the lower energy level E_1. If a photon of energy $hv = (E_2 - E_1)$ is incident on the atom, it imparts its energy to the atom and disappears. Then we say that the atom absorbed an incident photon. As a result of absorption of adequate energy, the atom jumps to the excited state E_2 (Fig. 4.4). The transition is called an *absorption transition*. It is also referred to as *induced absorption*. We may express the process as

$$A + hv = A^*$$

where A is an atom in the lower state and A* is an excited atom.

In each absorption transition event, an atom in the medium is excited and one photon is subtracted from the incident light beam, which results in attenuation of light in the medium.

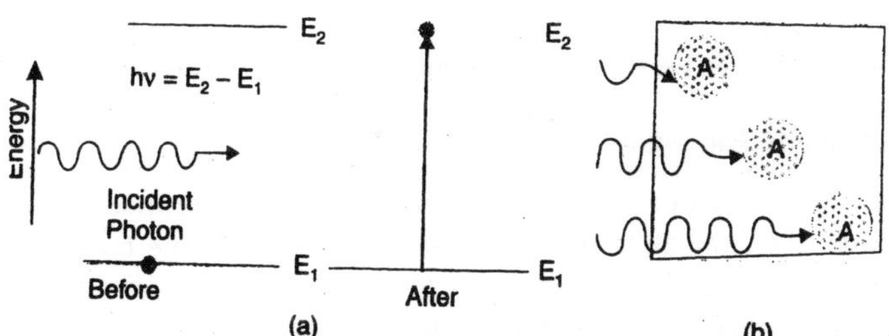

Fig. 4.3 : Absorption process (a) Induced absorption (b) Material absorbs photons.

The number of atoms per unit volume that undergo absorption transitions per second is called the *rate of absorption transition*. It is denoted by

$$R_{abs} = -\frac{dN_1}{dt} \quad ...(1)$$

where $\left(-\frac{dN_1}{dt}\right)$ represents the rate of decrease of population at the lower level E_1. The rate of absorption can also be represented by the rate of increase of population at the upper level E_2, as

$$R_{abs} = \frac{dN_2}{dt} \qquad ...(2)$$

The number of absorption transitions occurring in the material at any instant are proportional to the number of atoms in the lower level E_1 and the density of photons in the incident beam. When the atoms are more at the lower energy level, then more atoms can jump into the excited state.

Similarly, when more photons are incident on the assembly of atoms, then more atoms can get excited to the higher energy level. Then the rate of absorption transition is given by

$$R_{abs} = B_{12}\rho(v)N_1 \qquad ...(3)$$

where N_1 is the population of atoms at E_1, $\rho(v)$ the energy density of the incident beam and B_{12} is the constant of proportionality. B_{12} is known as the Einstein coefficient/or induced absorption. It indicates the probability of occurrence of an induced transition from level $1 \rightarrow 2$.

Induced absorption involves the excitation of atom to the fixed higher level only. As a result of this absorption N_1 decreases and N_2 increases. But under normal conditions N_2 cannot be greater than N_1. Therefore, as light propagates through the medium, it gets absorbed. However, N_2 can be made greater than N_1 using special techniques.

Spontaneous Emission

An atom cannot stay in the excited state for a longer time. In a time of about 10^{-8}s, the atom reverts to the lower energy state by releasing a photon of energy hv, where $hv = (E_2 - E_1)$. The emission of photon occurs on its own and without any external impetus given to the excited atom (Fig. 4.4). Emission of a photon by an atom *without any external impetus* is called *spontaneous emission.* We may write the process as

$$A^* \rightarrow A + hv$$

The number of spontaneous transitions depends only on the number of atoms N_2 at the excited state E_2. Therefore, the rate of spontaneous transitions is given by

$$R_{sp} = A_{21} N_2 \qquad ...(4)$$

where A_{21} is the proportionality constant and is called the *Einstein coefficient for spontaneous emission.* A_{21} represents the probability of a spontaneous transition from level 2 → 1. It is to be noted that the process of spontaneous emission is independent of the incident light energy.

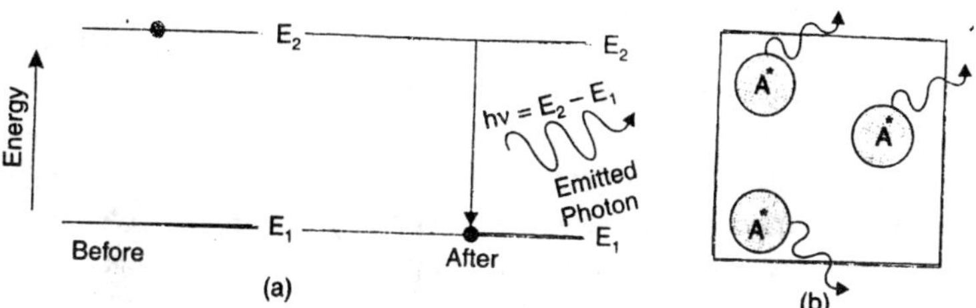

Fig. 4.4 : Spontaneous emission (a) emission process (b) Material emits photons haphazardly.

It follows from quantum mechanical considerations that spontaneous transition takes place from a given state to states lying lower in energy. Thus, spontaneous transition is not possible from level E_1 to level E_2. Therefore, the probability of spontaneous transition from E_1 to E_2 is zero.

$$\therefore \qquad A_{12} = 0 \qquad ...(5)$$

Let now look at the salient features of spontaneous emission of light.

Characteristics of spontaneous emission

(i) The process of spontaneous emission is essentially *probabilistic* in nature and is not amenable for control from outside.

(ii) The instant of transition, direction of propagation, the initial phase and the plane of polarisation of each photon are all *random*.

(iii) The light resulting through this process is *not monochromatic*.

(iv) As different atoms in the source emit photons in different directions, light spreads in all directions around the source. The light intensity goes on decreasing rapidly with distance from the source.

(v) Light emitted through this process is *incoherent*, as it results from a superposition of wave trains of random phases. The net intensity is proportional to the number of radiating atoms. Thus,

$$I_{total} = NI \qquad ...(6)$$

where N is the number of atoms and I is the intensity of light emitted by one atom. It is the process of spontaneous emission that dominates in conventional light sources.

Einstein's Prediction

It is seen that the rate of spontaneous transitions is determined only by the population N_2 at the higher energy level whereas the rate of absorption transitions is determined by the population N_1 at the lower energy level and the energy density $\rho(v)$ in the incident light. If absorption and spontaneous emission were the only processes operative, then obviously the number of atoms absorbing radiation per second would be more than the number of atoms emitting light per second. Eventually, we may end up with a non-equilibrium situation where all the atoms in the medium are excited. But this condition is not observed in practice. It means that equilibrium is maintained. Therefore, in order to account for the state of equilibrium between light and matter, Einstein pointed *out that if a photon can stimulate an atom to move from a lower energy level E_1 to a higher energy level E_2 by means of absorption transition, then a photon should also be able to stimulate an atom from the same upper level E_2 to the lower level E_1*. This alternative mechanism of photon emission depends on the photon density present in the medium and is known as *stimulated emission.*

Stimulated Emission

An atom in the excited state need not "wait" for spontaneous emission of photon. Well before the atom can make a spontaneous transition, it may interact with a photon with energy $hv = E_2 - E_1$ and make a downward transition. The photon is said to stimulate or induce the excited atom to emit a photon of energy $hv = (E_2 - E_1)$. The passing photon does not disappear and in addition to it there is a second photon which is emitted by the excited atom (Fig. 22.6). The phenomenon of forced photon emission by an excited atom *due to the action of an external agency* is called *stimulated emission or induced emission.* The process may be expressed as

$$A^* + hv \rightarrow A + 2hv$$

The rate of stimulated emission of photons is given by

$$R_{st} = B_{21}\,\rho(v)N_2 \qquad \text{...(7)}$$

where B_{21} is the *Einstein coefficient for stimulated emission* and represents the probability for induced transition from level $2 \rightarrow 1$.

In stimulated emission each incident photon encounters a previously excited atom, and the optical field of the photon interacts with the electron. The result of the interaction is a kind of resonance effect, which induces each atom to emit a second photon with the same frequency, direction, phase, and polarization as the incident photon.

Let us now look at the salient features of the stimulated emission of light.

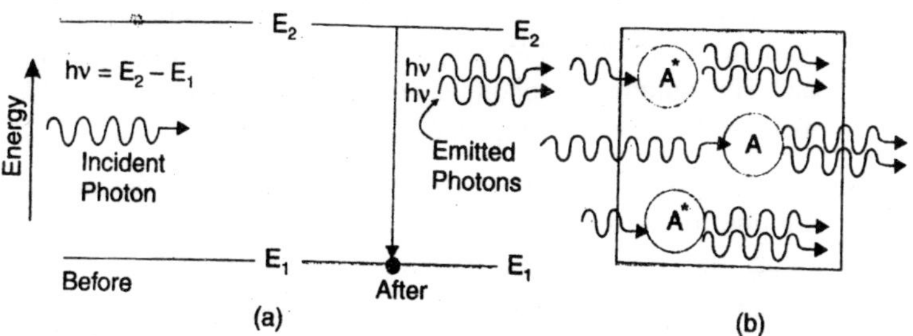

Fig. 4.5 : Stimulated emission (a) emission process. (b) Material emits photons in a coordinated manner.

Characteristics of Stimulated emission

(i) The process of stimulated emission is *controllable* from outside.

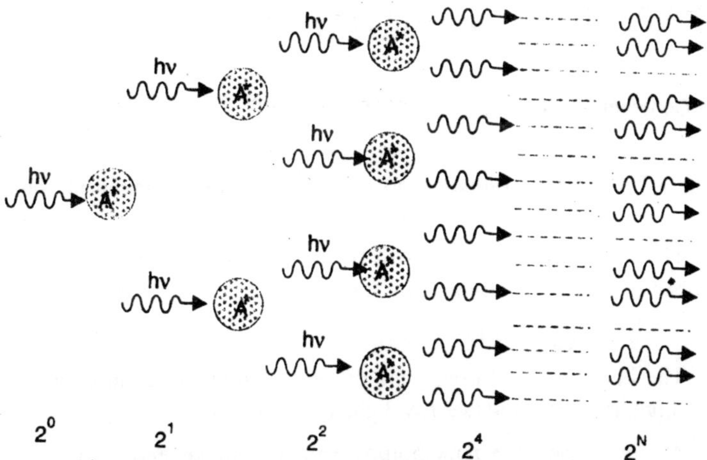

Fig. 4.6 : Multiplication of stimulated photons into an avalanche

(ii) The photon induced in this process propagates in the *same direction* as that of stimulating photon.

(iii) The induced photon has features identical to that of the inducing photon. It has the *same frequency, phase and plane of polarisation* as that of the stimulating photon.

(iv) *Multiplication of Photons* : The outstanding feature of this process is the *multiplication of photons*. For one photon interacting with an excited atom, there are two photons emerging. The two photons travelling in the same direction interact with two more excited atoms and generate two more photons and produce a total of four photons. These four photons in turn stimulate four excited atoms and generate eight photons, and so on. The number of photons builds up in an avalanche like manner, as shown in Fig. 4.6.

(v) Light amplification: All the light waves generated in the medium are due to one initial wave and all of the waves are in phase. Thus, the waves are coherent and interfere constructively (Fig. 4.7).

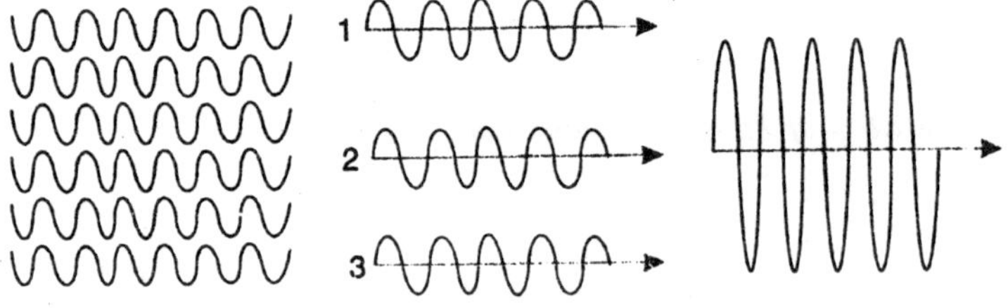

(a) Coherent radiation (b) Component waveform (c) Resultant waveform of waves 1, 2 & 3

Fig. 4.7 : Coherent radiation (a) coherent waves (b) component wave form (c) Resultant wave

The net intensity of light will be proportional to the square of the number of atoms radiating light. Thus,

$$I_{total} = N^2 I \qquad ...(8)$$

The light emitted through the process of stimulated emission is of very high intensity and we say light is *amplified.*

At any reasonable temperature there cannot be enough atoms in excited states for any appreciable amount of stimulated emission from these states to occur. Rather absorption is much more probable.

LIGHT AMPLIFICATION

If we consider a medium in thermal equilibrium, there would be more atoms in the lower level than at higher level. That is $N_1 >> N_2$.

As the probability for absorption transition is equal to the probability for stimulated transition, a photon travelling through the medium is more likely to get absorbed than to stimulate an excited atom to emit a photon. Therefore, usually the process of absorption dominates the process of stimulated emission. Similarly, an atom that is at the excited state is more likely to jump to the lower level on its own than being stimulated by a photon. Further, the photon density in the incident beam is not sufficient to interact with the excited atoms. Owing to this, the spontaneous emission dominates the stimulated emission.

Light amplification requires that stimulated emission occur almost exclusively. In practice, absorption and spontaneous emission always occur together with stimulated emission. The laser operation is achieved when stimulated emission exceeds in a large way the other two processes. Let us now look at the conditions under which the number of stimulated transitions can be made larger than the other two transitions.

Condition for Stimulated Emission to Dominate Spontaneous Emission

The ratio of eqn. (7) to eqn. (4) gives

$$R_1 = \frac{\text{Stimulated transitions}}{\text{Spontaneous transitions}} = \frac{B_{21}\rho(v)\,N_2}{A_{21}\,N_2} = \frac{B_{21}}{A_{21}}\rho(v)$$

The stimulated transitions will dominate the spontaneous transitions if the radiation density $\rho(v)$ is very large and the value of the ratio B_{21}/A_{21} is also large.

We have

$$R_1 = \left(\frac{B_{21}}{A_{21}}\right)\left[\frac{8\pi h v^3 \mu^2}{c^3}\cdot\frac{1}{e^{hv/kt}-1}\right]$$

But $$\frac{B_{21}}{A_{21}} = \frac{c^3}{8\pi h v^3 \mu^3}$$

$\therefore$ $$R_1 = \left(\frac{c^3}{8\pi h v^3 \mu^3}\right)\left[\frac{8\pi h v^3 \mu^3}{c^3}\cdot\frac{1}{e^{hv/kt}-1}\right]$$

or $$R_1 = \left[\frac{1}{e^{hv/kt}-1}\right] \qquad ...(9)$$

If we assume $v = 5 \times 10^{14}$Hz and $T = 300$ K, the value of R, comes to 10^{-58}

The above result shows that in the optical region stimulated emission is negligible compared to spontaneous emission.

Condition for Stimulated Emission to Dominate Absorption Transitions

It may be noted that the presence of a large number of photons will lead to more absorption transitions rather than stimulated emissions. Hence large photon density alone will not guarantee more stimulated emissions.

The ratio of eqn. (7) to eqn. (3) yields

$$R_2 = \frac{\text{Stimulated transition}}{\text{Absorption transition}} = \frac{B_{21}\,\rho(v)\,N_2}{B_{12}\,\rho(v)\,N_1} \qquad ...(10)$$

As $\quad B_{21} = B_{12},\; R_2 = \dfrac{N_2}{N_1} \qquad (11)$

At thermodynamic equilibrium, $N_2 << N_1$ and the population at the ground level far exceeds, that of the excited level. When light propagates through the medium, a photon may hit an excited atom leading to stimulated emission, or be absorbed on hitting an atom in the ground state. As more number of atoms are there at the ground level, a photon has a much higher probability of being absorbed than of stimulating an atom at the excited level. Therefore, the absorption transitions will be larger than stimulated transitions and the medium will absorb the incident light. If, on the other hand, the number of atoms are more in the excited state, the opposite situation prevails and photons are more likely to cause stimulated transitions than absorption transitions.

Two sum up, three conditions are to be satisfied to make stimulated transitions overwhelm the other transitions:

(i) the population at excited level should be greater than that at the lower energy level,

(ii) the ratio $\dfrac{B_{21}}{A_{21}}$ should be large and

(iii) a very high radiation density should be present in the medium. A medium amplifies light only when these three conditions are fulfilled.

To achieve high percentage of stimulated emissions:

(i) an artificial situation known as *population inversion* is to be created in the medium,

(ii) A larger value of $\frac{B_{21}}{A_{21}}$ is achieved by choosing a metastable energy level as the higher level. As spontaneous transitions are forbidden from a *metastable state* the coefficient A_{21} will be smaller and as a result the ratio $\frac{B_{21}}{A_{21}}$ will be larger,

(iii) $\rho(v)$ is made larger by enclosing the emitted radiation in an *optical resonant cavity* formed by two parallel mirrors. The radiation is reflected many times till the photon density reaches a very high value and a favourable condition is created for large stimulated emissions.

POPULATION INVERSION

When the material is in thermal equilibrium condition, the population ratio is governed by the Boltzmann distribution law according to the following equation:

$$\frac{N_2}{N_1} = e^{-[E_2 - E_1]/kT} \qquad \text{...(12)}$$

It means that the population N_2 at the excited level E_2 will be far smaller than the population N_1 at the level E_1. The condition in which there are more atoms in the lower energy level and relatively lesser number of atoms in the higher energy level is called *normal condition* or *thermal equilibrium* (Fig. 4.8a). Thus, under thermal equilibrium,

$$N_1 >> N_2.$$

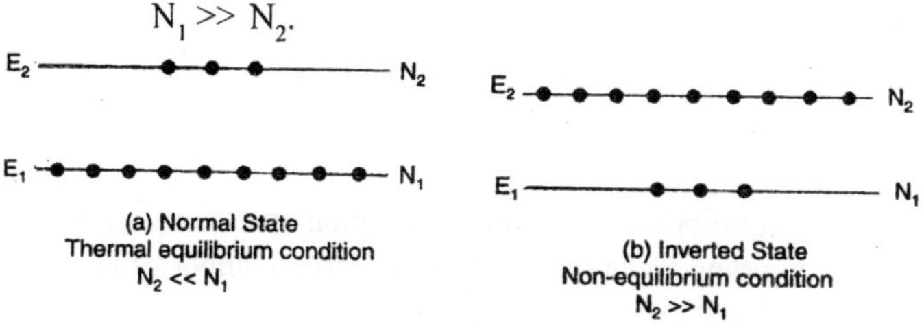

Fig. 4.8

To achieve a high percentage of stimulated emission, a majority of atoms should be at the higher energy level than at the lower level. We may somehow enhance the number of atoms in the excited level, such that the population ratio $\frac{N_2}{N_1}$ *momentarily* increases without change in temperature. This is a *non-equilibrium condition* and is known as *inverted population* condition. *Population inversion* is the non-equilibrium condition of the material in which population of the upper energy level N_2 momentarily exceeds the population of the lower energy level N_1 (Fig. 4.8b). That is,

$$N_2 >> N_1 \quad ...(13)$$

From eqn. (13) it is seen that N_2 can exceed N_1 only if the temperature were negative. In view of this, the state of population inversion is sometimes referred to as a *negative temperature state.* It does not mean that we can attain temperatures below absolute zero. The terminology underlines the fact that the state of population inversion is a non-equilibrium state. It should be borne in mind that the population inversion is attained at normal temperatures.

The system shown in Fig. 4.8 (a) has two energy levels. At thermal equilibrium, photon absorption and emission processes take place side by side, but because $N_1 > N_2$ the system absorbs photons rather than emits photons. Now suppose that the system is supplied with energy from an external source till N_2 exceeds N_1. Then, the system is said to have attained the state of population inversion. The population inversion has taken place between the levels E_2 and E_1.

When the system is in the population inversion condition, a few randomly emitted photons trigger stimulated emission of photons and those stimulated photons induce more stimulated emissions and so on. Consequently light gets amplified (Fig. 4.9) and a cascade of light is produced. However, in this process atoms from E_2 level make downward transitions and as soon as the population at lower level becomes equal or larger than that at the excited level, population inversion comes to an end. Energy is again to be supplied to the system to take it into the state of population inversion.

The non-equilibrium condition (population inversion) is attained by employing pumping techniques to transfer large number of atoms from lower energy level to higher energy level.

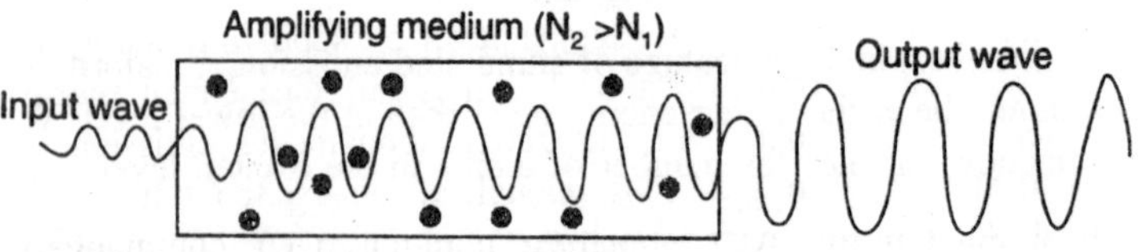

Fig. 4.9 : Amplification of a light wave in a medium with population inversion.

ACTIVE MEDIUM

Atoms are in general characterized by a large number of energy levels. However, all types of atoms are not suitable for laser operation. Even in a medium consisting of different species of atoms, only a small fraction of atoms of a particular type have energy level system suitable for achieving population inversion. Such atoms can produce more stimulated emission than spontaneous emission and cause amplification of light. Those atoms, which cause laser action, are called *active centers*. The rest of the medium acts as host and supports active centers. The medium hosting the active centers is called the *active medium*. An *active medium is a medium which when excited reaches the state of population inversion and promotes stimulated emissions leading to light amplification.*

PUMPING

For achieving and maintaining the condition of population inversion, we have to raise continuously the atoms in the lower energy level to the upper energy level. It requires energy to be supplied to the system. *Pumping* is the process of supplying energy to the laser medium with a view to transfer it into the state of population inversion. Because N_1 is originally very much larger than N_2, a large amount of input energy is required to momentarily increase N_2 to a value comparable to N_1.

There are a number of techniques for pumping a collection of atoms to an inverted state. Optical pumping, electrical discharge and direct conversion are some of the methods of pumping. In *optical pumping*, a light source such as a flash discharge tube is used to illuminate the active medium. This method is adopted in solid state lasers. In *electrical discharge* method, the electric field causes ionization of the medium and raises it to the excited state. In semiconductor diode lasers, a *direct conversion* of electrical energy into light energy takes place.

METASTABLE STATES

An atom can be excited to a higher level by supplying energy to it. Normally, excited atoms have short lifetimes and release their energy in a matter of nanoseconds (10^{-9}s) through spontaneous emission. It means that atoms do not stay long enough at the excited state to be stimulated. As a result, even though the pumping agent continuously raises the atoms to the excited level, they undergo spontaneous transitions and rapidly return to the lower energy level. Population inversion cannot be established under such circumstances. In order to establish the condition of population inversion, the excited atoms are required to 'wait' at the upper energy level till a large number of atoms accumulate at that level. In other words, it is necessary that the excited state has a longer lifetime. A metastable state is such a state. Because of restrictions imposed by conservation of angular momentum, an electron excited to a metastable state cannot return to the ground state by emitting a photon, as it is generally expected to do. Such a state in which single-photon emission is impossible, has an unusually long time and is called a *metastable state.* Atoms excited to the metastable states remain excited for an appreciable time, which is of the order of 10^{-6} to 10^{-3}s. This is 10^3 to 10^6 times the lifetimes of the ordinary energy levels.

Therefore, the metastable state allows accumulation of a large number of excited atoms at that level The metastable state population can exceed the population at a lower level and establish the condition of population inversion in the lasing medium. It would be impossible to create the state of population inversion without a metastable state. Metastable state can be readily obtained in a crystal system containing impurity atoms. These levels lie in the forbidden band gap of the host crystal. Population inversion readily takes place as the lifetimes of these levels are large, and secondly, there is no competition in filling these levels, as they are localized levels.

There could be no population inversion and hence no laser action, if metastable states do not exist.

PRINCIPAL PUMPING SCHEMES

Atoms in general are characterized by a large number of energy levels. Among them only three or four levels will be pertinent to the pumping process. Therefore, only those levels are depicted in the pumping scheme diagrams. Two important pumping schemes are widely employed. They are known as three-level and four-level pumping schemes.

Three-Level Pumping Scheme

A typical three-level pumping scheme is shown in Fig. 4.10. The state E_1 is the ground level; E_3 is the pump level and E_2 is the metastable upper lasing level. When the medium is exposed to pump frequency radiation, a large number of atoms will be excited to E_3 level. However, they do not stay at that level but rapidly undergo downward transitions to the metastable level E_2 through non-radiative transitions. The atoms are trapped at this level as spontaneous transition from the level E_2 to the level E_1 is forbidden. The pumping continues and after a short time there will be a large accumulation of atoms at the level E_2. When more than half of the ground level atoms accumulate at E_2, the population inversion condition is achieved between the two levels E_1 and E_2. Now a chance photon can trigger stimulated emission.

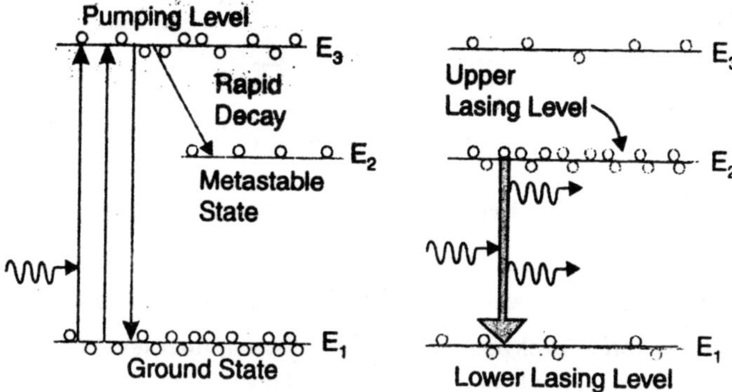

Fig. 4.10 : A typical three level pumping scheme—(a) optical pumping (fc) lasing action

Four-Level Pumping Scheme

A typical four-level pumping scheme is shown in Fig. 4.11. The level E_1 is the ground level, E_4 the pumping level, E_3 the metastable upper lasing level and E_2 the lower lasing level. E_2, E_3 and E_4 are the excited levels. When light of pump frequency v_p is incident on the lasing medium, the active centers are readily excited from the ground level to the pumping level E_4. The atoms stay at the E_4 level for only about 10^{-8}, and quickly drop down to the metastable level E_3. As spontaneous transitions from the level E_3 to level E_2 cannot take place, the atoms get trapped at the level E_3. The population at the level E_3 grows rapidly. The level E_2 is well above the ground level such that $(E_2 - E_1) > kT$. Therefore, at normal temperature atoms cannot jump to level E_2 from E_1 on the strength of thermal energy. As a result, the level E_2 is virtually empty. Therefore, population inversion is attained between the levels E_3 and E_2. A chance photon of energy $hv = (E_3 - E_2)$ emitted spontaneously can start a chain of stimulated emissions, bringing the atoms to the lower laser level E_2. From the level E_2 the atoms subsequently under go non-radiative transitions to the ground level E_1 and will be once again available for excitation.

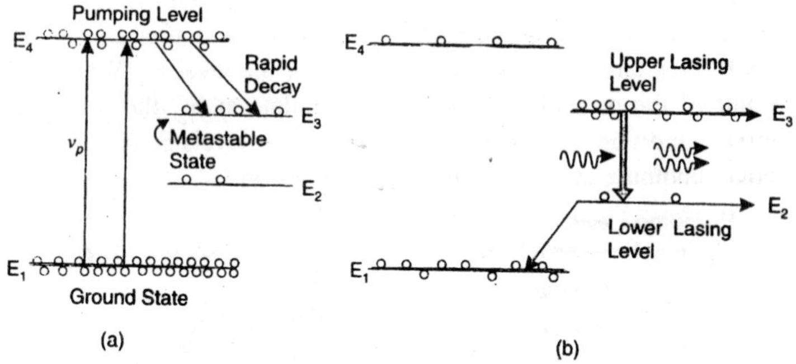

Fig. 4.11 : A typical four level pumping scheme (a) Pumping (b) Lasing action.

Comparison of Four-Level Laser with the Three-Level Laser

1. In the three-level pumping scheme, the terminal level of laser transition is simultaneously the ground level. Therefore, in order to achieve population inversion more than half of the ground level atoms have to be pumped up to the upper lasing level, such that $N_2 > N_1/2$. As the number of atoms in the ground level is very large, high pump power is required in order to promote $N_1/2$ atoms and establish the required population inversion.

   On the other hand, in the four-level pumping Scheme, the terminal level of laser transition is virtually empty and population inversion condition is readily established even if a smaller number of atoms arrive at the upper lasing level. Therefore, relatively small pumping power is required to establish population inversion in four level pumping schemes.

2. In case of three level pumping scheme, once stimulated emission commences, the population inversion condition reverts to normal population condition.. Lasing ceases as soon as the excited atoms drop to the ground level. Lasing occurs again only when the population inversion is re-established. The light output therefore is a *pulsed output.*

In case of four level scheme, the condition of population inversion can be held without interruption and light output is obtained continuously. Thus, the laser operates in *continuous wave* (cw) mode.

Necessity of Broad absorption Band at Pumping Level

In a laser medium, active centers are excited through absorption of energy from a pumping source. We desire that the energy given through the pumping agent is utilized to the largest possible extent in exciting the ground level atoms. This is denoted by *pumping efficiency*. In case of optical pumping, a flash discharge from a lamp serves as the pumping agent. Normally, light sources such as the flash discharge emit light over a wide frequency range. More number of atoms can be excited to the higher level if large number of the frequency components is utilized instead of a single frequency. That can happen *only* when there is a band of close spaced energy levels at the pumping level.

Thus, for a larger pumping efficiency, the pump level should be a broad band rather than a narrow discrete level.

OPTICAL RESONANT CAVITY

Laser is a light source and it is analogous to an electronic oscillator. An electronic oscillator (Fig. 4.12) is essentially an amplifier supplied' with a positive feed back. A part of the output of the amplifier is taken and fed back at its input. When the amplifier is switched on, electrical noise signal of appropriate frequency present at the input will be amplified; the output is fed back to the input and amplified again and so on. A stable output is quickly reached when the oscillator acts as a source of a particular frequency.

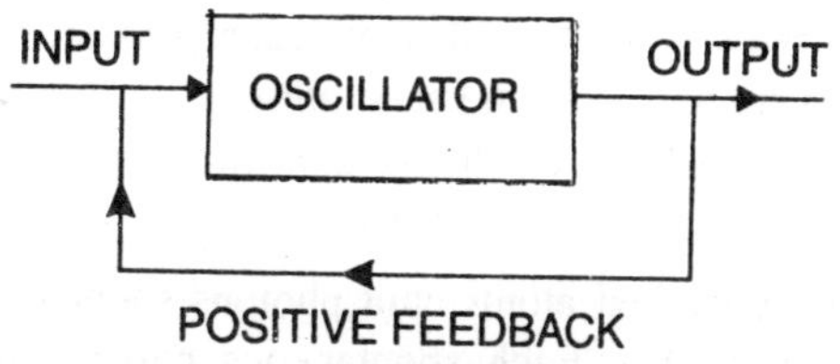

Fig. 4.12

In laser the active medium is the amplifying medium. It is converted into an oscillator through the feed back mechanism established by an optical resonator. A pair of optically plane parallel mirrors (Fig. 4.13) constitutes an *optical resonant cavity*. It is known as a *Fabry-Perot resonator*. One of these mirrors is fully reflecting and reflects all the light that is incident on it. The other mirror is made partially reflecting such that more than 90% of incident light is reflected from it and a small fraction is transmitted through it as the laser beam.

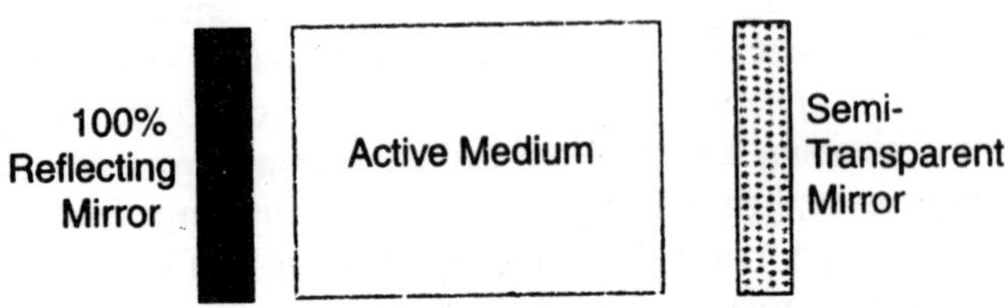

Fig. 4.13 : Fabry-Perot optical resonator

In laser, the role of noise is played by chance photons emitted spontaneously. The photons emitted along the optic axis of the resonant cavity travel through the medium and trigger stimulated emissions. They are reflected by the end mirror and reverse their path. The photons are thus fed back into the medium and travel toward the opposite end mirror causing more stimulated emissions. The photons are once more reflected at the mirror and travel toward the opposite mirror. Substantial light amplification takes place because the light beam is reflected several times at the mirrors and gains strength in each passage. Ultimately, when the amplification balances the losses in the cavity, the laser beam emerges out from the front—end mirror. *In the absence of resonator cavity, there would be no generation of light.*

Lasing Action

Fig. 4.14 shows the action of an optical resonator. The active centers in the medium are in the ground state initially, as shown in Fig. 4.14(a). Through suitable pumping mechanism, the medium is taken into the state of population inversion (Fig. 4.14b).

Some of the excited atoms emit photons spontaneously in various directions (Fig. 4.14c). Each spontaneous photon can trigger many stimulated transitions along the direction of its propagation.

As the initial spontaneous photons are moving in different directions, the photons stimulated by them also travel in different directions. Many of such photons leave the medium without reinforcing their strength. In the absence of the end mirrors, the net effect would have been the production of incoherent light. Now because of the end mirrors, a specific direction is imposed on photons. Photons travelling along the axis are amplified through stimulated emission while the photons emitted in any other direction will pass through the sides of the medium and are lost forever. Thus, a specific direction is selected for further amplification of light.

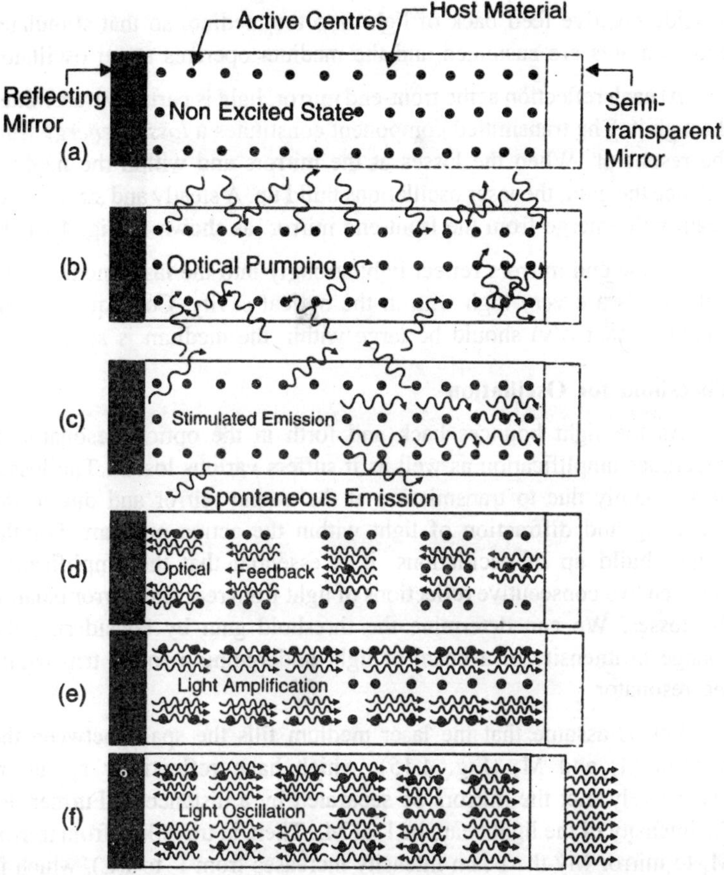

Fig. 4.14 : Light amplification and oscillations due to the action of optical resonator.

A majority of photons travelling along the axis are reflected back on reaching the end mirror. They travel towards the opposite mirror and on their way stimulate more and more atoms and build up the photon strength, as shown in Fig. 4.14 (d). The photons that strike the opposite mirror are reflected once more into the medium, as shown in Fig. 4.14 (e). The photons travel once more through the medium generating more photons and more amplification. The photons are then reflected again at the mirror and travel through the medium. As the photons are reflected back and forth between the mirrors, stimulated emission sharply increases and the amplification of light is augmented. The mirrors thus provide positive feed back of light into the medium so that stimulated emission acts are sustained and the medium operates as an oscillator.

At each reflection at the front-end mirror, light is partially transmitted through it. The transmitted component constitutes a *loss of energy* from the resonator. When the losses at the mirrors and within the medium balance the gain, the laser oscillations build up. A steady and strong laser beam will emerge from the front-end mirror, as shown in Fig. 4.14 (f).

As the end mirrors reflect light strongly into the laser medium, the light levels are very high with in the optical cavity. Consequently, the condition that ρ(v) should be large within the medium is satisfied.

Threshold for Oscillation

As the light bounces back and forth in the optical resonator, it undergoes amplification as well as it suffers various losses. The losses occur mainly due to transmission at the output mirror and due to the scattering and diffraction of light within the active medium. For the proper build up of oscillations, it is essential that the amplification between two consecutive reflections of light from rear end mirror balance the losses. We can determine the threshold gain by considering the change in intensity of a beam of light undergoing a round trip within the resonator.

Let us assume that the laser medium fills the space between the mirrors M_1 and M_2 (Fig. 4.15), which have reflectivity r_1 and r_2 respectively. Let the mirrors be separated by a distance L. Further, let the intensity of the light beam be I_o at M_1. Then, in travelling from mirror M_1 to mirror My the beam intensity increases from I_o to I(L), which is given by

$$I(L) = I_o e^{(\gamma-\alpha_s)L} \quad \text{...(14)}$$

After reflection at M_2 the beam intensity will be $r_2 I_o e^{(\gamma-\alpha_s)L}$ and after a complete round trip the final intensity will be

$$I\,(2L)\; r_1 r_2 I_o e^{(\gamma-\alpha_s)2L} \quad \text{...(15)}$$

The amplification obtained during the round trip is

$$G = \frac{I(2L)}{I_o} = r_1 r_2 e^{(\gamma - \alpha_s)2L} \qquad ...(16)$$

The product r_1r_2 represents the losses at the mirrors whereas α_s includes all the distributed losses such as scattering, diffraction and absorption occurring in the medium. The losses are balanced by gain, when $G \geq 1$ or $I(2L) = I_o$. It requires that

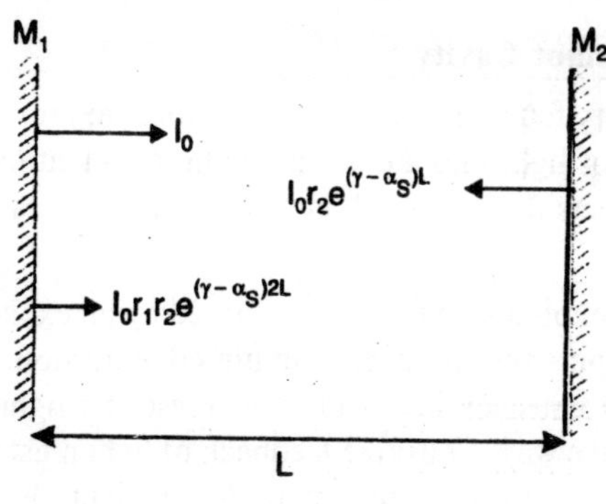

Fig. 4.15

$$r_1 r_2\, e^{2(\gamma - \alpha_s)L} \geq 1 \qquad ...(17)$$

or $$e^{2(\gamma - \alpha_s)L} \geq \frac{1}{r_1 r_2}$$

Taking logarithms on both sides, we get

$$2L(\gamma - \alpha_s) \geq \ln r_1 r_2$$

$$\gamma - \alpha_s \geq -\frac{1}{2L} \ln r_1 r_2$$

$$\gamma \geq \alpha_s - \frac{1}{2L} \ln r_1 r_2 \qquad ...(18)$$

or $$\gamma \geq \alpha_s + \frac{1}{2L} \ln \frac{1}{r_1 r_2} \qquad ...(19)$$

Eqn. (19) is known as the condition for lasing. It shows that the initial gain must exceed the sum of the losses in the cavity. This condition is used to determine the threshold value of pumping energy for lasing action.

γ, the amplification of the laser will be dependent on how hard the laser medium is pumped. As the pump power is slowly increased, a value of γ_{th} called threshold value is reached and the laser starts oscillating. The threshold value γ_{th} is given by

$$\gamma_{th} = \alpha_s + \frac{1}{2L} \ln \frac{1}{r_1 r_2} \quad ...(20)$$

Eqn. (20) states the condition when the net gain would be able to counteract the effect of losses in the cavity and is known as the *threshold condition for lasing*. The value of γ must be atleast γ_{th} for laser oscillations to commence.

Function of Resonant Cavity

(i) The primary function of optical resonator is to provide a positive feedback of light into the lasing medium so that the stimulated emission acts are sustained and the laser acts as a generator of light.

(ii) A chance photon spontaneously emitted by an excited atom acts as the input and induces stimulated emission. To sustain stimulated emission acts and to increase the light intensity in a cumulative way, a positive feedback of light must be provided. The mirrors by way of reflecting the incident photons provide the feedback.

(iii) Laser oscillation is initiated by the photons spontaneously radiated by some of the excited atoms.

Each spontaneous photon can trigger many stimulated transitions along the path of its travel. As the initial spontaneous photons are emitted in various directions, the secondary stimulated photons will also travel in various directions. The result would be production of incoherent light. In order to build up oscillations, a specific direction has to be defined for photon propagation in the lasing medium. The optical resonator sets its optic axis as the most favourable direction for build-up of light beam.

(iv) In order to make the stimulated emissions dominate spontaneous emissions, a high optical energy density $\rho(\nu)$ is necessary to be present in the active medium. The mirrors constituting the cavity confine more than 90% of the emitted photons to be within the laser medium such that a very high optical energy density is always present in the lasing medium.

(v) The optical cavity is very much similar to a resonating column. Just as standing waves form in a resonating column, standing waves of optical frequencies are formed in the optical cavity. If L is the length of the cavity, the longest wavelength that will produce a standing wave pattern is $\lambda = 2L$. In general the cavity supports the wavelengths

$$\lambda_m = \frac{2L}{m} \quad (m = 1,2,3,4 \ldots) \tag{21}$$

Waves of other wavelengths attenuate quickly. Thus, the optical cavity selects and amplifies only certain frequencies. Hence, the value of L should be properly selected.

(vi) Active centers may have a number of lasing transitions instead of only one transition. The mirrors of optical cavity suppress the undesired transitions. The reflectivity of the mirrors is further made less for the undesired photons, which therefore get absorbed at the mirrors.

AXIAL MODES

The wave properties of photons require that the waves be in phase and interfere constructively within the optical resonator such that progressive enhancement of light intensity takes place. The condition for constructive interference is that the path length travelled by a wave between two consecutive reflections at an end mirror should equal an integral multiple of the wavelength. It means that

$$2L = m\lambda \quad (m = 1, 2, 3 \ldots)$$

or

$$L = m\,(\lambda/2) \tag{22}$$

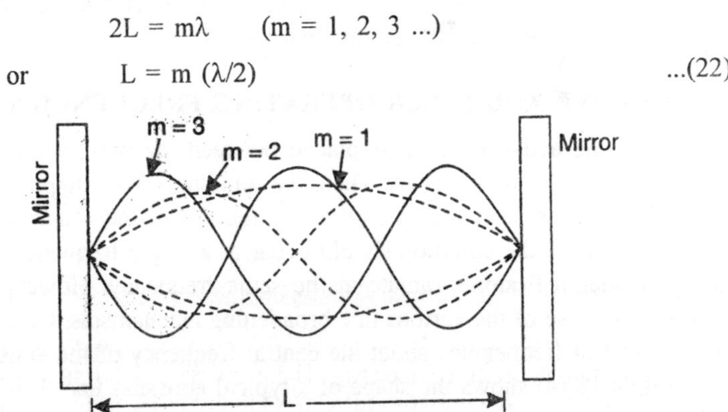

Fig. 4.16 : Standing wave pattern and axial modes in optical resonator.

The above equations indicate that only those waves, which can form standing wave pattern (Fig. 4.16), can exist inside the cavity in a steady state. Waves of other wavelengths interfere destructively and are quickly attenuated. Because of its length which is very large compared to light wavelength, optical resonator supports simultaneously several standing waves of multiple wavelengths. These wavelengths are called *longitudinal* or *axial modes*. Therefore, m in eqn. (22) is called the *mode number*.

The frequencies of the modes are given by

$$\nu_m = \frac{mc}{2L} \qquad \text{...(23)}$$

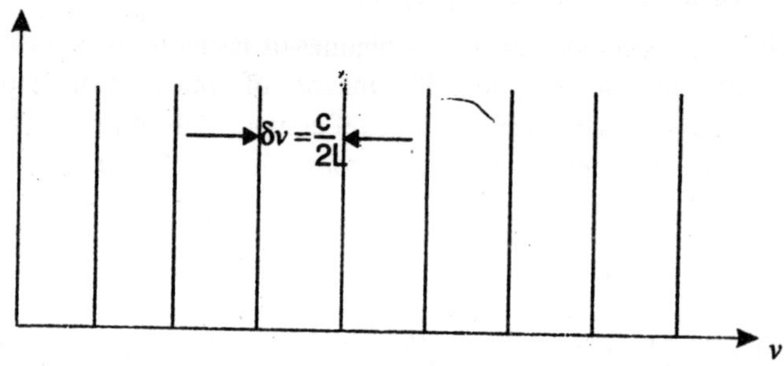

Fig. 4.17 : Cavity resonance frequencies representing the possible v longitudinal modes.

Theoretically, the cavity can resonate at a very large number of frequencies (Fig. 4.17) that satisfy the equation (23). For example, if we take L = 0.5 m, and λ, = 5000Å, we obtain m = 2 × 10^6. It means that the cavity supports 2 × 10^6 longitudinal modes.

GAIN CURVE AND LASER OPERATING FREQUENCIES

When the active medium is placed between the two mirrors, the cavity becomes an *active cavity*. Then, the standing waves supported by the cavity are the light waves emitted by the stimulated atoms of the medium. Ideally, the emission should occur at a single frequency, as a group of identical atoms radiate at the same frequency. However, in practice, because of the various line broadening mechanisms, there will be a spread of frequencies about the central frequency of the emission line. Fig. 4.18 (a) shows the shape of a typical emission line. It is also called gain *curve* or *gain profile* because it indicates the range of frequencies over which stimulated emission can provide sufficient gain.

The laser operating frequencies are determined together by the resonant frequencies of the cavity and by the laser emission line width. If an output has to exist at a particular frequency, the cavity must be resonant at that specific frequency and there must be sufficient gain at that frequency. Laser oscillation can take place only when the gain is large enough to maintain resonance. For example, two dashed lines are shown in Fig. 4.18 (a) which correspond to two threshold levels. If the threshold is at level (1), then only the central frequency v_0 is amplified. When the threshold is at (2), all frequencies between v_1 and v_2 can be amplified. However, there are only a few frequencies that resonate. As a result, the output of a laser consists of a few closely spaced frequencies as shown in Fig. 4.18 (b). Thus, the laser emission line transforms into a series of names spectral lines corresponding to cavity modes. If $\delta\lambda$ is the line width of the emission line, then me number of modes that would be ultimately present is given by

$$N = \frac{\delta\lambda}{\Delta\lambda} \qquad ...(24)$$

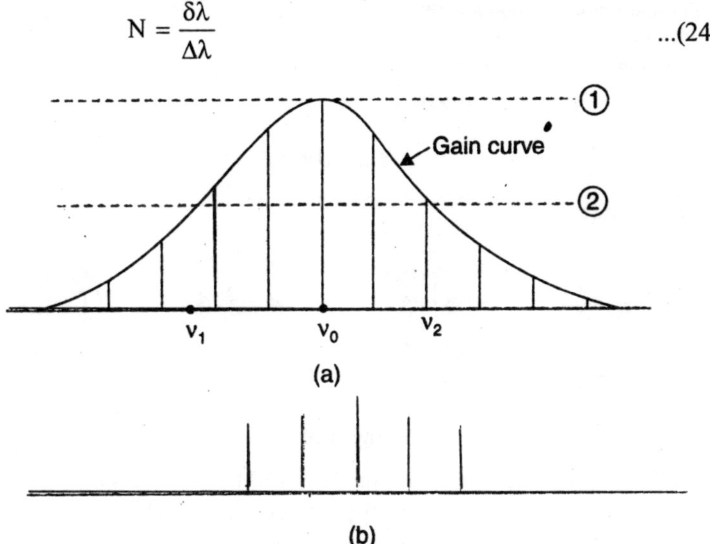

Fig. 4.18

TRANSVERSE MODES

The longitudinal dimension L of the resonant cavity governs the axial modes. The laser modes governed by the cross-sectional dimension of the optical cavity are called transverse *electromagnetic (TEM) modes* (Fig. 4.19). The TEM modes are generally few in numbers and they are easy to see. If the laser beam is spread out by a negative lens and focussed on to a screen, several bright patches are seen on the screen. The patches are separated by intervals called nodal lines. The transverse modes characterize the intensity distribution across the cross-section of the laser beam. In general, the allowed modes are designated as TEM_{mn}, where m and n are integers. The integers m and n represent the number of intensity minima in two orthogonal directions of the laser beam.

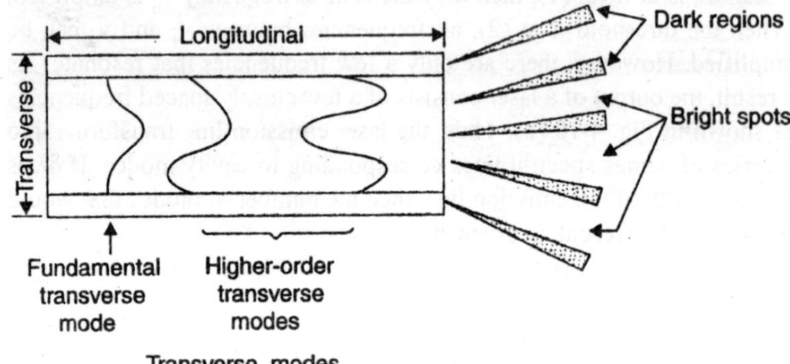

Fig. 4.19

BEAM PATTERN

$TEM_{0\,0}$

$TEM_{0\,1}$

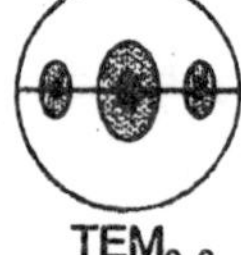

$TEM_{0\,2}$

$TEM_{1\,1}$

Fig. 4.20

The lowest order transverse mode is TEM_{00}. It is the simplest mode and has a smooth crosssection profile with a peak in the middle (Fig. 4.20). TEM_{01}. beam has a single minimum dividing the beam into two bright spots. A TEM_{11} beam has two perpendicular minima dividing the beam into four quadrants, and so on. Operation of a laser in multimode form provides considerably more power than in single mode operation.

TYPES OF LASERS

There are several ways in which we can classify lasers into different types. We prefer here to classify the lasers on the basis of the material used as active medium. Accordingly, they are broadly divided into four categories, namely solid state lasers, gas lasers, liquid lasers, and semiconductor diode lasers. Most lasers emit light in the red or IR regions. Lasers work in a continuous mode or in a pulsed mode.

Ruby Laser

Ruby laser belongs to the class of solid state lasers. The term solid state has different meanings in the field of electronics and lasers. A solid state laser is one in which the active centers are fixed in a crystal or glassy material. Solid state lasers are electrically nonconducting. They are also called *doped insulator lasers*.

Historically, the ruby laser was the first laser. It was invented in 1960 by Theodore Mail-nan, U.S.A. The ruby laser rod is in fact a synthetic ruby crystal. AlO_3 crystal, doped with chromium ions at a concentration of about 0.05% by weight. Cr^{3+} ions are the actual active centers and have a set of three energy levels suitable for realizing lasing action whereas aluminium and oxygen atoms are inert.

Construction : The schematic of a ruby laser is shown in Fig. 4.21. Ruby rod is taken in the form of a cylindrical rod of about 4 cm in length and 0.5 cm in diameter. Its ends are grounded and polished such that the end faces are exactly parallel and are also perpendicular the axis of the rod. One face is silvered to achieve 100% reflection while the other is silvered to give 10% transmission and 90% reflection. The silvered faces constitute the Fabry-Perot resonator. The laser rod is surrounded by a helical photographic flash lamp filled with xenon. Whenever activated by the power supply the lamp produces flashes of white light.

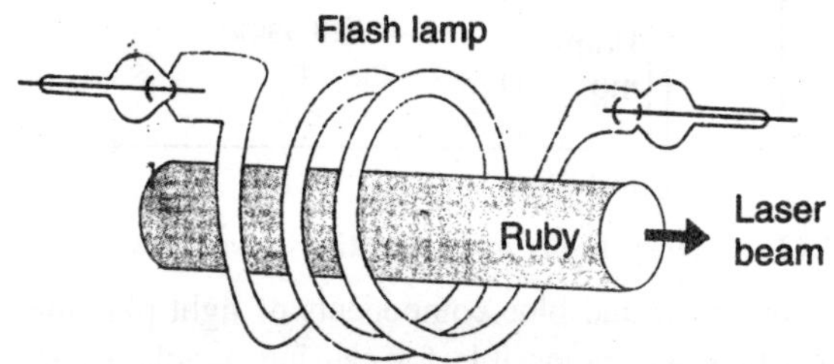

Fig. 4.21 : Schematic of a ruby laser.

Working : Ruby laser uses a three-level pumping scheme. The energy levels of Cr^{3+} ions in the crystal lattice are shown in Fig. 4.22. There are two wide energy bands E_3 and E'_3 and a pair of closely spaced levels at E_2. When the flash lamp is activated, the xenon discharge generates an intense burst of white light lasting for a few milliseconds. The Cr^{3+} ions are excited to the energy bands E_3 and E'_3 by the green and blue components of .white light. The energy levels in these bands have a very small lifetime ($\approx 10^{-9}$s). Hence the excited Cr^{3+} ions rapidly lose some of the energy to the crystal lattice and undergo non-radiative transitions. They quickly drop to the levels E_2. The pair of levels at E_2 are metastable states having a lifetime of approximately 1000 times more than the lifetime of E_3 level. Therefore, Cr^{3+} ions accumulate at E_2 level. When more than half of the Cr^{3+} ion population accumulates at E_2 level, the state of population inversion is established between E_2 and E_1 levels. A chance photon emitted spontaneously by a Cr^{3+} ion initiates a chain of stimulated emissions by other Cr^{3+} ions in the metastable state. Red photons of wavelength 6943Å travelling along the axis of the ruby rod are repeatedly reflected at the end mirrors and light amplification takes place. A strong intense beam of red light emerges out of the front-end mirror.

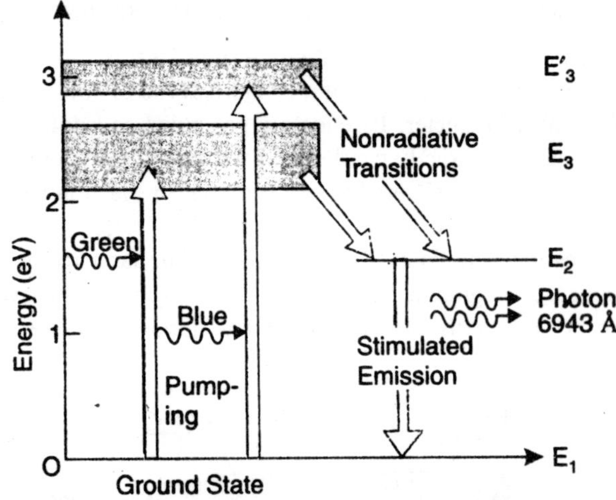

Fig. 4.22 : Energy levels and transitions in a ruby laser.

Note that the green and blue components of light play the role of pumping agents and are responsible for causing population inversion. The spontaneous photons of λ = 6943Å, corresponding to red colour, act as the input of the oscillator which actually gets amplified. The xenon flash lasts for a few milliseconds. However, the laser does not operate throughout this period. Its output occurs in the form of irregular pulses of microsecond duration. It is because the stimulated transitions occur faster than the rate at which population inversion is maintained in the crystal. Once stimulated transitions commence, the metastable state E_2 gets depopulated very rapidly and at the end of each small pulse, the population at E_2 has fallen blow the threshold value required for sustained emission of light. As a result the lasing ceases and laser becomes inactive.

The next pulse appears after the population inversion is once again restored. The process repeats.

Nd: Yag Laser

Nd : YAG laser is one of the most popular types of solid state laser. It is a four-level laser. Yttrium aluminium garnet, $Y_3Al_5O_{12}$, commonly called YAG is an optically isotropic crystal. Some of the Y^{3+} ions in the crystal are replaced by neodymium ions, Nd^{3+}. Doping concentrations are typically of the order of 0.725% by weight. The crystal atoms do not participate in the lasing action but serve as a host lattice in which the active centres, namely Nd^{3+} ions reside.

Construction : Fig. 4.23 illustrates a typical design of Nd: YAG laser. The system consists of an elliptically cylindrical reflector housing the laser rod along one of its focus line and a flash lamp along the other focus line. The light leaving one focus of the ellipse will pass through the other focus after reflection from the silvered surface of the reflector. Thus the entire flash lamp radiation gets focused on the laser rod. The YAG crystal rods are typically of 10 cm in length and 12 mm in diameter. The two ends of the laser rod are polished and silvered and constitute the optical resonator.

Working : A simplified energy level diagram for the neodymium ion in YAG crystal is shown in Fig. 4.24. The energy level structure of the free neodymium atom is preserved to a certain extend because of its relatively low concentration. However, the energy levels are split and the structure is complex. It is essentially a four-level system with the terminal laser level E_2 sufficiently far removed from the ground level. The pumping of the Nd^{3+} ions to upper states is done by a krypton arc lamp. The optical pumping with light of wavelength range of 5000 to 8000Å excites the ground state Nd^{3+} ions to the multiple energy levels at E_4. The metastable level E_3 is the upper laser level, while the E_2 forms the lower laser level. The upper laser level E_3 will be rapidly populated, as the excited Nd^{3+} ions quickly make downward transitions from the upper energy bands. The lower laser level E_2 is far above the ground level and hence it cannot be populated by Nd^{3+} ions through thermal transitions from the ground level. Therefore, the population inversion is readily achieved between the E_3 level and E_2 level. The laser emission occurs in infrared (IR) region at a wavelength of about 10,600Å (1.06 μm). As the laser is a four level laser, the population inversion can be maintained in the face of continuous laser emission.

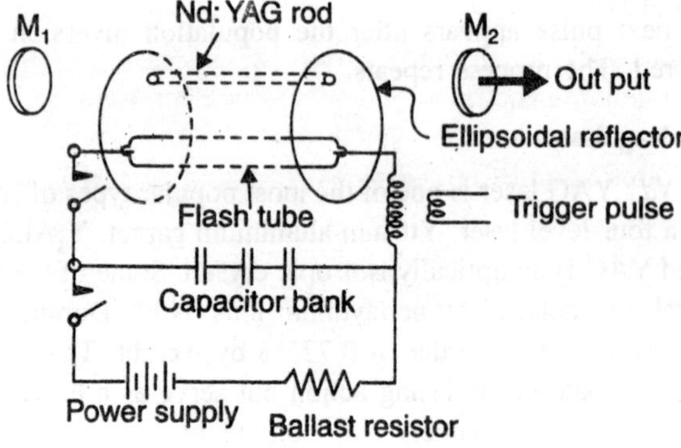

Fig. 4.23

Thus Nd: YAG laser can be operated in CW mode. An efficiency of better than 1 % is achieved. Nd: YAG lasers find many industrial applications such as resistor trimming, machining operations like welding, hole drilling etc. They are also used in surgery.

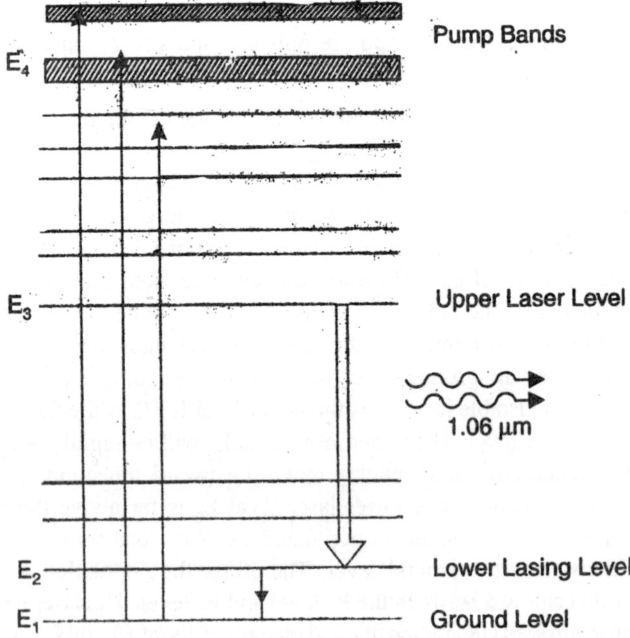

Fig. 4.24 : Energy levels and transitions in a Nd: YAG later.

Helium-Neon Laser

Gas lasers are the most widely used lasers. They range from the low power helium-neon laser used in college laboratories to very high power carbon dioxide laser used in industrial applications. These lasers operate with rarefied gases as the active media and are excited by an electric discharge. In gases, the energy levels of atoms involved in the lasing process are narrow and as such require sources with sharp wavelength to excite atoms. Finding an appropriate optical source for pumping poses a problem. Therefore optical pumping is not used in gas lasers. The most common method of exciting gas laser medium is by passing an electric discharge through the gas. Electrons present in the discharge transfer energy to atoms in the laser gas by collisions.

The first gas laser was He-Ne laser which was invented in 1961 by Ali Javan, William R. Bennett, Jr. and Donald R.Hemott.

Construction : The schematic of a He-Ne laser is shown in Fig. 4.25. Helium-Neon laser consists of a long discharge tube filled with a mixture of helium and neon gases in the ratio 10:1. Neon atoms are the active centers and have energy levels suitable for laser transitions while helium atoms help in exciting neon atoms. Electrodes are provided in the discharge tube to produce discharge in the gas. They are connected to a high voltage power supply. The tube is hermetically sealed by inclined windows arranged at its two ends. On the axis of the tube, two mirrors arc arranged externally which form the Fabry-Perot optical resonator. The distance between the mirrors is adjusted to be m λ/2 such that the resonator supports standing wave pattern.

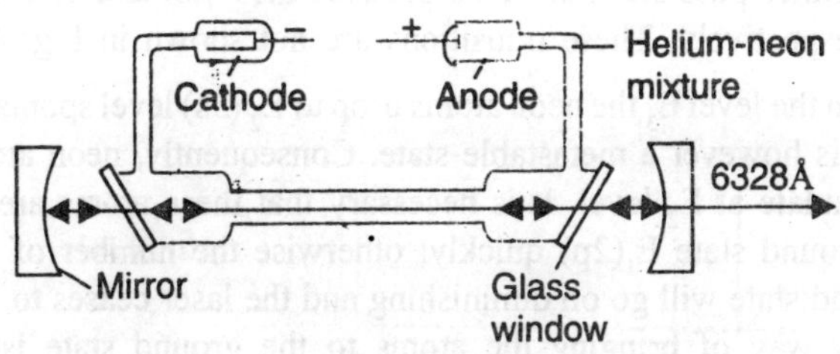

Fig. 4.25

Working : Helium-Neon laser employs a four-level pumping scheme. The energy levels of helium and neon are shown in Fig. 22.28. When the power is switched on, a high voltage of about 10 kV is applied across the gas. It is sufficient to ionize the gas. The electrons and ions produced in the process of discharge are accelerated towards the anode and cathode respectively. The energetic electrons excite helium atoms through collisions. One of the excited levels of helium F_3 (2s) is at 20.61 eV above the ground level. It is a metastable level and the excited helium atom cannot return to the ground level through spontaneous emission. However, it can return to the ground level by transferring its excess energy to a neon atom through collision. Such an energy transfer can take place when the two colliding atoms have identical energy levels. Such an energy transfer is known as *resonant energy transfer*. One of the excited levels of neon E_6 (5s) is at 20.66eV, which is nearly at the same level as F_3 of helium atom. Therefore, resonant transfer of energy can occur between the excited helium atom and ground level neon atom. The kinetic energy of helium atoms provides the additional 0.05 eV required for excitation of the neon atoms. Helium atoms drop to the ground state after exciting neon atoms. This is the pumping mechanism in He-Ne laser. The role of helium atoms is to excite neon atoms and to cause population inversion. The probability of energy transfer from helium atoms to neon atoms is more, as there are 10 helium atoms per 1 neon atom in the gas mixture. The probability of reverse transfer of energy from neon to helium atom is negligible.

The upper state of neon atom E_6 is a metastable state. Therefore, neon atoms accumulate in this upper state. The E_3 (3p) is sparsely populated at ordinary temperatures, and a state of population inversion is readily established between E_6 and E_3 levels. Random photons emitted spontaneously prompt stimulated emission and lasing occurs. The transition $E_6 \rightarrow E_3$ generates a laser beam of red colour of wavelength 6328Å. Other possible transitions produce 3.39 μm and 1.15 μm laser beams respectively. These transitions are not shown in Fig. 4.26.

From the level E_3 the neon atoms drop to E_2 (3s) level spontaneously. E_2 level is however a metastable state. Consequently, neon atoms tend to accumulate at E_2 level. It is necessary that these atoms are brought to the ground state E_1(2p) quickly; otherwise the number of atoms at the ground state will go on diminishing and the laser ceases to function. The only way of bringing the atoms to the ground state is through collisions. If the discharge tube is made narrow, the probability of atomic collisions with the tube walls increases. Because of frequent collisions with the walls, the neon atoms rapidly drop to the ground level and will be available for excitation once again.

If the diameter of the discharge tube is increased, the probability of collisions of atoms with the walls decreases and the neon atoms tend to accumulate at energy level E_2. In due course of time, the atoms are no more available at the ground level for further excitation. Therefore, the laser ceases to operate.

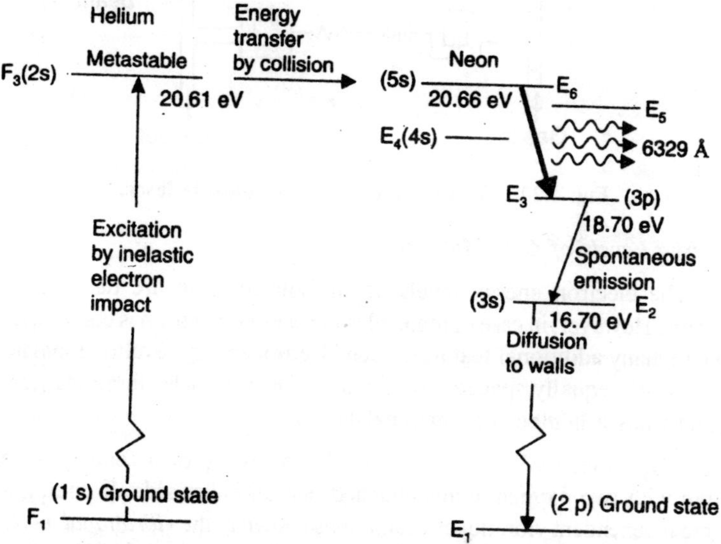

Fig. 4.26 : Energy level diagram for a helium-neon laser. Only the relevant energy levels are shown.

He-Ne laser operates in cw mode and is widely used in laboratories as a monochromatic source. It is also widely used in laser printing, bar code reading, etc.

Carbon Dioxide Laser

The carbon gas laser is a very useful and efficient laser. It is a four-level molecular laser and operates at 10.6 μm in far IR region.

Construction : The schematic of typical CO_2 laser is shown in Fig. 4.27. It is basically a discharge tube having a bore of cross section of about 1.5 mm^2 and a length of about 260 mm. The discharge tube is filled with a mixture of carbon dioxide, nitrogen and helium gases in 1:4:5 proportions respectively. Other additives such as water vapour are also added. The active centres are CO_2 molecules lasing on the transitions between the vibrational levels of the electronic ground state.

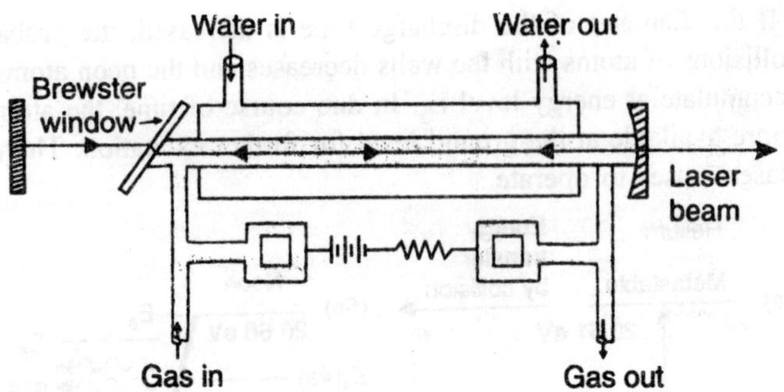

Fig. 4.27 : Schematic of a carbon dioxide laser.

Energy Levels of CO_2 Molecule

The electron energy levels of an isolated atom are discrete and narrow. However, in case of molecules the energy spectrum is complicated due to many additional features. Each electron energy level is associated with nearly equally spaced vibrational levels and each vibrational level in turn has a number of rotational levels.

CO_2 molecule is a linear molecule consisting of a central carbon atom with two oxygen atoms attached one on either side. It undergoes three independent vibrational-oscillations known as the *vibrational modes*. These vibrational degrees of freedom are quantized. At any one time, a CO_2 molecule can vibrate in a linear combination of three fundamental modes. The energy states of the molecule are then represented by three quantum numbers (m n q). These numbers represent the amount of energy associated with each mode. For example, the number (020)

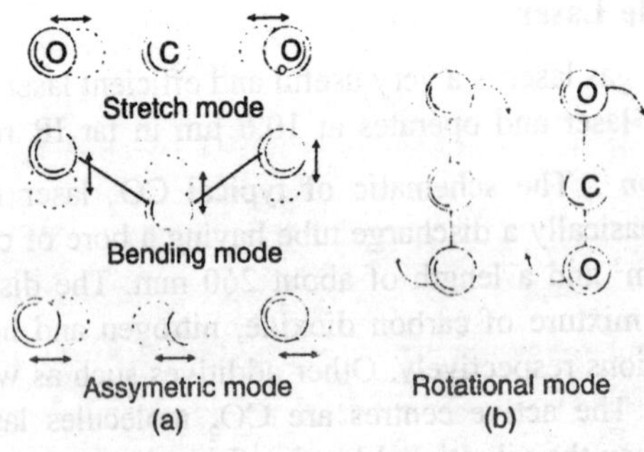

Fig. 4.28 : Vibrational modes of a CO_2 molecule.

indicates that the molecule in this energy state is in the pure bending mode with two units of energy, Each vibrational state is associated with rotational states corresponding to the rotation of CO_2 molecule about its centre of mass. The separations between vibrational-rotational states are much smaller on the energy scale compared to the separations between electron energy levels. The nitrogen molecule N_2 is also characterized by similar vibrational levels. Fig. 4.28 shows the vibrational modes and rotations of CO_2 molecule.

Working : Fig. 4.29 shows the lowest vibrational levels of the ground electron energy state of CO_2 molecule and an N_2 molecule. The excited state of an N_2 molecule is metastable and it is identical in energy to (001) vibrational level of CO_2 molecule, indicated as E_5 in Fig. 4.29. As current passes through the mixture of gases, the N3 molecules get excited to the metastable state. The excited N_2 molecules cannot spontaneously lose their energy and consequently, the number of N_2 molecules at the level keeps on increasing. The N_2 molecules return to ground state through inelastic collisions with ground state CO_2 molecules. Thereby the CO_2 molecules are excited to E_5 level. Some of the CO_2 molecules are also excited to the upper level E_5 through collisions with electrons. The excitation of CO_2 molecules through collisions with excited N_2 molecules is similar to that of helium atoms by neon atoms in helium-neon laser. The E_5 level is the upper lasing level while the (020) and (100) states marked as E_3 and E_4 levels act as the lower lasing levels. As the population of CO_2 molecules builds up at E_5 levels, population inversion is achieved between E_5 level and the levels at E_4 and E_3, The laser transition between $E_5 \rightarrow E_4$ levels produces far IR radiation at the wavelength 10.6 μm (1,06,000Å). The lasing transition between $E_5 \rightarrow E_3$ levels produces far IR radiation at 9.6 μm (96,000Å) wavelength. E_3 and E levels are also metastable states and the CO_2 molecules at these levels fail to the lower level E_2 through inelastic collisions with normal (unexcited) CO_2 molecules. This process leads to accumulation of population at E_2 level. And also, as the gaseous mixture heats up, the E_2 level, which is close to the ground state, E_1, tends to be populated through thermal excitations. Thus, the de-excitation of CO_2 molecules at the lower lasing level poses a problem and inhibits the laser action. The presence of helium along with CO_2 helps to decrease the population density at E_2 level. It de-exeites CO_2 molecules through inelastic collisions and aids cooling the gaseous mixture through heat conduction.

The CO_2 laser operates in CW mode and is capable of generating high powers of the order of several kilowatts at a relatively high efficiency of about 40%. Therefore, it is the most widely used laser. Its applications include use in communications, weaponry and laser fusion.

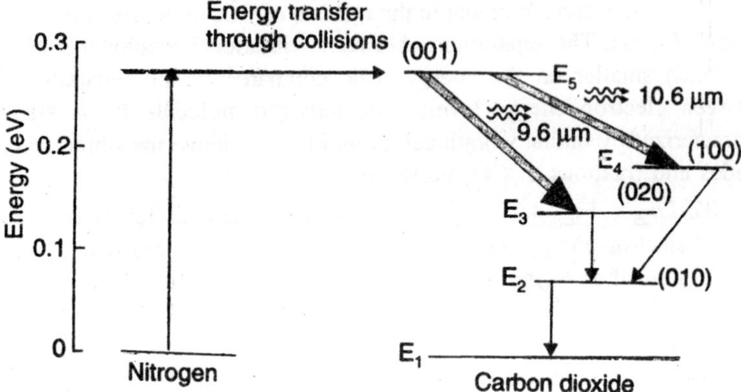

Fig. 4.29 : Energy levels of nitrogen and carbon dioxide molecules and transitions between the levels.

SEMICONDUCTOR LASER

A *semiconductor diode laser* is a specially fabricated pn junction device, which emits coherent light when it is forward biased. R.N. Hall and his coworkers made the first semiconductor laser in 1962. It is made from Gallium arsenide (GaAs) which operated at low temperatures and emitted light in the near IR region. Semiconductor lasers working at room temperature and in continuous wave mode are produced by 1970. Now pn-junction lasers are made to emit light almost anywhere in the spectrum from UV to IR. Diode lasers are remarkably small in size (0.1mm long). They have high efficiency of the order of 40%. Modulating the biasing current easily modulates the laser output. They operate at low powers. In spite of their small size and low power requirement, they produce power outputs equivalent to that of He-Ne lasers. The chief advantage of a diode laser is that it is portable. Because of the rapid advances in semiconductor technology, diode lasers are mass produced for use in optical fibre communications, in CD players, CD-ROM drives, optical reading, high speed laser printing etc wide variety of applications.

A *semiconductor* is a material with electrical properties intermediate to those of a conductor and an insulator. The allowed energy values of the valence electrons in semiconductors occur within two well-defined energy bands separated by an energy gap known as *band gap*. A pure semiconductor crystal has exactly enough electrons to fill all the states in the lower band, namely valence band. However, when a covalent bond is just broken, an electron is just set free. Then we say that the electron jumped into the upper band, namely *conduction band.* The electron jumping to the conduction band leaves behind a vacancy in the valence band. The vacancy is called a hole and is assigned a positive charge and a mass equivalent to that of an electron. In a pure semiconductor, for each covalent bond broken an electron and a hole are generated. Therefore, the number of electrons in the conduction band and the number of holes in the valence band are equal. When a conduction electron falls into the valence band, it recombines with a hole there. The electron rejoins the broken covalent bond and therefore both the electron and hole disappear. The recombination energy is released in the form of heat in silicon and germanium crystals. In some crystals it is released in the form of light.

Doping with small amounts of impurities can drastically increase the electrical conductivity of a pure semiconductor. When the dopant is a pentavalent element, each dopant atom contributes an electron to the conduction band without creating a hole simultaneously in the valence band. Hence the addition of the pentavalent element increases the number of conduction electrons which become the *majority carriers* in the silicon crystal. As negatively charged electrons are current carriers in this crystal, it is called a *n-type semiconductor*. On the other hand, a trivalent dopant atom produces a hole in the valence band without the simultaneous generation of electron in the conduction band. Hence the addition of the trivalent element increases the number of holes which become the *majority carriers* in the silicon crystal. As positively charged holes are current carriers in this crystal, it is called a *p-type semiconductor*.

There is a reference level in the energy band diagram of each type of semiconductor. The reference level is called the *Fermi level*. The Fermi level Ep is nearer to the top of the valence band in the p-type material and the Fermi level E_{Fn} is nearer to the bottom of the conduction band in the n-type material. When the p-type and n-type materials are joined at the atomic level to form a pn-junction device, equilibrium is attained only when equalization of Fermi levels takes place. The energy levels in p-region move up and those in n-region move down till the Fermi levels (E_{Fn} and E_{Fn}) in both the regions come to the same level. The mutual displacement of the energy levels on both sides of the junction causes a bending of the energy bands around the junction.

Achieving Population Inversion in a Semiconductor

Population inversion is required for producing stimulated emission. The way in which population inversion is achieved in semiconductors is very different from the way it is established in other types of lasers. A semiconductor cannot be regarded as two-level atomic system. It consists of electrons and holes distributed in the respective energy bands. Therefore, laser action in semiconductors involves energy bands rather than discrete levels. Secondly, in other types of lasers, population inversion is obtained by exciting electrons in spatially isolated atoms. In semiconductors, electrons are not associated with specific atoms but are injected into the conduction band from the external circuit. Therefore, the conduction band plays the role of excited level while the valence band plays the role of ground level. Population inversion requires the presence of a large concentration of electrons in the conduction band and a large concentration of holes in the valence band. A simple way to achieve population inversion is to use a semiconductor in the form of a pn-junction diode formed from heavily doped p- and n-type semiconductors.

PN-Junction Laser

Construction : Fig. 4.30 shows the schematic of a semiconductor laser. A simple diode makes use of the same semiconductor material, say, GaAs on both sides of the junction. Starting with a heavily doped n-type GaAs material, a p-region is formed on its top by diffusing zinc atoms into it. A heavily zinc doped layer constitutes the heavily doped p-region. The diode is extremely small in size. Typical diode chips are 500 μm long and about 100 μm wide and thick. The top and bottom faces are metallized and metal contacts are provided to pass current through the diode. The front and rear faces are polished parallel to each other and perpendicular to the plane of the junction. The polished faces constitute the Fabry-Perot resonator. In practice there is no necessity to polish the faces. A pair of parallel planes cleaved at the two ends of the pn-junction provides the required reflection to form the cavity. The two remaining sides of the diode are roughened to eliminate lasing action in that direction. The entire structure is packaged in small case which looks like the metal case used for discrete transistors.

Working : The energy band diagram of a heavily doped pn-junction is shown in Fig. 4.31 (a). Because of very high doping on n-side, the donor levels are broadened and extend into the conduction band. The

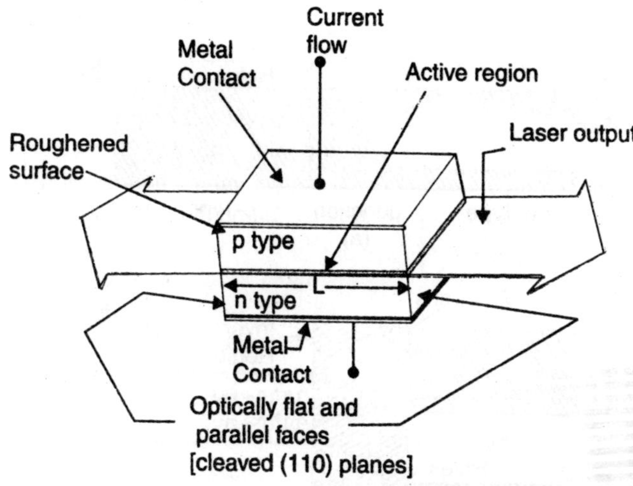

Fig. 4.30 : Schematic of a semiconductor diode laser.

Fermi level also is pushed into the conduction band. Electrons occupy the portion of the conduction band lying below the Fermi level. Similarly, on the heavily doped p-side the Fermi level lies within the valence band and holes occupy the portion of the valence band that lies above the Fermi level. At thermal equilibrium, the Fermi level is uniform across the junction.

When the junction is forward-biased, electrons and holes are injected into the junction region in high concentrations. In other words, carriers are *pumped* by the dc voltage source. At low forward current level, the electron-hole recombination causes spontaneous emission of photons and the junction acts as an LED. As the forward current through the junction is increased the intensity of the light increases linearly. However, when the current reaches a threshold value (Fig. 4.32), the carrier concentrations in the junction region will rise to a very high value. As a result, die junction region (Fig. 4.31 b) contains a large concentration of electrons within the conduction band and *simultaneously* a large number of holes within the valence band. Holes represent absence of electrons. Thus, the upper energy levels in the narrow region are having a high electron population while the lower energy levels in the same region are vacant. Therefore, the condition of population inversion is attained in the narrow junction region. This narrow zone in which population inversion occurs is called an *inversion region* or *active region.* Chance recombination acts of electron and hole pairs lead to emission

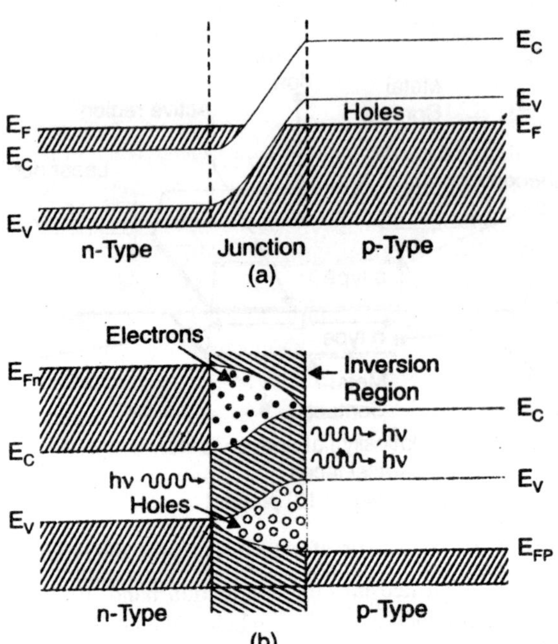

Fig. 4.31 : Energy band structure of a semiconductor diode (a) Heavily doped pn-junction without bias. (b) Heavily doped pn-junction forward biased above threshold value.

of spontaneous photons. The spontaneous photons propagating in the junction plane stimulate the conduction electrons to jump into the vacant states of valence band. This stimulated electron-hole recombination produces coherent radiation. GaAs laser emits light at a wavelength of 9000Å in IR region.

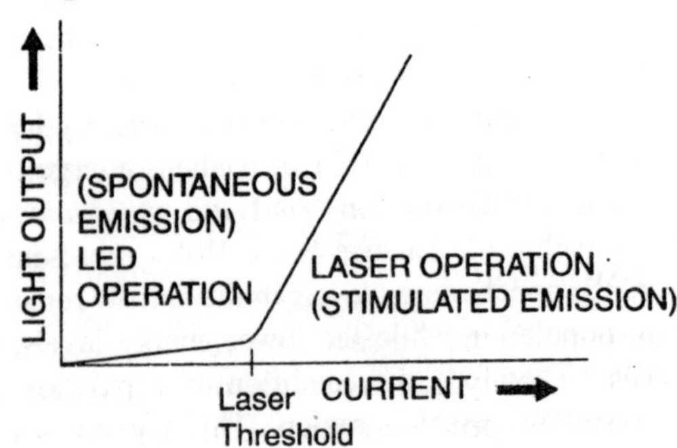

Fig. 4.32 : Light output-current characteristic of an ideal diode laser.

Q-SWITCHING

Since a laser is an oscillator, its resonator cavity is characterized by the quality factor, Q as is done in the case of an electronic oscillator. Q is defined as the ration of the energy W stored in the system to the energy losses AW per cycle. Thus,

$$Q = 2\pi \frac{\text{Energy stored in the resonator}}{\text{Energy lost in a cycle}}$$

In case of lasers Q is very high and is of the order of 108. Q-factor is a measure of the mirror losses.

The power of a laser beam can be drastically increased, provided a very large number of atoms in the active medium participates in stimulated emission. It can happen only when a very high population inversion density is established in the active medium. Normally when the laser starts oscillations, the high population inversion drops back to the threshold value of the steady state condition. It becomes therefore necessary that the onset of oscillations is delayed until largest possible number of atoms accumulates at the upper lasing level. The laser can be prevented from oscillating if, for example, the parallelism of the resonator mirrors is disturbed. If one of the end mirrors is misaligned, it cannot reflect incident photons into the active medium and therefore, stimulated emission cannot take place. Consequently, the pumping process can build up the population inversion to a very high value in the medium. In effect, Q-factor of resonator is spoiled and optical losses are increased to a high value. If now the end mirror is aligned suddenly, it reflects photons into the medium. The feed back of photons triggers a chain of stimulated emissions and builds up rapidly a photon avalanche. Thus, laser oscillations set in suddenly and the Q of the cavity is increased abruptly. All the energy stored in the cavity is emitted in a single giant pulse with peak power much higher than the laser could produce otherwise. The pulse lasts for a short time and depopulates the upper energy level quickly and the lasing action stops. This method of controlling the laser output power is called Q-switching method.

Method of Q-Switching

An electro-optic shutter can serve as a voltage controlled gate which rapidly switches the cavity from a passive to an active state. The shutter consists of a crystal that becomes double refracting when an electric field is applied across the crystal. The arrangement is shown in Fig. 4.33. A voltage is applied to the crystal during the pumping of the laser by the light from a flash lamp. The magnitude of the voltage is chosen such that it transforms the electro-optic crystal into a quarter- wave plate. Light emitted by the laser becomes linearly polarized on passing through the polarizer. The linearly polarized light is then incident on the electro-optic crystal and gets split into two mutually orthogonal components. As the two components travel through the electro-optic crystal, a relative phase retardation of 90° is produced between them. On emerging from the electro-optic crystal, the two components combine to produce circularly polarized light. The light beam reflects from the mirror and returns to the cavity travelling in the opposite direction. On reflection the sense of rotation of the circularly polarized light reverses and on repassing through the crystal, the two components of circularly polarized light experience a further retardation of 90° with respect to each other. Coming out of the crystal at the other end, the components recombine to produce linearly polarized light. The direction of polarization of the light is now al 90° with respect to its original direction of polarization and transmission axis of the polarizer. This light is not allowed to pass through the polarizer. Therefore, light does not come back to the laser rod and the cavity is switched off. Thus, the cavity Q is reduced to a low value. When the voltage applied to the electro-optic crystal is turned off, double refraction is absent in the crystal and the state of polarization

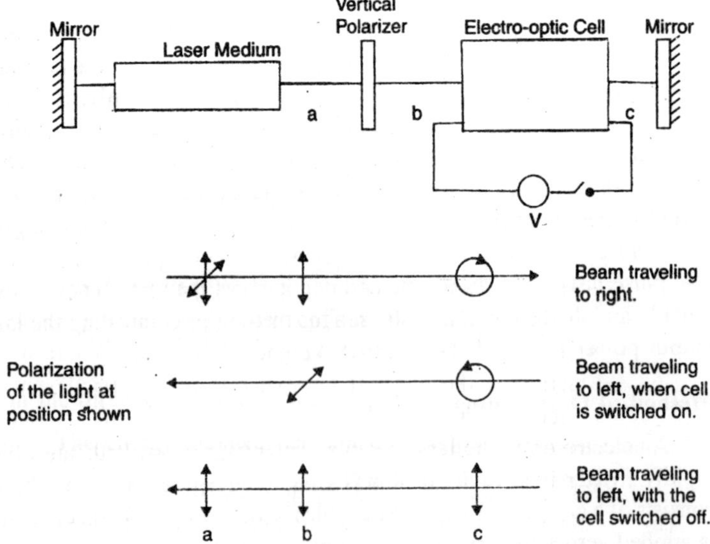

Fig. 4:33 : Electro-optic cell used as a Q-switch.

of the light passing through the crystal is unaffected. Light freely travels in both directions and returns to the laser rod to be further amplified. Now, the cavity is switched on and the Q regains its high value. The Q-switching is synchronized with pumping mechanism such at hat the voltage applied to the electro-optic crystal drops to zero value at the time when the population inversion in the laser medium attains its peak value.

LASER BEAM CHARACTERISTICS

The important characteristics of a laser beam are:

(i) directionality

(ii) negligible divergence

(iii) high intensity

(iv) high degree of coherence and

(v) high monochromaticity.

(i) *Directionality* : The conventional light sources emit light uniformly in all directions. When we need a narrow beam in a specific direction, we obtain it by placing a slit in front of the source of light.

In case of laser, the active material is in a cylindrical resonant cavity. Any light that is travelling in a direction other than parallel to the cavity axis is eliminated and only the light that is travelling parallel to the axis is selected and reinforced. Light propagating along the axial direction emerges from the cavity and becomes the laser beam. Thus, a laser emits light only in one direction.

(ii) *Divergence* : Light from conventional sources spreads out in the form of spherical wave fronts and hence it is highly divergent.

On the other hand, light from a laser propagates in the form of plane waves. The light beam remains essentially a bundle of parallel rays. The small divergence that exists is due to the diffraction of the beam at the exit mirror. A typical value of divergence of a He-Ne laser is 10^{-3} radians. It means that the diameter of the laser beam increases by about 1 mm for every meter it travels.

(iii) *Intensity* : The intensity of light from a conventional source decreases rapidly with distance as it spreads out in space. Laser emits light in the form of a narrow beam with its energy concentrated in a small region of space. Therefore, the beam intensity would be tremendously large and stays constant with distance. The intensity of a laser beam is approximately given by

$$I = \left[\frac{10}{\lambda}\right]^2 P \qquad ...(25)$$

where P is the power radiated by the laser.

To obtain light of same intensity from a tungsten bulb, it would have to be raised to a temperature of 4.6×10^6K.

(iv) *Coherence* : The light that emerges from a conventional light source is a jumble of short wave trains which combine with each other in a random manner.

The resultant light is incoherent. Coherence length is one of the parameters used as a measure of coherence. In case of a laser a large number of identical photons are emitted through stimulated emissions and therefore they will be in phase with each other. The resultant light exhibits a high degree of coherence.

The coherence length of light from a sodium lamp, which is a traditional monochromatic source, is of the order of 0.3 mm. On the other hand the coherence length of light emitted by an ordinary helium-neon laser is about 100 m.

(v) *Monochromaticity* : If light coming from a source has only one frequency (single wavelength) of oscillation, the light is said to be *monochromatic* and the source a *monochromatic source.*

Light from traditional monochromatic sources spreads over a wavelength range of 100Å to 1000Å. On the other hand, the light from lasers is highly monochromatic and contains a very narrow range of a few angstroms (< 10Å).

APPLICATIONS

Lasers find application in almost every field. They are used in mechanical working, industrial electronics, entertainment electronics, communications, information processing, and even in wars to guide missiles to the target. Lasers are used in CD players, laser printers, laser copiers, optical floppy discs, optical memory cards etc.

SOLVED EXAMPLES

Example 1:

Deduce the de Broglie wavelength of :

(i) neutrons of kinetic energy 1 ev,

(ii) thermal neutrons.

Solution:

(i) Since p = $\sqrt{2mE}$, we get (using 1 ev = 1.6 × 10^{-19} joule).

$$\lambda = \frac{h}{\sqrt{2mE}} = \frac{6.6\times10^{-34}}{(2\times1.67\times10^{-27}\times1.6\times10^{-19})^{1/2}}$$

$$= 2 \times 10^{-11} \text{ m}$$

(ii) For *thermal neutrons, energy* E = kT. We get

$$E = 1.38 \times 10^{-23} \times 300 = 4 \times 10^{-21} \text{ joule}$$

$$\lambda = \frac{6.6\times10^{-34}}{(2\times1.67\times10^{-27}\times4\times10^{-21})^{1/2}} = 2 \times 10^{-10} \text{ m.}$$

Example 2:

Make an order of magnitude calculation of the de Broglie wavelengths of a 100-volt electron (i.e., an electron accelerated through 100 volt potential difference).

Solution:

The momentum p is given by the relation $p^2 = 2mE$,

$$p = (2 \times 9.1 \times 10^{-31} \times 1.6 \times 10^{-19} \times 100)^{1/2}$$

$$= 5 \times 10^{-24} \text{ kg ms}^{-1}$$

Hence, de Broglie wavelength is

$$\lambda = \frac{h}{p} = \frac{6.6\times10^{-34}}{5\times10^{-24}} = 1.3 \times 10^{-10} \text{ m.}$$

Example 3:

In a long chain organic molecule of length 5Å electrons may bet created as free to move along the length. Deduce the zero point energy, the energy gap between the first two energy states of the electron, and also the wavelength of absorption line arising from this transition.

Solution:

$$E_1 = \frac{h^2}{8ma^2} = \frac{(6.6\times10^{-34})^2}{8\times9.1\times10^{-31}\times(5\times10^{-10})^2} = 2.4 \times 10^{-19}J$$

This is the zero-point energy. Its value *per mole* comes to

$NE_1 = 1.5 \times 15^5$ J/gm mole.

$E_2 = 4.\ E_1 = 9.6 \times 10^{-19}$ J

$\therefore$ Energy gap $\Delta E = 7.2 \times 10^{-19}\ J = \frac{7.2\times10^{-19}}{1.6\times10^{-19}} eV = 4.5\ eV$

Absorption wavelength l is given by

$\Delta E = hv = hc/\lambda$

$$\lambda = \frac{hc}{\Delta E} = \frac{6.6\times10^{-34}\times3.0\times10^{8}}{7.2\times10^{-19}} m = 2.8 \times 10^4 Å$$

[Note results have been checked quantitatively in the absorption spectra of several organic molecules].

Example 4:

Harmonic 2ω is to generated out of a laser beam ω = 4 × 10¹⁴ Hz. If the medium has refractive index 1.528 at ω and 1.542 at 2ω, deduce the momentum mismatch due to colour dispersion. Also estimate the length of the medium for which the 2ω beam will remain coherent if the mismatch is not compensated.

Solution:

We have $k = \frac{2\pi}{\lambda} = \frac{2\pi v}{c'} = \frac{\omega}{c} n$

where c is speed of light in the medium and n is the appropriate refractive index. Substitution leads to

$$\Delta k = \frac{8\times10^{14}\times1.542}{3\times10^{8}} - 2\,\frac{4\times10^{14}\times1.528}{3\times10^{8}}$$
$$= 4 \times 10^4 m^{-1}$$

$$\Rightarrow \qquad L = (\Delta k)^{-1} = 2 \times 10^{-5} m.$$

Example 5:

Find the ratio of populations of the two states in a He-Ne laser that produces light of wavelength 6328 Å At 27°C.

Solution:

The ratio of population is given by $\frac{N_2}{N_1} = e^{-(E_2-E_1)/KT}$

$$E_2 - E_1 = \frac{12400}{6328} eV$$

$$\therefore \frac{N_2}{N_1} = \exp\left[\frac{-1.96 eV}{(8.61 \times 10^{-5} eV)(300K)}\right] = e^{-75.88} = 1.1 \times 10^{-33}.$$

Example 6:

The wavelength of emission is 6000 Å and the coefficient of spontaneous emission is 10^6/s. Determine the coefficient for the stimulated emissions.

Solution:

The coefficient for stimulated emission is given by

$$E_{21} = \frac{c^3}{8\pi hv^3\mu^3} A_{21} = \frac{\lambda^3}{8\pi h} A_{21} \quad \text{(Taking } \mu = 1\text{)}$$

$$\therefore B_{21} = \frac{(6000 \times 10^{-10})m^3(10^6/s)}{8\pi \times 6.626 \times 10^{-34} Js} = 1.3 \times 10^{19}/kg.$$

Example 7:

A plane diffraction grating with 5000 lines/cm is used in Littrow mounting with angle of incidence 20° and focal length of collimator/ camera lens is 50 cm. Deduce for the second order spectrum :

(i) The wavelength falling at the centre of the field of view,

(ii) Linear dispersion in the spectrum (in Angstrom/mm).

Solution:

(i) grating element e = 1/5000 cm = 2×10^{-4} cm. In the Littrow mount $\theta = i = 20°$ (given). Hence, for second order,

$$2\lambda = e(\sin i + \sin\theta) = 2 \times 10^{-4}(0.34 + 0.34)$$

$$\therefore \lambda = 6.8 \times 10^{-5} \text{ cm} = 6800\text{Å}.$$

(ii) Angular dispersion $d\theta/d\lambda$ for Littrow mount is given by

$$e \cos\theta . d\theta = n d\lambda$$

$\therefore$ Linear dispersion dx/dl is given by

$$\frac{dx}{d\lambda} = f.\frac{d\theta}{d\lambda} = f\frac{n}{e\cos\theta}$$

$$= \frac{50\times2}{2\times10^{-4}\times0.94} x10^{-8} \text{ cm/angstrom}$$

$$= 5.3 \times 10^{-2} \text{ mm/A}$$

$\therefore d\lambda/dx = 19$ A/mm.

Example 8:

Deduce Δv values per angstrom difference of wavelength at $\lambda = 2500\mathring{A}$, and at $\lambda = 5000\mathring{A}$.

Solution:

$$\bar{v} = 1/\lambda \Rightarrow \Delta\bar{v} = -\Delta\lambda/\lambda^2$$

For $\Delta\lambda = 1 \times 10^{-8}$ cm at $\lambda = 2500 \times 10^{-8}$ cm

$$\Delta\bar{v} = (1 \times 10^{-8})/(2.5 \times 10^{-5})^2 = -16 \text{ cm}^{-1}$$

For $\Delta\lambda = \lambda \times 10^{-8}$ cm at $\lambda = 5000 \times 10^{-8}$ cm

$$\Delta\bar{v} = -(1 \times 10^{-8})/(5 \times 10^{-5})^2 = -4 \text{ cm}^{-1}$$

(The minus sign only means that for greater l the $\bar{v}$ value is smaller).

Example 9:

Excited by 5460.7Å radiation a sample gives a Raman line at 5542.6Å. Compute:

(i) the Raman frequency,

(ii) position of the corresponding anti-stokes line.

Solution:

$$v_o = \frac{10^8}{5460.7} = 18312 \text{ cm}^{-1}$$

$$v_R = \frac{10^8}{5542.6} = 18042 \text{ cm}^{-1}$$

$\therefore$ Raman frequency $\Delta v_R = v_o - v_R = 270 \text{ cm}^{-1}$

anti-stokes frequency $v'R = v_o + \Delta v_R$

$$= 18312 + 270 = 18582 \text{ cm}^{-1}$$

$\therefore$ anti-stokes wavelength $\lambda' = \dfrac{10^8}{v'R} = 5381.5\text{Å}.$

Example 10:

At what temperature are the rates of spontaneous and stimulated emission equal? Assume λ = 5000 Å.

Solution:

If the ratio of spontaneous and stimulated emission are equal, the

$$R_1 = \left[\frac{1}{e^{h\nu/kT} - 1}\right] = 1 \quad \text{or} \quad e^{h\nu/kT} = 2$$

As λ = 5000 Å, $\nu = c/\lambda = 6 \times 10^{14}$Hz and

$$\frac{h\nu}{kT} = \frac{6.626\times10^{-34}\,\text{Js}(6\times19^{14}/\text{s})}{(1.38\times10^{-23}\,\text{J/K})T} = \frac{28.8\times10^3}{T}\text{K}$$

$$e^{h\nu/kT} = \exp\left[\frac{28.8\times10^3}{T}\text{K}\right] = 2$$

or $$\frac{28.8\times10^3}{T}\text{K} = \text{In } 2 = 0.693$$

$\therefore$ $$T = \frac{28.8\times10^3}{0.693}\text{K} = 41{,}558 \text{ K}.$$

Example 11:

The length of a laser tube is 150 mm and the gain factor of the laser material is 0.0005/cm. If one of the cavity mirrors reflects 100% light that is incident on it, what is the required reflectance of the other cavity mirror?

Solution:

$$\gamma_{th} = \frac{1}{2L}\text{In}\frac{1}{r_1 r_2}$$

$$\therefore r_2 = \frac{1}{r_1 e^{2L\gamma}} = \frac{1}{1\times e^{2\times15\times0.0005}} = 0.985$$

It means that the second mirror should have a reflectance of 98.5%.

Example 12:

The half-width of the gain profile of a He-Ne laser material is 2×10^{-3} nm. If the length of the cavity is 30 cm, how many longitudinal modes can be excited? The emission wavelength of He-Ne laser is 6328 Å.

Solution:

The separation between successive longitudinal modes is given by

$$\Delta\lambda = \frac{\lambda^2}{2L} = \frac{(6328\times10^{-10}\,m)^2}{2(30\times10^{-2}\,m)}$$

$$= 0.66 \times 10^{-3}\ nm$$

$$\text{Number of modes } N = \frac{\delta\lambda}{\Delta\lambda}$$

$$= \frac{2\times10^{-3}\,nm}{0.66\times10^{-3}\,nm} = 3.$$

Example 13:

The coherence length for sodium D_2 line is 2.5 cm. Deduce :

(i) the coherence time τ,

(ii) the spectral width of the line,

(iii) the purity factor (Q).

Solution:

We have

$$t = \frac{L}{c} = \frac{2.5cm}{3.0\times10^{10}\,cm/s} = 0.8 \times 10^{-10}s$$

From Eq. (21.3b), taking $\lambda = 6 \times 10^{-5}$ cm

$$\Delta\lambda = \frac{\lambda^2}{L} = \frac{36\times10^{-10}\,cm^2}{2.5cm} = 14 \times 10^{-10}\ cm = 0.14\ A$$

We have,

$$Q = \frac{\lambda}{\Delta\lambda} = \frac{6\times10^{-5}\,cm}{14\times10^{-10}\,cm} = 0.4 \times 10^5$$

Example 14:

With a He-Ne laser, Michelson interferometer fringes remained clearly visible when the path difference was increased upto 8 m. Deduce the lower limits for :

(i) the coherence length,

(ii) coherence time,

(iii) spectral half width, and

(iv) Q of the line, (Given $l = 11.5 \times 10^{-5}$ cm).

Solution:

The experimental fact of observing interference for path difference upto 8 m directly gives–coherence length L > 8m

Then give–coherence time $\tau = \frac{L}{c} > 2.7 \times 10^{-8}$ sec

Spectral half-width $\Delta\lambda = \frac{\lambda^2}{L} = 1.6 \times 10^{-3}$Å

$$Q = \frac{L}{\lambda} = 7 \times 10^6$$

Example 15:

A laser source of λ = 6000Å, coherence width 4 mm and power 10mW shines on a surface 100m away. Deduce the illumination. Compare it with that due to a focussed beam from a torch of filament diameter 1mm, power 10W, lens focal length 10 cm and aperture 4 cm dia.

Solution:

For the laser all 100mW power goes in a cone of semi angle $\theta = \lambda/a = 6 \times 10^{-5}/0.4 = 15 \times 10^{-5}$ rad. Hence a real spread at 10^4 cm distance (besides the initial area $\pi \times 0.2^2$ cm^2) is –

$$A = D^2\pi\,\lambda\theta^2 = 10^8 \times 2.14 \times (15 \times 10^{-5})^2 = 7\text{cm}^2$$

Illumination = Power/Area = 10mW/7 cm^2 = 1.4 mW/cm^2

For the torch the 10W power spreads isotropically, and the lens collects a fraction π (2 cm)^2 ÷ 4π (10 cm)^2 = 1/100. The light spreads over a cone of semiangle determined by the filament size ÷ focal length of the lens = 0.1/10 = 1/100 rad. Hence a real spread at distance 10^4 cm is

$$A' = D^2\pi\,\theta^2 = 10^8 \times 3.14 \times (10^{-2})^2 = 3.1 \times 10^4 \text{ cm}^2$$

Illumination = Power/Area

$$= \frac{1}{100} \times 10\text{W}/3.1 \times 10^4\text{cm}^2 = 0.003 \text{ mW/cm}^2.$$

Example 16:

If the laser beam of Ex. 21.3 is focussed by a lens of focal length 10 cm, deduce (i) the radius, (ii) area, and (iii) power density of the image.

Solution:

Angular spread of the laser beam is – l/d, due to diffraction alone.

Hence $\theta = 6 \times 10^{-5}/0.4 = 1.5 \times 10^{-4}$ rad

$\therefore$ radius of the image $= f\theta = 1.5 \times 10^{-3}$ cm

area of the image $= \pi\ (1.5 \times 10^{-3})^2 = 8 \times 10^{-6}$ cm^2

$$\text{power density} = \frac{10\times10^{-3}\,\text{W}}{8\times10^{-6}\,\text{cm}^2} = 1.2\ \text{KW/cm}^2.$$

Example 17:

With certain units for P and E, let the three χ coefficients of the order of 10^{-4}, 10–9 and 10^{-16} respectively. evaluate the relative contributions of the three terms for (a) E = 1 unit, (b) E = 104 units, (c) E = 107 units.

Solution:

Simple substitution gives for the three cases:

(a) $1 : 10^{-5} : 10^{-12}$

(b) $1 : 10^{-1} : 10^{-4}$

(c) $1 : 10^{+2} : 10^{+2}$

Note: $\chi^{(1)}, \chi^{(2)}, \chi^{(3)}$, are dimensionally different. Hence their relative numerical values have no meaning by themselves; they depend on the units for E and P. Thus, if the unit for E were taken 10^6 – fold larger (numerical E values 10^{-6} fold), then in the same example $\chi^{(1)}, \chi^{(2)}, \chi^{(3)}$, would be 10^{+2}, 10^{+3} 10^{+5} respectively.

EXERCISES

1. What do you understand by an optical resonant cavity? Explain.
2. Why is the optical resonator required in lasers? Illustrate your answer with neat sketches.
3. What are the essential components of a laser? Explain their functions briefly.
4. Describe the working of solid state ruby laser.
9. Explain the principle and working of a He-Ne laser.
5. In helium-neon laser lasing is through neon gas. What is then the role of helium gas?
6. In helium -neon laser why is it necessary to use narrow tubes?
7. What is the reason for monochromaticity of laser beam?

8. With the help of energy band diagram discuss the working of a semiconductor laser.
9. Explain in brief the characteristics of a laser beam.
10. A pulsed laser is constructed with a ruby crystal as the active element. The ruby rod contains typically a total of 3×10^{19} Cr^{3+}ions. If the laser emits light at 6943Å wavelength, find
    (a) the energy of emitted photon (in eV)
    (b) the total energy available per laser pulse (assuming total population inversion)
11. The He-Ne system is capable of lasing at several different IR wavelengths, the prominent one being 3.3913μm. Determine the energy difference (in eV) between the upper and lower levels for this wavelength.
12. Explain with neat diagram absorption, spontaneous emission and stimulated emission of radiation.
13. What is population inversion? Explain why laser action cannot occur without population inversion between atomic levels.
143. What do you understand by a negative temperature state? How can it be achieved?
15. Discuss the four-level (pumping) scheme for laser action.
16. The CO_2 laser is one of the most powerful lasers. The energy difference between the two laser levels is 0.117 eV. Determine the frequency and wavelength of the radiation.
17. A laser beam can be focused on an area equal to the square of its wavelength. For a He-Ne laser, the wavelength of emitted light is 6328Å. If the laser radiates energy at the rate of 1 mW, find out the intensity of the focused beam.
18. Find the relative populations of the two states in a ruby laser that produces a light beam of wavelength 6943Å at 300K.
19. Find the ratio of populations of the two states in a He-Ne laser that produces light of wavelength 6328Å at 27°C.

5

HOLOGRAPHY

INTRODUCTION

Images of objects are generally obtained using photographic method. In this method a lens focuses the light reflected from a three-dimensional object onto a photographic film where a two-dimensional image of the object is formed. A negative is first obtained by developing the film and then a positive is obtained through printing. The positive print is a two information about the square of the amplitude of the light wave that produced the image but information about the phase of the wave is not recorded and is lost.

In 1948 Dennis Gabor outlined a two-step lensless imaging process. It is radically a new technique of photographing the objects and is known as wave front *reconstruction*. The technique is also called *holography*. The word 'holography' is formed by combining parts of two Greek words: 'holos', meaning ''whole'', and 'graphein,' meaning "to write". Thus holography means writing the complete image. Holography is actually according of interference pattern formed between two beams of coherent light coining from the same source. In this process both the amplitude and phase components of light wave are recorded on a light sensitive medium such as a photographic plate. The recording is known as a *hologram*. Holography required an intense coherent light source. Laser was not available when Gabor formulated the idea of holography. Holographic technique became a practical proposition only after the invention of lasers. Leith and Upatnicks prepared laser holograms for the first time. In this chapter we discuss the fundamental concept of holography.

PRINCIPLE OF HOLOGRAPHY

Holography is a two-step process. First step is the *recording* of hologram where the object is transformed into a photographic record and the second step is the *reconstruction* in which the hologram is transformed into the image. Unlike in the conventional photography, lens is not required in either of the steps. A hologram is the result of interference occurring between two waves, an object beam which is the light scattered off the object and a coherent background, the reference beam, which is the light reaching the photographic plate directly. In Gabor's original experiments, the reference beam and object beams were coaxial. Further advance was made by Leith and Upatnieks, who used the reference beam at an offset angle. That made possible the recording of holograms of three-dimensional objects.

The off-axis arrangement for generating and viewing holograms is described here.

Recording of the Hologram

In the off-axis arrangement a broad laser beam is divided into two beams, namely a reference beam and an object beam by a beam splitter. The reference beam goes directly to the photographic plate. The second beam of light is directed onto the object to be photographed. Each point of the object scatters the incident light and acts as the source of spherical waves. Part of the light, scattered by the object, travels towards the photographic plate. At the photographic plate the innumerable spherical waves from the object combine with the plane light wave from the reference beam. The sets of light waves are coherent because they are from the same laser. They interfere and form interference fringes on the

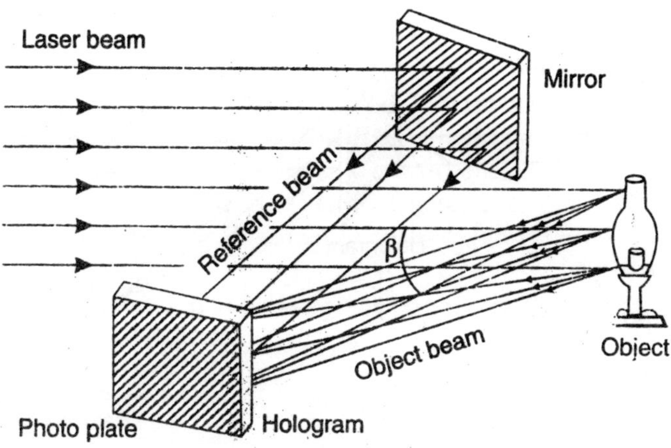

Fig. 5.1 : Generation of a hologram.

plane of the photographic plate. These interference fringes are a series of zone-plate like rings, but these rings are also superimposed, making a complex pattern of lines and swirls. The developed negative of these interference fringe-patterns is a hologram. Thus, the hologram does not contain a distinct image of the object but carries a record of both the intensity and the relative phase of the light waves at each point.

Reconstruction of the Image

Whenever required, the object can be viewed. For reconstruction of the image, the hologram is illuminated by a parallel beam of light from the laser. Most of the light passes straight through, but the complex of fine fringes acts as an elaborate diffraction grating. Light is diffracted at a fairly wide angle. The diffracted rays form two images: a virtual image and a real image. The virtual image appears at the location formerly occupied by the object and is sometimes called as the true image. The real image is formed in front of the hologram. Since the light rays pass through the point where the real image is, it can be photographed. The virtual image of the hologram is only for viewing. Observer can move to different positions and look around the image to the same extent that he would be able to, were he looking directly at the real object. This type of hologram is known as a transmission hologram since the image is seen by looking through it.

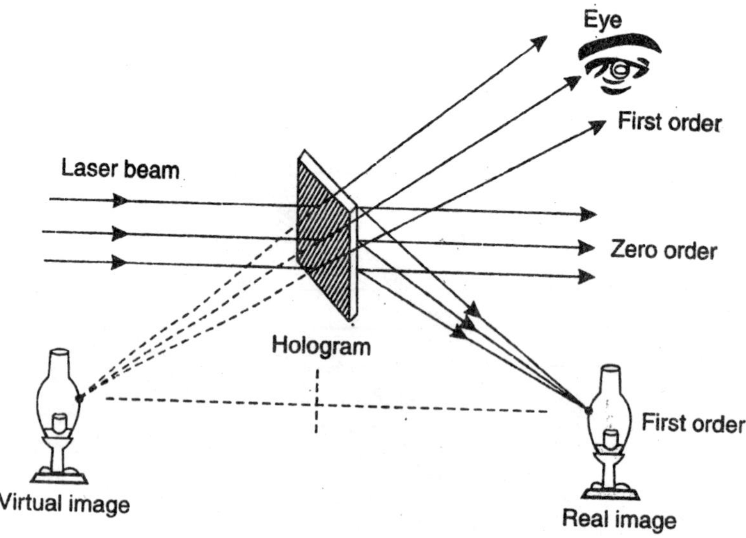

Fig. 5.2 : Viewing of hologram.

The original configuration adopted by Gabor for recording hologram was a coaxial arrangement, as illustrated in Fig. 5.3. In this arrangement the real image is located in front of the virtual image and is inconvenient for viewing or photographing. The advantage of the off-axis configuration is that the two images are separated.

The fundamental difference between a hologram and an ordinary photograph is like this. In a photograph the information is stored in an orderly fashion: each point in the object relates to a conjugate point in the image. In a hologram there is no such relationship; light from every object point goes to the entire hologram. This has two main advantages:

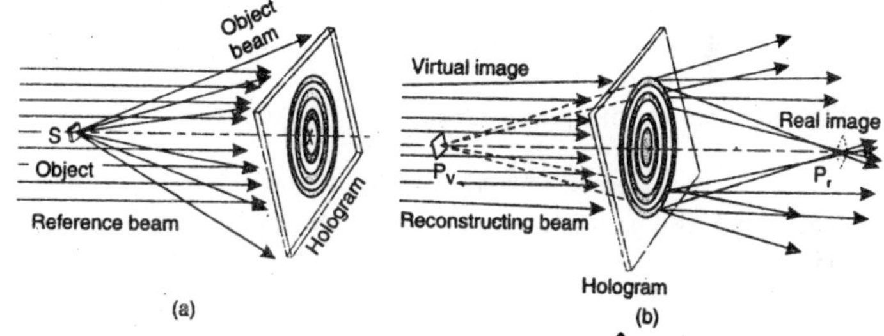

Fig. 5.3 : In-line holography (a) Generation (b) viewing

1. As the observer moves sideways in viewing the hologram, the image is seen in three dimensions.
2. If the hologram were shattered or cut into small pieces, each fragment would still reconstruct the whole object, not just part of the object.

THEORY

The general theory of holography is much involved and cumbersome. We illustrate here it by taking the simple example of a point object in a coaxial configuration.

Let the light beam from a coherent source illuminate a point object P (Fig. 5.4). The beam consists of plane waves. Most of the plane waves reach the photographic plate directly. Part of the light is scattered by the point object and spherical waves are produced. They also reach the photographic plate. The plane waves of the reference beam and spherical

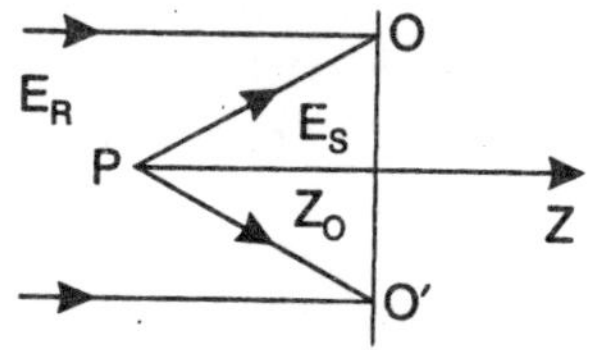

Fig. 5.4 : Hologram of a point object

waves of the object beam superpose at the plane of the photographic plate and produce interference.

We may write the optical field arriving at point O on the photographic plate as

$$E = E_R + E_S \quad ...(1)$$

where E_R is the field due to the reference beam and Eg is the field scattered from the object. The scattered field Eg is not simple, both amplitude and phase vary greatly with position. The reflected wave fronts are spherical and concentric around the point of origin. We represent the field of the scattered wave front by

$$E_S = \frac{E_o}{r_o} \exp\left[i(kr_o - \omega t)\right] \quad ...(2)$$

and the field E_R by the plane wave

$$E_R = E_r \exp\left[i(k z_o - \omega t)\right] \quad ...(3)$$

where r_o = PO and z_o is the distance from P to the plate. The intensity at O is

$$I = |E + E_S|^2$$

$$= |E_R| + \frac{|E_S|^2}{r_0^2} + \frac{E_S E_r^*}{r_o} \exp\left[ik(r_o - r_o)\right]$$

$$+ \frac{E_S^* E_r}{r_o} \exp\left[ik(z_o - r_o\right] \quad ...(4)$$

We combine the last two terms of the above equation and write it as

$$I = |E_R|^2 + \frac{|E_S|^2}{r_0^2} + K\frac{\cos\left[k(r_o - z_o) + \phi\right]}{r_o} \quad ...(5)$$

The total intensity I is a function of cosine term and shows a series of maxima and minima. Thus, the interference of the spherical wave E_S and plane wave E_R produces a set of circular interference fringes. If we assume that the plate response is proportional to the intensity I, the power transmission of the plate, T^2 is given by

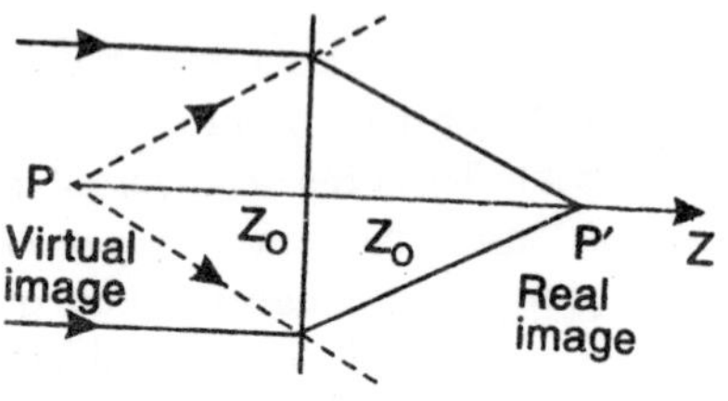

Fig. 5.5

$$T^2 = 1 - \alpha I \qquad ...(6)$$

where a is a constant. Eqn. (6) can be approximated as

$$T \cong 1 = 1/2\ \alpha\ I \qquad ...(7)$$

When the hologram is illuminated by the reference beam, the field of the transmitted wave may be written as

$$E = TE_R = \left[1 - \frac{\alpha}{2} I\right] E_r \exp\left[i\left(kz_o - \omega t\right]\right.$$

$$= \left[1 - \frac{\alpha}{2}|E_R|^2 - \frac{\alpha}{2}\frac{|E_S|^2}{r_o^2}\right] E_R \exp\left[i(kz_o - \omega t)\right]$$

$$= \frac{\alpha}{2}\frac{E_S E_R^*}{r_o} \exp\left[ik(r_o - z_o)\right] E_R \exp\left[i(kz_o - \omega t)\right]$$

$$- \frac{\alpha}{2}\frac{E_S^* E_R}{r_o} \exp\left[ik(r_o - z_o)\right] E_R \exp\left[i(kz_o - \omega t)\right]$$

$$= \left[1 = \frac{\alpha}{2}|E_R|^2 - \frac{\alpha}{2}\frac{|E_S|^2}{r_o^2}\right] E_R \exp\left[i(kz_o - \omega t)\right]$$

$$- \frac{\alpha E_S\ |E_S|^2}{2r_o} \exp\left[i(kr_o - \omega f)\right]$$

$$- \frac{\alpha}{2}\frac{E_S^*\ E_R^2}{} \exp\left[i(2kz_o - kr_o - \omega t)\right] \qquad ...(8)$$

The first term in (8) represents the incident plane wave with some attenuation.

The second term represents a spherical wave identical with that emitted by the object except for a constant factor.

The wave surface when projected back appears to have come from an apparent object located at the place where the original object was located. This is the *virtual image* of the object.

The third term represents also a spherical wave, which is identical to the original wave but converges at a point P'.

A *real image* is produced at P' which can be photographed without a lens. The hologram thus produces both a real image P and a virtual image P'.

IMPORTANT PROPERTIES OF A HOLOGRAM

1. In an ordinary photograph each region contains separate and individual part of the original object. Therefore, destruction of a portion of a photographic image leads to an irreparable loss of information corresponding to the destroyed part. On the other hand, in a hologram each part contains information about the entire object. From even a small part of the hologram the entire image can be reconstructed if only with a reduced clarity and definition of the image. Therefore, a hologram is a reliable medium for data storage.

2. It is not useful to record several images on a single photographic film. Such a record cannot give information about any of the individual images. On the other hand, several images can be recorded on a hologram. Therefore the information holding capacity of a hologram is extremely high. While a 6 × 9 mm photograph can hold one printed page, a hologram of the same size can store up to 300 such pages.

3. On a hologram information is recorded in the form of interference pattern. The type of the pattern obtained depends on the reference beam used to record the hologram. The information can be decoded only by a coherent wave identical to that of the reference wave. The reference wave can be chosen appropriately. Consequently without the knowledge of the shape of the reference wave front the information encoded in the form of interference pattern on the hologram cannot be deciphered.

4. The reconstruction of the image of the hologram can be done with reference beam of any wavelength if it is coherent and identical to the original reference beam. If the wavelength λ of the reconstructing beam is greater than that λ_0 of the reference beam, the reconstructed image will be a magnified image. The magnification will be proportional to the ratio of the two wavelengths.

ADVANCES

Reflection Holography

In the recording process of reflection holography, the object beam and the reference beam reach the recording medium from opposite directions. The beams combine within the medium and generate interference fringes which are recorded in the thick emulsion. During reconstruction the hologram scatters the illuminating beam back towards the viewer. One sees a virtual image behind the hologram, as if looking into a mirror. Since this technique utilizes a thick emulsion on the photographic plate, *reflection holography* is also known as *thick emulsion holography*.

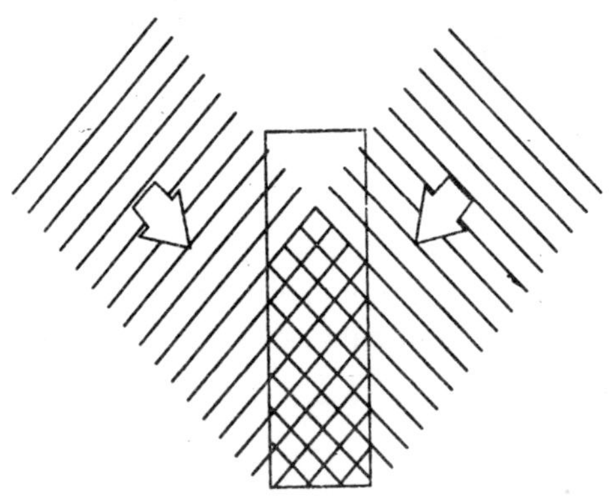

Fig. 5.6

Volume Holography

In 1962 Yuri Denisyuk used a process similar to Lippmann colour photography. In this method the object wave is reflected from the object and propagates backward and overlaps the incoming reference wave. The two waves form standing wave pattern. The fringes are recorded by the photo-emulsion throughout its entire thickness to form a *volume hologram*. The hologram may be regarded as a three-dimensional grating. In the volume holograms there is an interdependence of the wavelength and the scattering angle, because the scattering follows Bragg's law, Id sin $\theta = m\lambda$. Therefore, by successively changing the incident angle or wavelength, a number of holograms can be stored in the medium. Different and mutually incoherent laser beams may be used to produce different component holograms of the object and when they are illuminated a multicolored image is seen.

White-light Reflection Holography

This method is developed by Stroke and Labeyrie. In this scheme the hologram is generated using coherent light but in the reconstruction process an ordinary white-light beam having a wave front similar to the original coherent waves is used. Using coherent sources at different wavelengths, several holograms are stored in a single film. When the hologram is illuminated by ordinary white light, a multicolored image is seen in reflection.

Rainbow Holography

In 1969 another type of color holography was developed by Stephen Benton. The hologram produced has been called a white-light transmission hologram, but it is best known as a rainbow hologram. Benton technique for constructing a true-color rainbow hologram is cumbersome and not easily implemented. When the hologram is viewed with a white-light point source, a very bright color image can be reconstructed. A true-color hologram image can be observed when the hologram is viewed in the correct plane. If the viewer moves -off this plane, different shades of color can still be seen, but the color will be different from that of the original object.

APPLICATIONS

Holography has many fascinating applications. It is not possible to describe all of them here. Only some typical applications are discussed here.

Holographic Interferometery

Holography is widely used in non-destructive testing to study distortions resulting from stresses, strains, heat and vibrations etc. In the *double exposure technique* two exposures are made of the object, one before processing and the other after. The original object and the object after deformation are recorded on the same hologram. When the hologram is illuminated with the reference wave, both the images are viewed simultaneously. Since the waves are slightly different, the two images form interference fringe pattern indicative of the changes suffered by the object.

Acoustic Holography

It is easy to produce coherent sound waves. Sound waves readily propagate in solids. Therefore, a three dimensional acoustical hologram of an opaque object can be made. By viewing such hologram in visible light the internal structure of the object can be observed. Such techniques will be highly useful in the fields of medicine and technology. In one of the techniques, two submerged coherent sound generators emit the reference and the signal, scattered by an object, respectively (Fig. 5.7). On a calm surface of water, these two contributions produce ripples. The ripple pattern is the hologram. The pattern may be photographed and then reconstructed. As sound waves can propagate through, dense liquids and solids, acoustical holography has an advantage in locating underwater submarines etc., and internal body organs.

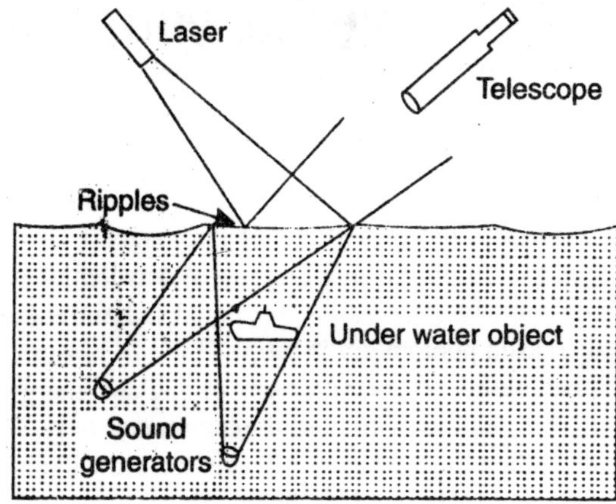

Fig. 5.7

Holographic Optical Elements

Holographic optical elements (HOE) are diffractive devices which are holograms of typical optical components such as lenses or mirrors. HOEs perform precisely the same type of operation as the lenses or mirrors, but they do it by diffraction and not by refraction or reflection. They are used in barcode scanners, office copy machines, solar concentrators etc.

EXERCISES

1. What is the advantage of off-axis configuration over the coaxial configuration?
2. Explain the principle of holography.
3. Explain some of the important properties of hologram.
4. What is meant by holography? Why is it called wave front reconstruction?
5. Discuss any two important applications of holography.
6. Describe how hologram is generated and image is reconstructed using off-axis configuration?

6

VISUAL PHOTOMETRY

INTRODUCTION

'Photometry' in its general meaning involves the measurement of radiant energy. However, out of the total electromagnetic spectrum the human eye is sensitive to only a small region of wavelengths and even within this region the visual sensation for equal radiant energy is not uniform. For this reason photometry in terms of *visual sensation* is very different from that based on measurement of radiant energy. In this Chapter we will be dealing with visual photometry only. The final measuring device is then the human eye. Therefore, we will also include a brief description of the eye and its functioning.

QUANTITIES RELATED WITH VISUAL PHOTOMETRY

The basic measure of visual sensation created by the total radiation emanating from a source or incident on a receiver is called the luminous flux, expressed by the symbol ϕ, and measured in a unit *lumen* (lm). We talk of a given area S of a source giving out luminous flux ϕ lm in a specified solid angle Ω. We also talk of a screen of area S receiving luminous flux ϕ lm from one or many sources.

Out of the quantities defined, brightness (B) and, intensity (I) are direction dependent. In contrast, luminance is the integrated effect of B summed for all directions. Also, for a source of finite dimension, I is the integrated effect of B over the entire area of the source. Finally ϕ is the integrated effect of I over any given solid angle or say receiving area.

It may be noted that if solid angle is taken to be dimensionless, then there can be considerable mix-up in units. For this reason, we take the dimension of solid angle as $\left(L_y^2\ L_x^{-2}\right)$, where L_x and L_y are lengths along and transverse to any ray, respectively.

Candela, the Fundamental SI Unit of Intensity

A 'standard candle' was used early to define intensity. But the latest fundamental unit is *Candela defined as the luminous flux emitted per unit solid angle along normal to the surface by one-sixtieth cm2 of a black-body radiator kept at the temperature of solidification of platinum* (2,046 K).

Thus 1 cm^2 of this blackbody surface at 2,046 K will have intensity 60 cos θ cd in a direction Q from the normal. The brightness of a blackbody is the same in all directions, and its value for the 2,046 K surface would be 60×10^4 cd/m^2.

Some Theorems in Photometry

In simple situations several inter-relations hold good between the quantities defined above. We will see some of them which are used widely.

(a) Relation $E = I \cos\theta/r^2$. If the dimensions of the source are much smaller than the distance of the observer, then the source is called a 'point source.' In such cases, the solid angle $\Delta\Omega$ subtended by a given surface of area ΔS at the source is defineable: We have $\Delta\Omega = \Delta S \cos\theta/r^2$, where θ is the angle between the ray and the normal to the surface (Fig. 6.1). If the source has intensity led in the concerned direction, the flux received is $Id\Omega$ lm, and illumination is

$$E = \frac{Id\Omega}{dS} = \frac{IdS\cos\theta}{r^2 dS} = \frac{I\cos\theta}{r^2} \qquad ...(1)$$

This is one of the Lambert relations. It combines the inverse square law with cos θ dependence. *But it is useful only if the source can be treated as a point source.*

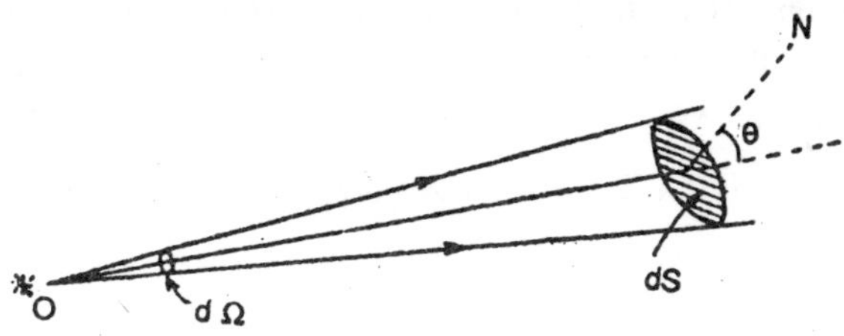

Fig. 6.1 : Illumination by a point source.

(b) Relation $L = \pi B$. Fora surface of uniform brightness B, the luminance L is given by πB. The elementary solid angle for the hollow cone between θ and $\theta + d\theta$ from the normal to a surface is $2\pi r \sin\theta . r d\theta \div r^2 = 2\pi \sin\theta\, d\theta$. The brightness is defined by

$$B = \frac{\Delta I}{\Delta S \cos\theta} \quad ...(2)$$

Hence the luminous flux in the hollow cone due to area ΔS is B $\Delta S \cos\theta$. $2\theta \sin\theta\, d\theta$. Integrating this over the whole available solid angle gives $L.\Delta S$.

Thus

$$L\Delta S = \int_0^{\pi/2} B\ \Delta S \cos\theta\ 2\pi \sin\theta\ d\theta$$

If B is taken as independent of θ, we get the result

$$L = \pi B \quad ...(3)$$

This is another *Lambert relation.* One consequence of this is that luminosity of the blackbody surface used for defining the Candela is $6\pi \times 10^5$ *l*m/m^2 [check it].

(c) *Brightness of a surface is independent of the distance of the viewer*. This ts because the illuminance falls as $1/r^2$, and the solid angle subtended at the eye by a given area of the surface also falls as $1/r^2$. Hence the (lux received *per unit solid angle* remains unchanged. Of course, this assumes that there is no attenuation in the medium.

(d) Relation $E = \pi L$: The illuminance at the centre of a flat surface under a hemispherical cover" of uniform luminance L is given by πL. An elementary ring on the hemisphere defined by θ and $\theta + d\theta$ has area $2\pi r \sin\theta\, r d\theta$ and hence intensity L. $2\pi r^2 \sin\theta\, d\theta$. Its contribution to the illuminance at the centre is $\cos\theta/r^2$ times this (Eqn. 6.1)

That gives

$$dE = L.2\ \pi \sin\theta \cos\theta\ d\theta$$

Integration from $\theta = 0$ to $\theta = \pi/2$ gives ...(4)

$$E = \theta L$$

The result is independent of r.

The illuminance over the surface due to a parallel infinite surface of luminance L is also πL. The proof of this we will leave for the reader. The result is independent of the distance between the two surfaces.

(e) *Relation* I = BA For a surface of uniform brightness B, the intensity for a distant observer is B.A, where A is the area of the source projected on a plane transverse to the line of sight. This follows directly from Eqn. (2).

$$dI = BdS \cos\theta = BdS'$$

$$I = B\int dS' = B.A. \qquad ...(5)$$

The relation has special interest in the viewing of stars.

(f) *No image can have brightness greater than that of the source.* Fig. 6.2 shows the geometry in image formation, with areas S, S' and solid angles Ω, Ω' shown, besides u and v.

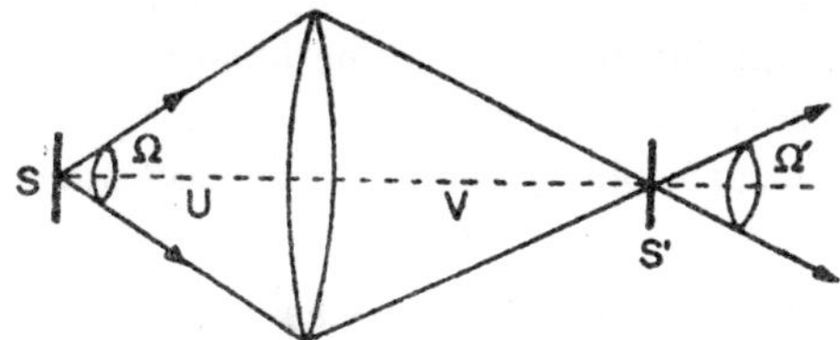

Fig. 6.2 : To discuss brightness of images.

Then we have from geometry

$$S' = \frac{\upsilon^2}{u^2}S;\ \Omega' = \frac{u^2}{\upsilon^2}\Omega$$

If we consider a source of uniform brightness B, then the luminous flux collected by the lens is B.S.Ω and this spreads over area S', so that *illuminance* E on the image plane is

$$E = B\,\Omega\,\frac{S}{S'} = B\frac{A}{\upsilon^2} = B\,\frac{A}{\upsilon^2} \qquad ...(6)$$

where A is the area of the lens. Thus E can be increased by having larger lens area and smaller focal length (to give lower υ).

But consider *brightness* B' of the image. The flux collected by the lens emerges in a solid angle Ω'. Hence the flux per unit area of the image per unit solid angle (which is B') comes to

$$B' = \frac{BS\Omega}{S'\Omega'} = B\left(\frac{S}{S'}\right)\left(\frac{\Omega}{\Omega'}\right) = B\,\frac{u^2}{\upsilon^2}.\frac{\upsilon^2}{u^2} = B \qquad ...(7)$$

Thus the best we can get is B' = B; in practice losses due to various effects give B' < B.

SOME TYPICAL INTENSITY AND ILLUMINANCE VALUES

For comfortable work in different situations the illuminations recommended are as follows:

Home, libraries, office, etc.	:	100-300 lx
Sewing and finer work	:	500-1000 lx
Operation theatre table	:	5000-8000 lx

The accommodation provided by nature in the eye is able to take care if there is excess illumination; but in poor illumination one has less than the optimum efficiency of work.

A typical 100 W bulb of incandescent filament type has intensity ~200 cd in the forward direction, so that the illuminance 2m below it would be ~50 lx due to direct light Multiple scattering from walls of the room raise this to 2 to 5-fold, depending on the wall colour and room size.

Sunlight at mid-day produces an illuminance about 10^5 lx. In fullmoon at mid-night it is about 0.25 lx. That we can at least see, if not read, under such a large range of illuminations is remarkable.

A sodium vapour lamp has much larger intensity than an incandescent filament lamp; but the filament could be much brighter than the sodium vapour lamp, because the area is very much smaller in the second case.

Any illuminated surface acts as a secondary source of luminous flux, defined by brightness B and luminance L, just as for a primary source. Typically, a matt surface coated with MgO will have luminance L about 95% of the illuminance—and also have B equal in all directions. A mirror reflector, on the other hand, will have B. zero in all directions except the range of specular reflection and L around 80-98% of E, depending on the quality of the reflecting polish.

LUMINOSITY AND LUMINOUS EFFICIENCY OF A SOURCE

A primary source of light consumes energy at the rate of X wait. Out of this a certain part, X' watt, goes out as radiant energy, the rest is consumed otherwise. This X' watt creates a total luminous flux ϕ lm.

Then the ratio (ϕ/X') lm/W is called the luminosity of the source, and the ratio (ϕ/X) lm/W is called the *luminous efficiency* of the source. Not unoften, the distinction between X and X' is omitted and one writes

$$\eta = \frac{\text{Lu min ous flux (lm)}}{\text{Power concerned (W)}} \quad \text{...(8)}$$

where η is used for both luminosity and luminous efficiency.

The value of η defined as ϕ/X', depends on the spectral distribution of the radiation. Obviously, if all the radiation falls outside the visible range (~4000 to ~7000^1), then $\phi = 0$ and hence $\eta = 0$. Within the visible range, ϕ/X' is wavelength-dependent, being the maximum for the green region λ~5550 A°, and failing off with λ both ways. This maximum value is

$$\eta_{max} = 683 \text{ lmlW} \qquad \text{at } \lambda = 5550\text{A}$$

The number 683 comes from the way 1 candela is defined. Fig. 6.3 shows the relative spectral sensitivity, K_λ of the human eye.

It will be seen that if a source emits all its radiation at λ = 5550Å, it will have η = 683 lm/W. If the spectral distribution (intensitywise) matches the curve of Fig. 6.3, then η = 683 × averaged K_λ, + 200 lm/W. For a general case it would be

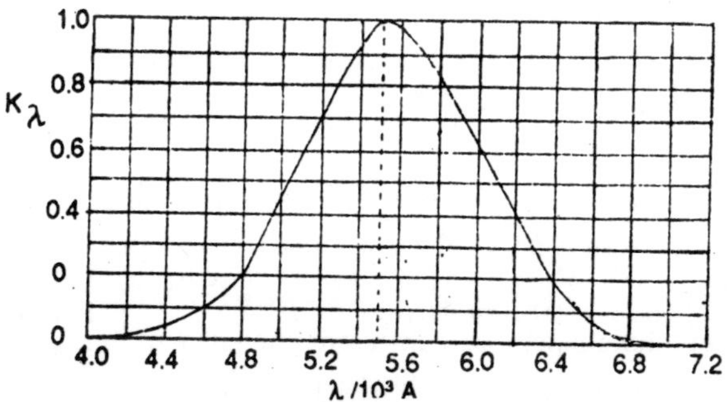

Fig. 6.3 : Relative spectral sensitivity of the human eye.

$$\eta = \frac{\int K_\lambda P_\lambda d_\lambda}{\int P_\lambda d_\lambda} \times 683 \text{ lm/W} \quad \text{... (9)}$$

where P_λ is the radiated power at λ. This is for η defined by ϕlX'. The realistic η is defined by ϕ/X and it will be lower, depending upon the power (X–X') consumed otherwise by the concerned system.

Sunlight happens to have its radiant energy maximum close to 5550A°, and it appears that nature has adapted the human eye to match its sensitivity curve to secure the best from sunlight. For other sources the typical η values are as follows:

Incandescent lamp	:	14 lm/W
Fluorescent lamp	:	43 lm/W
Halogen lamp	:	58 lm/W

The values mean that a 100 TV incandescent lamp would emit ~1400 lm flux. Considered isotropic, its intensity would be 1400/4π 120 cd. With a 'shade,' the I value in the front direction may reach ~400 cd; from a 2m distance it would create illuminance of ~ 100 lx.

PHOTOMETERS

In general a photometer compares the luminous intensities of two sources by matching the brightness produced by them on a surface. The steps involved are (a) illumination produced = I cos $\frac{\theta}{r^2}$; (b) luminance- = f.I cos $\frac{\theta}{r^2}$; (c) brightness = ff' I cos θ/(r^2 cos θ'). Here f and f' are appropriate fractions, and θ and θ' are angles of incidence and viewing. If θ is kept zero and f,f', θ' are identical for the two cases, then equality of brightness means

$$\frac{I_1}{r_1^2} = \frac{I_2}{r_2^2} \Rightarrow \frac{I_1}{I_2} = \left(\frac{r_1}{r_2}\right)^2 \qquad ...(9a)$$

An assumption involved is that the concerned sources are 'point sources,' which means that r values are much greater than the size of either source.

Different photometers use different methods of matching the brightness. We describe some of them.

(a) *Bunsen 's Grease-Spot Photometer :* It uses a mat-white paper (like a filter paper) with a central greased spot. The ungreased part gives diffuse scattering of light falling on the same face, while the greased part gives diffuse scattering of light falling on the opposite face. Fig. 6.4 shows the arrangement. If the distances of the sources are adjusted to make the grease spot vanish in the background of the ungreased part, Eqn. (9a) gives I_1/I_2.

If we assume that the ungreased part scatters a fraction a, and the greased part scatters fraction b and c of the light falling on the same and the opposite face, respectively, then the brightness of face 1 is

$$\left(\frac{I_1}{r_1^2}\right)a \quad \text{for the ungreased part}$$

$$\left(\frac{I_1}{r_1^2}\right)b + \left(\frac{I_2}{r_2^2}\right)c \quad \text{for the greased part} \qquad ...(10)$$

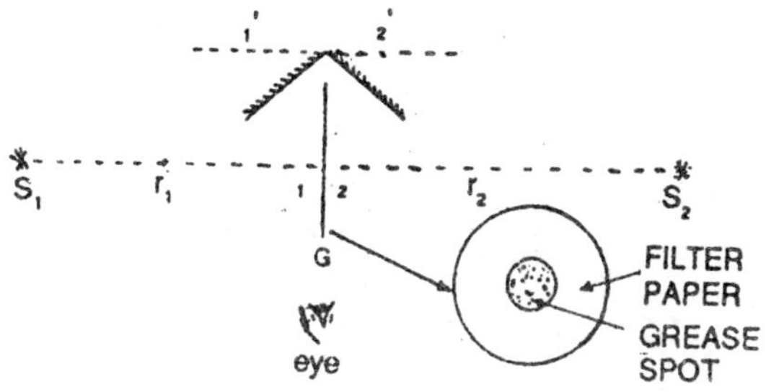

Fig. 6.4 : The grease-spot photometer.

Equality of brightness will still lead to Eqn. 9a if b + c = a (*i.e.*, there is no loss by absorption). In practice, the grease spot never fully vanishes in the background, and what one has to adjust is the 'identical view' of the two faces of the detector paper.

(b) *Lummer-Brodhun Photometer :* In this also the two sides of a mat white surface (a thick one, this time) are illuminated by light from the two sources under study (Fig. 6.5), but comparison of brightness of the two faces is made with the help of a 'photometer cube' C. In this 'cube' an air space separates the hypotenuse faces of two prisms, but a central portion is sealed with Canada balsam, which is transparent, like glass. Thus an observer views face 1 through the central portion of C and face 2 through the outer portion of C. In effect, this is like setting b = 0 and c = a in relations 10. Because total internal reflection gives 100% reflection, these conditions hold well, assuming that the faces of all the right-angled prisms are kept very clean.

(c) *Flicker Photometers :* If the sources under comparison give out light of different colours, then the aforesaid methods fail. A flicker photometer is then used. In principle, the two surfaces, illuminated separately by the two sources, are kept in the same line of sight, but an arrangement is made that these two are seen alternately at a fairly large frequency. The result is that what one sees is neither colour of source 1 nor that of source 2, but a mix. What is important is that if the visual sensations of the two are equal, then no flicker is observed. A crucial fact is that at low frequency of the alternating view, the colon r mixing (in the sensation at the eye) does not take place, and at very high frequency of alternating view the flicker is not observable even when brightnesses are unequal (due to persistence of vision). So a careful balance is needed. One starts from low rate and goes on to higher ones to see if the colour distinction becomes absent; and then one adjusts the distances till the flicker is eliminated.

kept very clean.

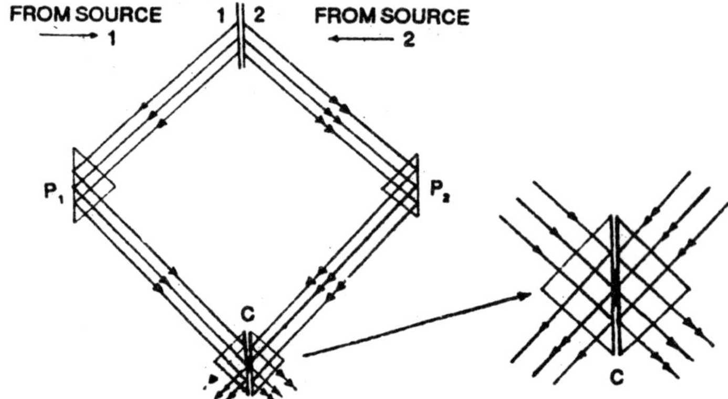

Fig. 6.5 : Lummer-Brodhun photometer, using total reflection prisms P_1 and P_2 and the Lummer-Brodhun 'cube' C. Inset shows how the rays from sources 1 and 2 pass through the central and outer parts of C.

Fig. 6.6 shows the schematic of *Guild's flicker photometer*. Surfaced is illuminated with source 1 and seen at 45° from its normal. Surface B is a sector disc illuminated with source 2 and seen again at 45° from the normal. As the sector disc is rotated, the eye alternately views A and B. If both the surfaces have identical diffuse coating (say, of MgO), which is neutral for colours, then absence of flicker means $\frac{I_1}{r_1^2} = \frac{I_2}{r_2^2}$.

The sectoral disc B is rotated so that alternately surfaces A and B are seen by the eye. The viewing chamber has a diaphragm D painted white and kept illuminated to serve as a comfortable background against which A/B is seen ($r_1 = a + b$).

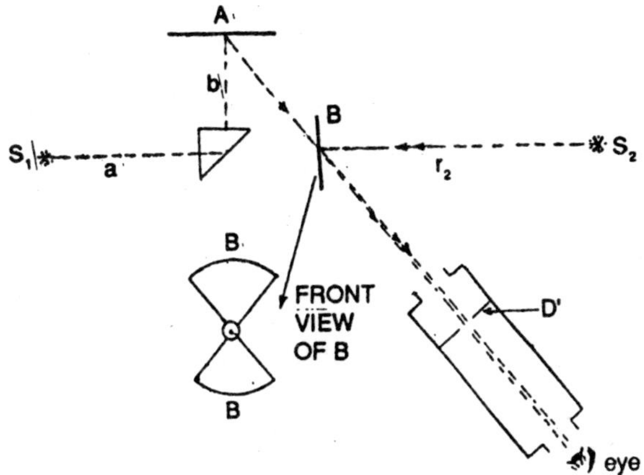

Fig. 6.6 : Guild's Flicker Photometer.

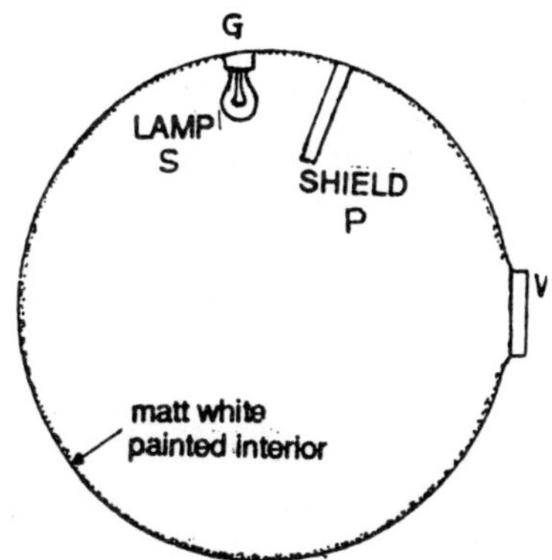

Fig, 6.7 : Ulbright's integral photometer. G is gate for introducing the source S; P is shield to prevent direct light from reaching W.

(d) *Ulbright's Integral Photometer :* The luminous intensity of a source is often direction-dependent. One may therefore measure I in different orientations and then find the averaged I over the total solid angle. But Ulbright's photometer does the integration in another way. Fig. 6.7 shows the source S placed inside a hollow sphere, whose inside walls are perfectly diffusing, mat painted typically with MgO. Due to multiple reflections within, the radiation inside becomes isotropic. If a window W is placed on the wall, the output flux there is equal to the total flux emitted by the source. Comparing the brightness of an opal glass placed at this window with that produced by a standard source from a measured distance gives the desired result.

THE HUMAN EYE

A sketch of the human eye is shown in Fig. 25.8. The different optical parts are as follows:

Cornea : 1, a thin transparent front opening.

Aqueous Humour : 2, a watery fluid between the cornea and the crystalline lens.

Pupil : 3, the central opening of the crystalline lens front.

Iris : 4, a pigmented muscularating which can expand or contract to control the size of the pupil and hence the amount of light entering the eye.

Crystalline lens : 5, the image forming lens.

Ciliary Muscle : 6, which can stretch or slacken to vary the focal power of the crystalline lens.

Vitreous Humour : 7, a kind of transparent jelly.

Retina : 8, the sensory layer on which images are formed.

The system made up of the cornea, the aqueous humour, the crystalline lens, and the vitreous humour add up to an effective focal length 17 mm; the ciliary muscle can change it by a small margin.

The major mechanical parts of the eye are:

Scalera : 9, which encloses the eyeball, protects the inside, and gives it the needed stiffness.

Vascular Layer : 10, which supports the rods and cones behind the retina.

Finally, the parts which convert the physical image into the visual sensations of intensity and colour. The *retina*, the shape of a hemisphere, contains light receptors, called rods and cones. These are light-sensitive cells. In the human eye there are ~125 million rods and 6.5 million cones. The cones are responsible for colour vision and are concentrated at the depression in the retina, called *fovca centralis*, 11, which falls on the optic axis. The remaining areas are occupied mainly by the rods, with the exception of the junction of cells of the retina with the optic nerve, 12; there are no rods or cones here, so that this junction forms the blind spot.

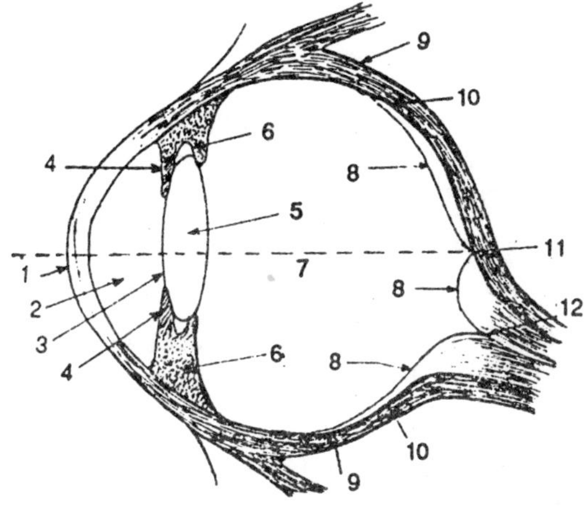

Fig. 6.8 : The human eye.

In rods the light causes the rhodosin [a compound of a form of vitamin A (retinen) and the protein of the retina (opsin)] to change from the cis-to trans-isomer. Thereby each rod generates a *nerve impulse*, which is conveyed by the *optic nerve* to the brain. In the process, the energy is drawn from that stored already by the receptors; the light only triggers the reaction. That is the reason why each rod is capable of responding to a photon.

Rods do not perceive colour. That is the function of the cones. Hence colours are perceived only if the image falls into the central area of the retina. Probably there are at least three kinds of cones, perceiving green, red and blue. Intermediate colours are seen depending on the relative stimulation. The brain perceives the colours on the basis of different trains of pulses from the cones. However, there are many unexplained features and a consistent theory of colour vision is still lacking.

SOLVED EXAMPLES

Example 1:

The angular diameter of a star is negligibly small compared to the diffraction spread λ/a due to viewing with an instrument of aperture diameter a. If the effective brightness of the star is 10^{-4} times that of the background sky in viewing with the naked eye, what will the relation be in viewing through a telescope of 200 times the diameter of the pupil of the eye?

Solution:

We will use subscripts s and o for the star and background sky, and primes' and" for viewing with the eye and the telescope respectively. We have the general relations.

$$I = BA = B'A' = B''A''$$

and $$A = r^2\Omega$$

where B is the brightness, A is the transverse area, r is the distance, and W the solid angle. For the star, Ω arises due primarily to diffraction, so that

$$\Omega'_s = \pi\ (\lambda/a')^2;\ \Omega_s'' = \pi\ (\lambda/a')^2$$

Substitutions lead to

$$B_s'' = B_s'\ (a''/a')^2$$

In the case of the sky, the intrinsic Ω_o far exceeds the contributions due to diffraction, and hence $\Omega_o'' = \Omega_o' = \Omega_o$. That means

$$B_o'' = B_o'$$

Hence $$(B_s''B_o'') = (B_s'/B_o')\ (a''/a')^2 = (10^{-4})\ (200)^2 = 4$$

Thus, in the telescope the star brightness is 4 times the sky brightness.

(*Note:* It is incorrect to take the effect as due to larger light gathering by a telescope lens. That effect is common for the star and the background light. The effect is because of smaller diffraction spread in star image as a increases).

Example 2:

A 100 W lamp may be treated as emitting radiations at different wavelengths as listed below:

λ/1000 A = 4.8	*5.2*	*5.6*	*6.0*	*6.4*	*6.8*	*7.2*
Pλ/Watt = 0.4	*0.8*	*1.2*	*1.6*	*2.0*	*2.8*	*3.0*

The rest of the power is spent otherwise. Using the sensitivity curve of the human eye, and given $\mu_{max} = 683$ *lm/W, deduce the* μ *of the lamp.*

Solution:

We list the corresponding Kλ values and hence PλKλ

$K_\lambda = 0.20$	0.71	0.95	0.60	0.16	0.04	0.00
$P_\lambda K_\lambda = 0.0.8$	0.57	1.14	0.96	0.32	0.11	0.00

That gives $\Sigma P_\lambda K_\lambda = 3.2W$. Hence

$$\mu = \frac{3.2W \times 6.83 lm/W}{100W} = 22 \text{ lm/W}.$$

Example 3:

A typical coiled-coil filament lamp of 100 cm intensity has a filament of apparent length 2.5 cm and diameter 0.30 cm. Deduce the total flux emission, luminosity and brightness, assuming isotropic character of emission.

Solution:

Total flux $= Id\Omega = I.4\pi = 400\pi$ lm

Area of cross-section $= 2.5 \times 0.30 \times 10^{-4}m^2 = 7.5 \times 10^{-5}m^2$

Luminosity $(L_{average})$

$$= \frac{\phi}{S} = \frac{400\pi \text{ lm}}{7.5 \times 10^{-5} m^2} = 5.3\pi \times 10^6 \text{ lm/m}^2$$

Brightness $(B_{average})$

$$= \frac{I}{S} = \frac{400 cd}{7.5 \times 10^{-5}} = 5.3 \times 10^6 \text{ cd/m}^2$$

(*Note:* L here is –9 times that for the surface defining the candela. This is not because of higher temperature – T is of same order – but is be cause of coiled coil. The actual coil is much longer than that given as apparent length here).

Example 4:

A lamp of I = 400.cd hangs 1.5 m above the centre of a table of dimensions 2.0m × 2.0m. Deduce the illumination at a diagonal end of the table. If a paper placed there gives diffuse scattering with 60% efficiency, deduce the luminance of the paper and also the brightness.

Solution:

For the diagonal end,

$r^2 = [(1.5)^2 + (1^2 + 1^2)] = 4.25 \text{ m}^2;$

$\cos\theta = \frac{1.5}{r}$

$$\text{Illuminance } E = \frac{I\cos\theta}{r^2} = \frac{400\times 1.5}{(4.25)^{3/2}} = 70\text{lx}$$

Luminance L = 0.60 × E = 42 lm/m^2

Brightness B = L/π = 14 cd/m^2.

Example 5:

A source of light has intensity depending on q according to $I = I_o \cos(\theta/2)$. Deduce the total flux emitted by it, and the averaged intensity as a point source.

Solution:

The solid angle between θ and θ + dθ is 2π sin θ dθ.

Hence $d\phi = 12\pi \sin\theta\, d\theta = I_o 2\pi \sin\theta \cos 1/2\, \theta\, d\theta$

Integrating from θ = 0 to θ = π (point source)

$$\phi = 2\pi I_o \left[-\frac{4}{3}\cos^3\frac{\theta}{2}\right]_\theta^\pi = \frac{8\pi}{3} I_o$$

$$I_{av} = \frac{\phi}{4\pi} = \frac{2}{3} I_o.$$

Example 6:

A source P and an observer Q are placed at angles 3 0° and 60° respectively from the normal to a surface. The source has intensity 1400 cd in the concerned direction, and its distance from the surface is 2.5m. The surface gives diffuse scattering of the incident light with 80% efficiency. Deduce the brightness of the surface for the viewer.

Solution:

The illuminance is

$$E = \frac{I\cos\theta}{r^2}$$

$$= \frac{1400\times\cos 30}{(2.5)^2}\,\text{lm/m}^2$$

With 80% efficiency, the luminance must be 0.8 E. Also, for an ideal diffuse scatterer, B is the same in all directions, with the relation L = πB. That gives

$$B = \frac{L}{\pi} = \frac{0.8E}{\pi}$$

$$= \frac{0.8\times 1400\times\cos 30}{\pi(2.5)^2}\,\text{cd/m}^2.$$

(Notice that angle 60° irrelevant for an ideal diffuse scatterer).

EXERCISES

1. (a) Phot, is a unit for illuminance equal to 1 lm/cm^2. Deduce the relation between a phot and a lux.

   (b) Lambert is a unit of brightness equal to 1 cd/cm^2 . Deduce the relation between a lambert and 1 cd/m^2.

   (c) Metre-candle is a unit for illuminance. Deduce in meter candles the illuminance for a surface placed 1m from a source of intensity 100 erf and normal to the beam.

2. (a) A chamber encloses *diffuse* radiations. If the flux across unit area anywhere, considered in all directions, is ϕ, show that the flux component along one particular direction is $(1/4)\phi$.

   (b) If the above chamber has a long cylindrical window of cross-sectional area S' and a source of average spherical intensity I cd, deduce the value of ϕ. How will the result change if a window of same area S' is cut right on the chamber face?

3. A collimated beam is created by placing a source at the focus of a convergent lens of f = 25 cm and aperture diameter 4.0 cm. If the source is taken as a sphere of diameter 3.0 mm, deduce the cross-sectional diameter of the beam at distances (i) 10 cm, (ii) 1.0 m, (iii) 10 m from the lens, ignoring diffraction effect. If the source has intensity 60 cd in the concerned range, deduce the illuminance in the beam -at the three given distances.

4. Comment critically on the following:

   (a) The contrast of star brightness relative to the background sky brightness improves when telescopes of large apertures are used; but this is not because of larger collection of luminous flux.

   (b) In a given image-forming system, the narrowing of tens aperture does not affect brightness of the image.

   (c) A sodium vapour lamp has much larger *intensity* than a typical in candescent filament lamp, but in *brightness* the latter could be greater.

5. Two if sources of light of intensities f and 6I are placed 200cm apart. At what place should a screen be placed on the line joining the sources, so that illumination due to the two sources may be equal on the screen.

6. For the-human eye, consider the pupil diameter ~3 mm, mean wavelength ~600A; and focal length ~20 mm. Compute the area of the image of an ideal point source at infinity as determined by diffraction. If the real density of the cones in the retina is designed by nature to use the optimum available resolution, estimate the area! density of the cones on the retina.

7. A 60W lamp has intensity 100 cd and is isotropic. Deduce :

   (a) the luminous efficiency,

   (b) illuminance on a surface placed 4.0 m away with inclination $\theta = 30°$. If the surface is a diffuse scatterer with 60% efficiency, deduce its luminance, and brightness.

8. The angular diameter of the moon, as seen from f the earth, is 0.008 radian, and the moon's surface re-emits 20% of the light incident on it from the sun. Deduce the ratio of illumination on the earth in full moon light to that in full sunlight.

9. Deduce the luminance of the sun, given that its disc subtends an angle 30' arc and the illuminance due to it at earth's surface at mid-day is 10^5 lux. An image of the sun is formed by a lens of diameter 8 cm and focal length 30 cm; deduce the illuminance of the image.

10. In a street lighting system, the poles are 12 m apart and the lamps 5.0m above the road. The lamps are designed to give uniform

intensity over a solid angle 1.5π sr and zero outside. If the averaged intensity of the lamps is 600 cd each, deduce :

(a) the intensity within the 15π sr solid angle cone,

(b) the illuminance at the bottom of each lamp, and

(c) the illuminance at the mid-point between the bases of adjacent poles.

11. A 100 W lamp has uniform intensity 250 cd in a cone of semiangle 60° and zero intensity outside. Deduce :

(a) the luminous flux,

(b) the luminous efficiency, and

(c) brightness of a perfect (Effuse scatterer surface placed 5 m away normal to the incident beam.

12. A photometric balance is obtained between two sources P and Q when P is 58.0 cm from the photometer. When a plane glass plate is introduced in the beam from P, with its face normal to the beam, the balance is restored when P is shifted by 2.6 cm. Deduce the percentage to is of light by reflection at each face of the plate. What do you expect if the plate were inclined from the beam at an increasing angle?

7

FIBRE OPTICS

INTRODUCTION

In photonics photons play the same role as that played by electrons in electronics. The subject of photonics is therefore also referred to as *optoelectronics* or *optronics*. Photons travel with a speed far larger than that of electrons, and light beams can cross each other without affecting each other. Optical signals have large bandwidth and can accommodate a large number of channels per a given volume. Photonics is mainly concerned with lasers, fibre-optics, harmonic generation using nonlinear media, electro- optic devices, imaging, optical computing etc.

In 1870 John Tyndall, a British physicist demonstrated that light can be guided along a curved stream of water. Owing to total internal reflections light gets confined to the water stream and the stream appears luminous. A luminous water stream is the precursor of an optical fibre.

Electronic communications use radio and microwaves to carry information over copper wires and coaxial cables. The information carrying capacity of the wires is highly restricted because of their limited bandwidths and does not suffice for the modem needs. The use of optical fibres in the place wires enhances tremendously the number of signals that can be transmitted simultaneously. Historically, the first effort to use light as a vehicle of communication was in 1880 due to Alexander graham Bell. By 1960 it had been established that light could be guided by a glass fibre but the fibres available at that time heavily attenuated light waves propagating through them. It was the fabrication of low-loss glass fibres by Coming Glass Works and the invention of solid state lasers in 1970 that made optical communications practicable. The overwhelming advantages of optical fibres over traditional wires and coaxial cables were soon recognised. Commercial communication systems based on optical fibres made their appearance by 1977.

Apart from the use as communicational channel, optical fibres are widely used in other areas. Fibroscopes made of optical fibres are widely used in a variety of forms in medical diagnostics. Sensors for detecting electrical, mechanical, thermal energies are made using optical fibres.

OPTICAL FIBRE

An optical fibre is a transparent conduit as thin as human hair, made of glass or clear plastic, designed to guide light waves along its length. An optical fibre works on the principle of total internal reflection. When light enters one end of the fibre, it undergoes successive total internal reflections from sidewalls and travels down the length of the fibre along a zigzag path, as shown in Fig. 7.1. A small fraction of light may escape through sidewalls but a major fraction emerges out from the other end of the fibre.

A practical optical fibre has in general three coaxial regions (Fig. 72). The inner most region is the light guiding region known as the core. It is surrounded by a coaxial middle region known as the cladding. The outermost region is called the sheath. The refractive index of cladding always lower than that of the core. The purpose of the cladding is to make the light to be confined to the core. Light launched into the core and striking the core-to-cladding interface at an angle greater than critical angle will be reflected back into the core. Since the angles of incidence and reflection are equal, the light will continue to rebound and propagate through the fibre. The sheath protects the cladding and the core from abrasions, contamination and the harmful influence of moisture. In addition, it increases the mechanical strength of the fibre.

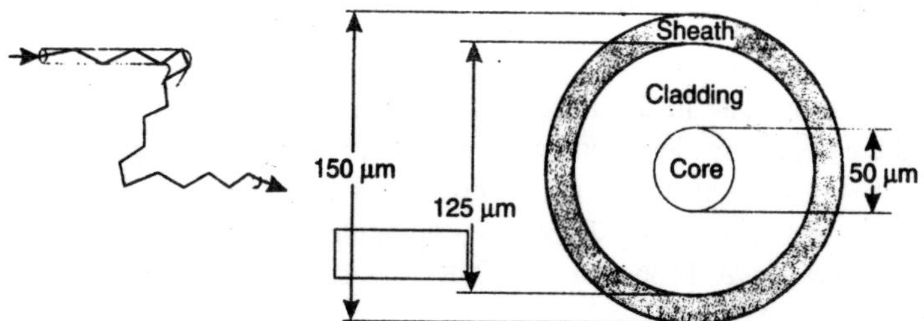

Fig. 7.1 **Fig. 7.2 : Cross-sectional view of an optical fibre.**

CRITICAL ANGLE OF PROPAGATION

Let us consider a step index optical fibre into which light is launched at one end. The end at which light enters the fibre is called the *launching* end. Fig. 7.3 depicts the conditions at the launching end. In a step-index fibre, the refractive index changes abruptly form the core to the cladding. Now, we consider two rays entering the fibre at two different angles of incidence. The ray shown by the broken line is incident at an angle θ_2 with respect to the axis of the fibre. This ray undergoes refraction at point A on the interface between air and the core. The ray refracts into the fibre at an angle θ_1 ($\theta_1 < \theta_2$). The ray reaches the core-cladding interface at point B. At point B, refraction takes place again and the ray travels in the cladding. Finally, at point C, the ray refracts once again and emerges out of fibre into the air. It means that the ray does not propagate through the fibre.

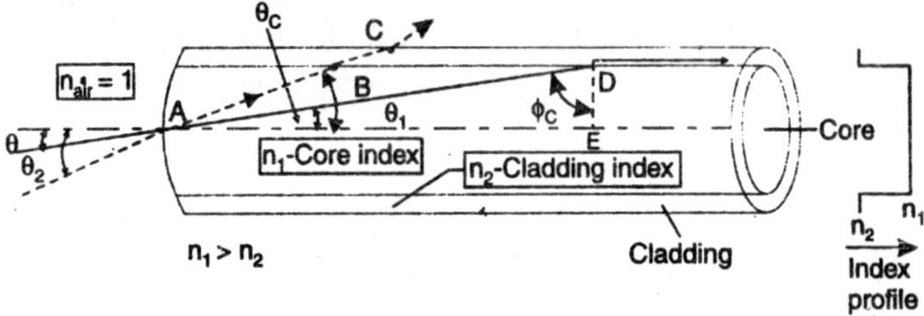

Fig. 7.3 : Light propagation in a step-index fibre.

Let us next consider the ray shown by the solid line in Fig. 7.3. The ray incident at an angle θ undergoes refraction at point A on the interface and propagates at an angle θ_c in the fibre. At point D on the core-cladding interface, the ray undergoes total internal reflection, since $n_1 > n_2$. Let us assume that the angle of incidence at the core-cladding interface is the *critical angle* ϕ_c where ϕ_c is given by

$$\phi_c = \sin^{-1}\left(\frac{n_2}{n_1}\right) \qquad ...(1)$$

A ray incident with an angle larger than ϕ_c will be confined to the fibre and propagate in the fibre. A ray incident at the critical angle is called a *critical ray*. The critical ray makes an angle θ_c with axis of the fibre. It is obvious that rays with propagation angles larger than θ_c will not propagate in the fibre. Therefore, the angle θ_c is called the *critical propagation angle*. From the Δ^{le} ADE, it is seen that

$$\frac{AE}{AD} = \sin\phi_c.$$

Also $$\frac{AE}{AD} = \cos\theta_c$$

From the relation (1), $\sin \phi_c = \frac{n_2}{n_1}$.

$$\cos \theta_c = \frac{n_2}{n_1} \qquad ...(2)$$

$$\therefore \qquad \theta_c = \cos^{-1}\left(\frac{n_2}{n_1}\right) \qquad ...(3)$$

MODES OF PROPAGATION

When light is launched into an optical fibre, waves having ray directions less than the critical angle θ_c, will be trapped within the fibre due to total internal reflection. But all such waves do not propagate along the fibre. In reality, only certain ray directions are allowed to propagate. The allowed directions correspond to the *modes* of the fibre. In simple terms *modes can be visualised as the possible number of paths of light* in an optical fibre. The paths are all zigzag paths, excepting the axial direction. As a zigzag ray gets repeatedly reflected at the walls of the fibre, phase shift occurs. Consequently, the waves travelling along certain zigzag paths will be in phase and intensified while the waves coursing along certain other paths will be out of phase and diminish due to destructive interference. The light ray paths along which the waves are in phase inside the fibre are known as modes. The number of modes that a fibre will support depends on the ratio d/λ, where a is the diameter of the core and λ, is the wavelength of the wave being transmitted.

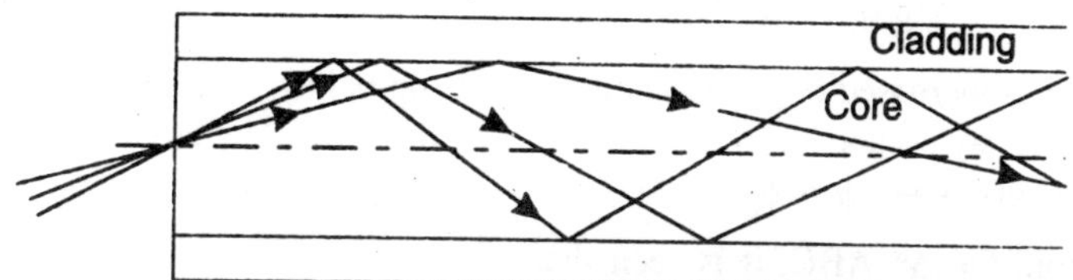

Fig. 7.4 : Propagation modes.

Modes are designated by an 'order' number 'm'. In a fibre of fixed thickness, the higher order modes propagate at angles close to the critical angle θ_c, and lower order modes propagate with angles much lower than the critical angle. The zero order ray travels along the axis and is known as the *axial ray*. The higher order modes tend to send light energy into the cladding. This energy is lost ultimately.

ACCEPTANCE ANGLE

Consider a step index optical fibre into which light is launched at one end, as shown in Fig. 7.5. Let the refractive index of the core be n_1 and the refractive index of the cladding be n_2 ($n_2 < n_1$). Let n be the refractive index of the medium from which light is launched into the fibre. Assume that a light ray enters the fibre at an angle θ_i to the axis of the fibre. The ray refracts at an angle θ_r and strikes the core-cladding interface at an angle ϕ. If ϕ is greater than critical angle ϕ_c, the ray undergoes total internal reflection at the interface, since $n_1 > n_2$. As long as the angle ϕ is greater than ϕ_c, the light will stay within the fibre.

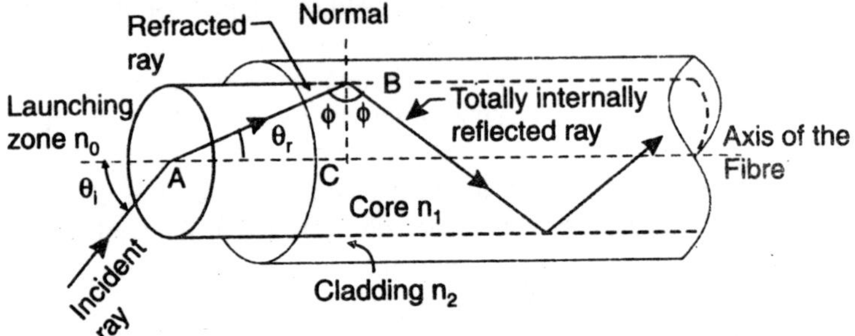

Fig. 7.5 : Illustration of the path of a light ray incident on the end of an optical fibre at an angle to the fibre axis.

Applying Snell's law to the launching face of the fibre, we get

$$\frac{\sin \theta_i}{\sin \theta_r} = \frac{n_1}{n_o} \qquad ...(4)$$

If θ_i is increased beyond a limit, ϕ will drop below the critical value ϕ_c and the ray escapes from the sidewalls of the fibre. The largest value of θ_i occurs when $\phi = \phi_c$.

From the Δ^{le} ABC, it is seen that

$$\sin \theta_r = \sin (90° - \phi) = \cos \phi \qquad ...(5)$$

Using equation (5) into equation (4), we obtain

$$\sin \theta_i = \frac{n_1}{n_o} \cos \phi$$

When $\phi = \phi_c$, $\sin \left[\theta_{i_{max}}\right] = \frac{n_1}{n_o} \cos \phi_c$ ...(6)

But $\sin \phi_c \frac{n_2}{n_1}$

$$\therefore \qquad \cos \phi_c \ \frac{\sqrt{n_1^2 - n_2^2}}{n_1} \qquad \text{...(7)}$$

Substituting the expression (7) into (6), we get

$$\sin\left[\theta_i \ (\max)\right] \frac{\sqrt{n_1^2 - n_2^2}}{n_o} \qquad \text{...(8)}$$

Quite often the incident ray is launched from air medium, for which $n_o = 1$.

Designating θ_i (max) = θ_o, equation (8) may be simplified to

$$\sin \theta_o = \sqrt{n_1^2 - n_2^2}$$

$$\therefore \qquad \theta_o = \sin^{-1}\left[\sqrt{n_1^2 - n_2^2}\right] \qquad \text{...(9)}$$

The angle θ_o is called the *acceptance angle* of the fibre. *Acceptance angle is the maximum angle that a light ray can have relative to the axis of the fibre and propagate down the fibre.*

In three dimensions, the light rays contained within the cone having a full angle $2\theta_o$ are accepted and transmitted along the fibre. Therefore, the cone is called the acceptance cone.

Light incident at an angle beyond θ_o refracts through the cladding and the corresponding optical energy is lost. It is obvious that the larger the diameter of the core, the larger the acceptance angle.

FRACTIONAL REFRACTIVE INDEX CHANGE

The fractional difference A between the refractive indices of the core and the cladding is known as *fractional refractive index change.* It is expressed as

$$\Delta = \frac{n_1 - n_2}{n_1} \qquad \text{...(10)}$$

This parameter is always positive because n_1 must be larger than n_2 for the total internal reflection condition. In order to guide light rays effectively through a fibre, $\Delta << 1$. Typically, Δ is of the order of 0.01.

NUMERICAL APERTURE

The main function of an optical fibre is to accept and transmit as much light from the source as possible. The light gathering ability of a fibre depends on two factors, namely core size and the numerical aperture. The acceptance angle and the fractional refractive index change determine the numerical aperture of fibre.

The numerical aperture (NA) is defined as the sine of the acceptance angle. Thus,

$$NA = \sin \theta_o$$

where θ_o is the acceptance angle.

But $$\sin \theta_o = \sqrt{n_1^2 - n_2^2}$$

$$\therefore \quad NA = \sqrt{n_1^2 - n_2^2} \qquad \text{...(11)}$$

$$n_1^2 - n_2^2 = (n_1 + n_2)(n_1 - n_2) = \left(\frac{n_1 + n_2}{2}\right)\left(\frac{n_1 - n_2}{n_1}\right) 2n_1$$

Approximating $\frac{n_1 + n_2}{2} \approx n_1$, we can express the above relation as $\left(n_1^2 - n_2^2\right) = 2n_1^2 \, \Delta$.

It gives $$NA = \sqrt{2n_1^2 \, \Delta}$$

$$\therefore \quad NA = n_1 \sqrt{2\Delta} \qquad \text{...(12)}$$

Numerical aperture determines the light gathering ability of the fibre. It is a measure of the amount of light that can be accepted by a fibre. It is seen from eqn. (11) that NA is dependent only on the refractive indices of the core and cladding materials and does not depend on the physical dimensions of the fibre. The value of NA ranges from 0.13 to 0.50. A large NA implies that a fibre will accept large amount of light from the source.

TYPES OF OPTICAL FIBRES

Optical fibres are in general of two types: namely the single mode fibre (SMF) and the multimode fibre (MMF). A single mode fibre has a smaller core diameter and can support only one mode of propagation. On the other hand, a multimode fibre has a larger core diameter and supports a number of modes. Multimode fibres are further distinguished on the basis of index profile. Index profile is a plot of refractive index drawn on horizontal axis versus the distance from the core axis drawn on the vertical axis. The index profile of a MMF can be either a step index (SI) type or a graded index (GRIN) type. The index profile of a SMF is usually a step index type.

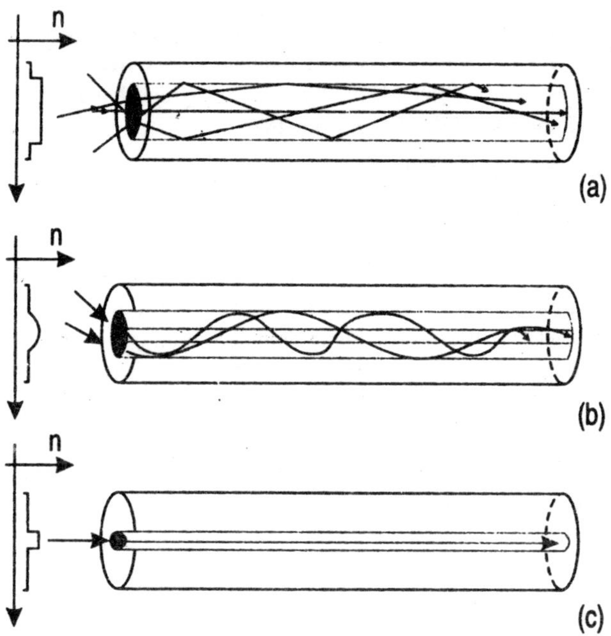

Fig. 7.6 : The three major fiberoptic configurations and their index profiles, (a) Multimode step-index fibre. (b) Multimode graded-index fibre, (c) Single-mode step-index fibre.

Single Mode Step Index Fibre

A typical SMF has a core diameter of 4μm which is of the order of a few wavelengths of light. The fibre is surrounded by an opaque protective sheath. The refractive index of the fibre changes abruptly at the core-cladding boundary, as shown in Fig. 7.6. Light travels in SMF along a single path that is along the axis. Obviously, it is the zero order mode that is supported by a SMF. A single mode step index fibre is designed to have a V number between 0 and 2.4. This relatively small value is obtained by reducing the fibre radius and by making Δ the relative refractive index change, to be small. Both Δ and NA are very small for single mode fibres. The low NA means a low acceptance angle. Therefore, light coupling into the fibre becomes slightly difficult.

Intermodal dispersion does not exist in single mode fibres because only one mode exists. With careful choice of material, dimensions, and wavelength, the total dispersion can be made extremely small. Low dispersion makes the fibre suitable for use with high data rates.

In these fibres part of the light propagates in the cladding. Therefore, the cladding must have a low loss and be relatively thick. Typically, for a core diameter of 10 μm, the cladding diameter is 120 μm. Thus, overall fibre size is as that of other fibres. Manufacturing and handling is more difficult and therefore, the fibre is costlier.

Multimode Step Index Fibre

A multimode step index fibre is very much similar to the single mode step index fibre except that its core is of larger diameter. A typical fibre has a core diameter of 100 μm which is very large compared to the wavelength of light. Light follows zigzag paths inside the fibre. Many such zigzag paths of propagation are permitted in a MMF. Typical structure and index profiles of a step index MMF are shown in Fig. 7.6(a). The NA of a MMF is larger as the core diameter of the fibre is larger and it is of the order of 0.3.

Larger NA leads to more modes which also mean higher dispersion. Higher dispersion means lower data rate and a less efficient transmission. In MMF, the dispersion is mostly intermodal. The multimode step index fibre is relatively easy to manufacture and is less costly.

Graded Index (Grain) Fibre

A graded index fibre is a multimode fibre with a core consisting of concentric layers of different refractive indices. Therefore, the refractive index of the core varies with distance from the fibre axis. It has a high value at the centre and falls of with increasing radial distance from the axis. A typical structure and its index profile are shown in Fig. 7.6(b). Such a profile causes a periodic focussing of light propagating through the fibre.

In case of GRIN fibres, the acceptance angle and numerical aperture decrease with radial distance from the axis (Fig. 7.6b). The number of modes in a graded index fibre is about half that in a similar multimode step-index fibre. The lower number of modes in the GRIN fibre results in lower dispersion than is found in the MMF. The size of the graded index fibre is about the same as the step index fibre. The manufacture of graded index fibre is more complex.

The effective acceptance angle of the GRIN fibre is somewhat less than that of an equivalent SI fibre. It makes coupling fibre to the light source more difficult.

NORMALIZED FREQUENCY

An optical fibre is characterised by one more important parameter. known as V-number, which is more generally called *normalized frequency* of the fibre. The normalised frequency is a relation among the fibre size, the refractive indices, and the wavelength. It is given by

$$V = \frac{2\pi a}{\lambda}\sqrt{n_1^2 - n_2^2} \qquad ...(13)$$

where 'a' is the radius of the core and λ is the free space wavelength.

As $\sqrt{n_1^2 - n_2^2}$ = N.A., we can write (13) as

$$V = \frac{2\pi a}{\lambda}(NA) \qquad ...(14)$$

We may also write it as $V = \frac{2\pi a}{\lambda} n_1 \sqrt{2\Delta}$ ...(15)

The V-number determines the number of modes that can propagate through a fibre. According to the relation (15), it is seen that the number of modes that propagate through a fibre increases with increase in the numerical aperture. However, the intermodal dispersion is proportional to the square of the N.A., and therefore, more modes imply more dispersion.

The maximum number of modes M_N supported by a multimode SI fibre is given by

$$M_N \cong \frac{1}{2}V^2 \text{ .....MMF} \qquad ...(16)$$

while the number of modes in a GRIN fibre is about half that in a similar step-index fibre.

Thus, $M_N \cong \frac{1}{4}V^2$ ....GRIN fibre ...(17)

For $V < 2.405$, the fibre can support only one mode and is classified as a SMF.

It is seen from eqn. (15) that single mode properties can be realised by decreasing the core diameter and/or decreasing Δ such that $V < 2.405$. Either a large core and/or a larger A will result in multimode properties.

PULSE DISPERSION

The term dispersion is used to describe pulse-broadening effect by fibres. The pulse that appears at the output of the fibre is wider than the input pulse (for example, Fig. 7.8). As the signal, which is a pulse of light, travels along the fibre, it becomes wider because of various propagation phenomena. Dispersion is measured in units of time, typically, nanoseconds or picoseconds. In practical terms, if we have an input pulse of width t_{p1} and an output pulse t_{p2}, then the dispersion At is defined as

$$\Delta t = \sqrt{t_{p2}^2 - t_{p1}^2} \qquad ...(18)$$

The total dispersion of a fibre depends on its length. A longer fibre causes more pulse broadening; it has a large dispersion. Usually, dispersion of a particular fibre is specified per unit length, from which the total dispersion for a given length of the fibre is calculated. Thus,

$$\Delta t = L \times (\text{dispersion/km}) \qquad ...(19)$$

There are three mechanisms, which contribute to the distortion of light pulse in a fibre. They are known as intermodal dispersion, intramodal dispersion, and wave guide dispersion.

Intermodal Dispersion

Intermodal dispersion results from the fact that the wave propagates in modes. It is dispersion between the modes caused by the difference in propagation time for the different modes.

When numerous modes are propagating in a fibre, they travel with different net velocities with respect to the fibre axis. Parts of the wave arrive at the output before other parts (Fig. 7.7), leading to a spread of the input pulse (Fig. 7.8). This is known as intermodal dispersion.

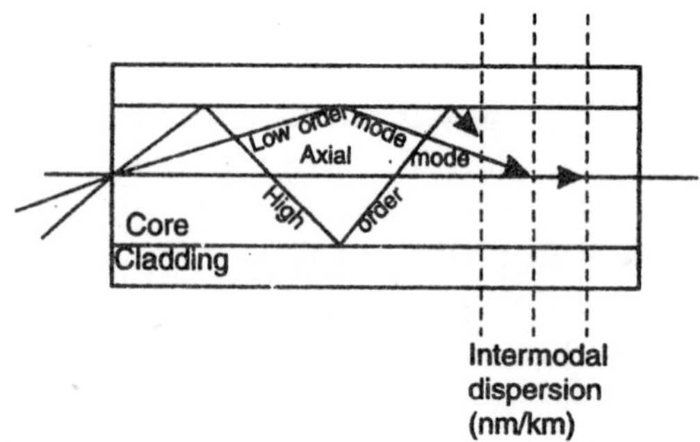

Fig. 7.7 : Intermodal dispersion in a stepped-index multimode fibre.

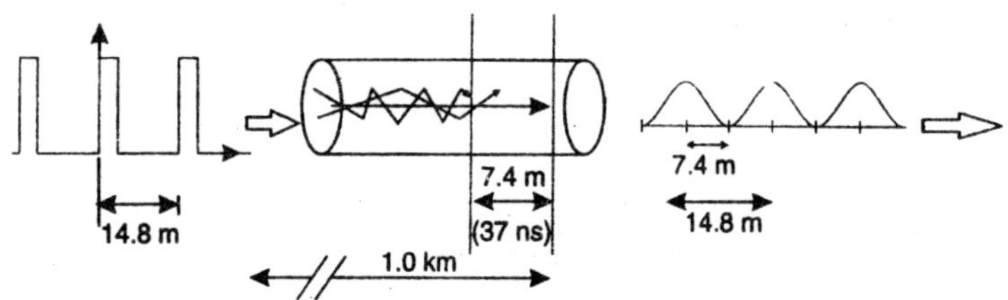

Fig. 7.8 : The spreading of an input/signal due to intermodal dispersion.

Intramodal Dispersion

Intramodal dispersion is a direct result of the fact that the light in the fibre consists of a group of wavelengths. Light waves of different wavelength travel at different speeds in a medium.

The short wavelength waves travel slower than long wavelength waves. Consequently, narrow pulses of light tend to broaden as they travel down the fibre. This is also known as *material dispersion.* The spectral width of the light source determines the extent of material dispersion.

Wave Guide Dispersion

Wave-guide dispersion arises from the guiding properties of the fibre. The effective refractive index for any mode varies with wavelength, which causes pulse spreading just like the variation in refractive index does. This is known as wave guide dispersion.

In fibres with large numerical aperture (NA) more modes exist leading to larger dispersion.

Therefore, dispersion may be restricted by selecting low NA fibre and a narrow spectral width source. Another solution to this problem is to use GRIN fibres, which produce less distortion, compared to SI fibres. However, GRIN fibres are more expensive.

Dispersion limits the bandwidth of a fibre. The information flow must be slow enough that dispersion does not cause adjacent pulses to overlap. If the distortion due to dispersion is large, the pulses may overlap and appear at the output as a single pulse with a number of humps. The detector cannot distinguish between the individual pulses.

ATTENUATION

An optical signal propagating through a fibre will get progressively attenuated. The signal attenuation is defined as *the ratio of the optical output power from a fibre of length L to the input optical power.*

$$\alpha = \frac{10}{L} \log \frac{P_i}{P_o} \quad \text{...(20)}$$

where P_i is the power of optical signal launched at one end of the fibre and P_o is the power of the optical signal emerging from the other end of the fibre. In case of an ideal fibre, $P_o = P_i$ and the attenuation would be zero. The unit of measurement of attenuation is decibel/kilometre (dB/km).

Different Mechanisms of Attenuation

There are three fundamental mechanisms responsible for attenuation in optical fibres.

1. *Absorption by material :* This includes absorption due to the light interacting with the molecular structure of the material, as well as loss because of material impurities. Even a highly pure glass absorbs light in specific wavelength regions. Strong electronic absorption occurs at UV wavelengths, while vibrational absorption occurs at IR wavelengths. These absorption losses are inherent property of the glass itself and are known as *intrinsic absorption.* Intrinsic losses are insignificant where fibre systems operate at present.

   Impurities are a major source of losses in fibres. Hydroxyl radical ions (OH), and transition metals such as copper, nickel, chromium, vanadium and manganese have electronic absorption in and near visible part of the spectrum. Their presence causes heavy losses. Losses due to impurities can be reduced by better manufacturing processes. In improved fibres, metal ions are practically negligible. The largest loss is caused by OH ions. These cannot be sufficiently reduced. The absorption of light either through intrinsic or impurity process constitutes a transmission loss because that much energy is subtracted from the light propagating through the fibre. The absorption losses are found to be at minimum at around 1.3 μm.

2. *Scattering :* When light is scattered by an obstruction, the result is power loss. The local microscopic density variations in glass

cause local variations in refractive index. These variations, which are inherent in the manufacturing process and cannot be eliminated, act as obstructions and scatter light in all directions. This is known as Rayleigh scattering. The *Rayleigh scattering* loss greatly depends on the wavelength. It varies as $1/\lambda^4$ and becomes important at lower wavelengths. Thus, Rayleigh scattering sets a lower limit, on the wavelengths that can be transmitted by a glass fibre at 0.8 urn, below which the scattering loss is very high.

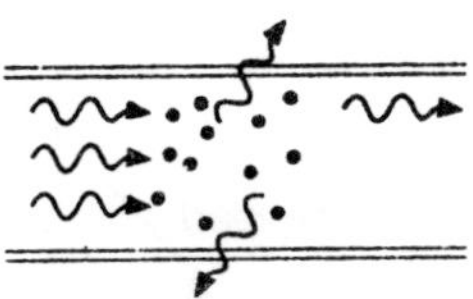

Fig. 7.9 : Rayleigh scarring, showing attenuation of an incident stream of photons due to localized variations in refractive index

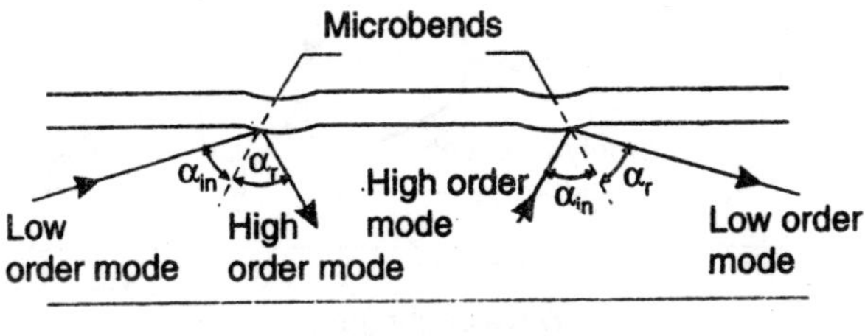

Fig. 24.10

3. *Waveguide and microbend losses :* These are the fibre losses introduced during manufacturing or installation processes. Structural variations in the fibre, or fibre deformation, cause radiation of light away from the fibre. Microbends, very minute disturbances in core size, also cause radiation of light.

An attenuation curve combining the losses due to various effects is shown in Fig. 7.11. For better performance, the choice of wavelength must be based on minimising loss and minimising dispersion. Such windows are selected for communication purposes. It is seen from the attenuation curve that it has a minimum at around a particular band of optical wavelengths. The band of wavelengths at which the attenuation is a minimum is called *optical window* or *transmission window* or *low-loss window*. There are three *low-loss windows* where losses from the various causes are relatively low.

λ (nm)	Approx. loss (dB/km)
820–880	2.2
1200–1320	0.6
1550-1610	0.2

From the above data it is seen that the range 1550 to 1610 is most preferable. From the point of view of dispersion, the low intramodal dispersion wavelength of about 1300nm is most suitable.

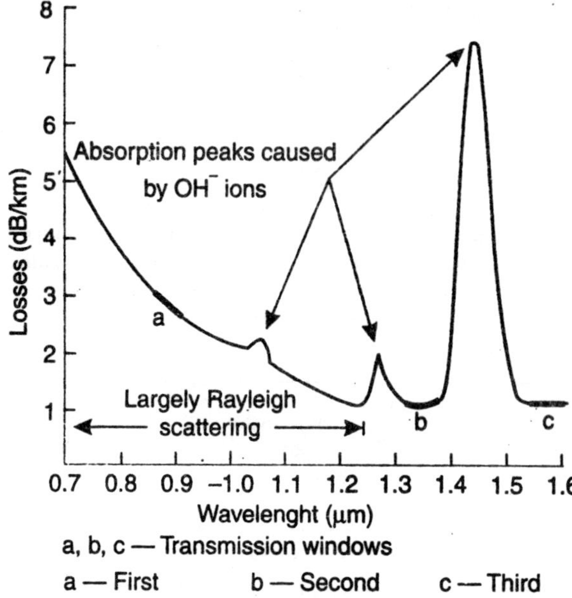

Fig. 7.11 : Fibre losses over the 0.7 to 1.6 μm spectrum.

APPLICATIONS

Transmission of light via an optical fibre has a wide variety of applications. We discuss here some of the applications. Broadly, optical fibres have three different applications, apart from other miscellaneous applications.

They are used for illumination and short distance transmission of images.

They are used as waveguides in telecommunications.

They are used in fabricating a new family of sensors.

Illumination and Image Transmission

Bundles of free fibres whose ends are bound together, ground and polished form flexible light guides. If no attempt is made to align the fibres in an ordinary array, they form an incoherent bundle in the sense that there is no correlation in the positions of the fibre terminations at one end of the bundle with that at the other end of the bundle. The primary function of such bundles is simply to conduct light from one region to another. Such *flexible light carriers* are relatively easy to make and inexpensive. They are used for illumination purpose.

When the fibres are carefully arranged so that their terminations occupy the same relative positions in both of the bound ends of the bundle, the bundle is said to be coherent. Such a bundle is capable of transmitting images. When one end of such a *flexible image carrier* is placed face down flat on an illuminated surface, a point-by-point image of the surface appear at the other end. Endoscopes are such coherent bundles.

By allowing direct viewing of what was formerly hidden, it has become a vital diagnostic tool for industry and medicine.

Optical Communications

Traditionally, electronic communications were carried out by sending electrical signals through copper cables, coaxial cables or waveguides. In recent years optical fibres are being used, where light signals replace electrical signals. A basic communications system consists of a transmitter, a receiver and an information pathway. Normally, the information to be communicated is a non-electrical message, which is to be converted first into an electrical form. The conversion is done by a transducer. For example, a microphone converts sound waves into currents. Similarly, a video camera converts images into currents. These electrical messages are of low frequency and cannot be transmitted directly. Therefore, they are superposed on a carrier wave of very high frequency. The process of imposing a message signal on a carrier wave is called modulation. Two different techniques of modulation are available. In analog modulation a continuous wave carries the message. In digital modulation message is transmitted in discrete form using binary digits. The message travels along the transmission channel and is received at the receiver. The receiver demodulates the modulated wave and separates out the message and feeds to a transducer such a loud speaker. The bandwidth requirement of the message and the bandwidth of the carrier determine the number of messages that can be simultaneously transmitted on an information channel. For example, a bandwidth of 4 kHz is required for voice transmission while 6 MHz bandwidth is required for TV signal transmission. When signals are transmitted in analog form the carrier should have double the above bandwidth. The normal TV communications has a bandwidth of 250 MHz and therefore, it can simultaneously transmit 20 TV programmes. However, instead of microwaves if light waves are used as carrier wave, the bandwidth will be about 10^8 MHz and can therefore transmit about 10^6 TV programs at a time. Thus, the use of the light waves expands our communication capabilities tremendously.

Optical Fibre Sensors

The variation of refractive index of the optical fibre under the influence of external forces is utilised in fabrication of optical fibre sensors.

If the optical fibre is subjected to heating, the temperature causes a change in the refractive index of the optical fibre. As temperature increases, the difference between the refractive indices of the core and cladding materials reduces and light tends to leak away from the cladding. A simple thermometer can be built using LED as a light source, a coil of optical fibre as heat sensing element and a photo-detector to measure the intensity of light. Temperatures in the range from 80° to 700° are measured using this technique.

A smoke and pollution detector can be built using optical fibres. A beam of light radiating from one end of a fibre can be collected by another optical fibre. If foreign particles are present in the region between the two fibres, they scatter light. The variation in intensity of the light collected by the second optical fibre reveals the presence of foreign particles.

A loop of optical fibre can be used to determine the level of the liquid in a container. A part of the cladding is scraped and the fibre loop is suspended above the liquid level. Light is directed to pass through one end of the fibre and its intensity is measured at the other end. A bare core loses more light when it is immersed in liquid than when it is in air. Therefore, a sudden change of outcoming light intensity indicates the liquid level. An LED, a MMF and a photo-detector are used in fabricating such a liquid level sensor. Such sensors are used to monitor the filling of petroleum tanks.

Double refraction is induced in the optical fibre when external forces are applied, the property of induced double refraction is exploited in fabrication of some of the sensors.

Medical Applications

Fibre optic technology is used in medical diagnostics as well as in medical procedures. The fibre optic endoscope is used to inspect internal organs for diagnostic purposes.

In ophthalmology, a laser beam guided by optical fibres is used to reattach detached retina and to correct defective vision.

In cardiology, optical energy transmitted through a optical fibre is used to evaporate built-up plaque that is blocking an artery.

In the treatment of cancer also the optical fibre technology is used. The process involves injection of special chemicals that penetrate only the cancerous cells. Infrared energy is transmitted via the fibre illuminates the affected area and is absorbed by the special chemical in the cancerous cells. The heat generated destroys the cancerous cells.

Military Applications

An aircraft, a ship or a tank needs tons of copper wire for wiring of the communication equipment, control mechanisms, instrument panel illumination etc. Use of optical fibre in place of copper reduces weight and further maintains true communication silence to the enemy. For example, a shipboard radar system requires about 250 m of coaxial cable, with a weight of 7 tons and a diameter of 45 cm. These cables can be replaced by optical fibre weighing 20 kg and measuring 2.5 cm in diameter.

Fibre guided missiles are used in recent wars. Sensors mounted on the missile transmit video information through the optical fibre to a ground control van and receive commands from the van again. The control van continuously monitors the course of the missile and if necessary corrects its course to ensure that the missile precisely hits the target.

OPTIC COMMUNICATION SYSTEM

A fibre optic communication system is very much similar to a traditional communications system and has three major components. A transmitter converts electrical signal to light signals, an optical fibre transmits the signals and a receiver captures the signals at the other end of the fibre and converts them to electrical signals.

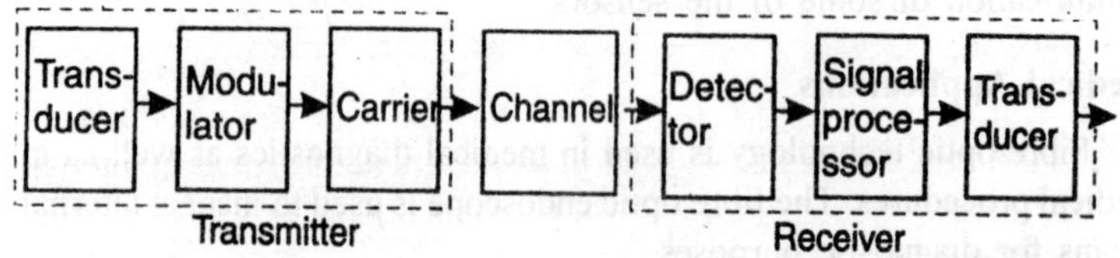

Fig. 7.12 : A generalized communications system.

The block diagram Fig. 7.12 illustrates a typical communications system. The transmitter consists of a light source supported by necessary drive circuits. A transducer converts a non-electrical message into an electrical signal and is fed to a light source. The light source is a miniature source, either a light emitting diode or a semiconductor laser. In either case, light is emitted in the IR range with a wavelength of 850 nm (0.85 μm), 1300 nm (1.3 μm) or 1550 nm (1.55 μm). The light waves are modulated with the signal. By varying the intensity of the light beam from the laser diode or LED, analog modulation is achieved. By flashing the laser diode or LED on and off at an extremely fast rate, digital modulation is achieved. A pulse of light represents the number 1 and the absence of light at a specified time represents zero. A message can be transmitted by a particular sequence of these Is or Os. If the receiver is programmed to recognise such digital patterns, it can reconstruct the original message. Though the digital modulation requires more complicated equipment such as encoders and decoders and also more bandwidth than analog modulation, it allows greater transmission distance with the same power. This is a great advantage and hence digital modulation has become popular and widely used nowadays. The transmitter feeds the analog or digitally modulated light wave to the transmission channel, namely optical fibre link. The optical signal travelling through the fibre will get attenuated progressively and distorted due to dispersion effects. Therefore, repeaters are to be used at specific intervals to regenerate the signal. At the end of the fibre, an output coupler directs the light from the fibre onto a semiconductor photodiode, which converts the light signals to electrical signals. The photodetector converts the light waves into electrical signals which are then amplified and decoded to obtain the message. The output is fed to a suitable transducer to convert it into an audio or video form.

Applications

Optical fibre communications systems can be broadly classified into two groups: (i) local and intermediate range systems where the distances involved are small and (ii) long-haul systems where cables span large distances.

(i) *Local area networks* : The local area network (LAN) is a computer oriented communication system. LAN operates over short distances of about 1 to 2 km. It is multiuser oriented system. In LAN, a number of computer terminals are interconnected over a common channel allowing each computer to use data and programs from any other. An optical data bus offers a great reduction in cost and increases enormously the information handling capacity.

(ii) *Long-haul communication* : On of the most important applications of fibre optic communication is long-haul communication. Long-haul communication systems are used for long distances, 10 km or more. Telephone cables connecting various countries come under this category. A rather sophisticated long-haul network is the NSFNET which links six supercomputer centres throughout U.S.A.

ADVANTAGES

Optical fibres have many advantageous features that are not found in conducting wires. Some of the important advantages are given here.

1. *Cheaper* : Optical fibres are made from silica (SiO_2) which is one of the most abundant materials on the earth. The overall cost of a fibre optic communication is lower than that of an equivalent cable communication system.
2. Smaller in size, lighter in weight, flexible yet strong: The cross section of an optical fibre is about a few hundred microns. Hence, the fibres are less bulky. Typically, a RG-19/U coaxial cable weighs about 1100 kg/km whereas a PCS fibre cable weighs 6 kg/km only. Optical fibres are quite flexible and strong.
3. *Not hazardous* : A wire communication link could accidentally short circuit high voltage lines and the sparking occurring thereby could ignite combustible gases in the area leading to a great damage. Such accidents cannot occur with fibre links since fibres are made of insulating materials.

5. *Immune to EMI and RJFI* : In optical fibres, information is carried by photons. Photons are electrically neutral and cannot be disturbed by high voltage fields, lightening, etc. Therefore, fibres are immune to externally caused background noise generated through electromagnetic interference (EMI) and radiofrequency interference (RFI).
6. *No cross talk* : The light waves propagating along the optical fibre are completely trapped within the fibre and cannot leak out. Further, light cannot couple into the fibre from sides. In view of these features, possibility of cross talk is minimised when optical fibre is used. Therefore, transmission is more secure and private.
7. Wider bandwidth: Optical fibres have ability to carry large amounts of information. While a telephone cable composed of 900 pairs of wire can handle 10,000 calls, a 1mm optical fibre can transmit 50,000 calls.
8. Low loss per unit length: The transmission loss per unit length of an optical fibre is about 4 dB/km. Therefore, longer cable-runs between repeaters are feasible. If copper cables are used, the repeaters are to be spaced at intervals of about 2 km. In case of optical fibres, the interval can be as large as 100 km and above.

SOLVED EXAMPLES

Example 1:

A step index fibre in air has a NA of 0.16 a core refractive index of 1.45 and a core diameter of 60 μm. Determine the normalized frequency for the fibre when light at a wavelength of 0.9 μm is transmitted.

Solution:

$$V = \frac{\pi d}{\lambda}(NA) = \frac{(3.143)(60\times10^{-6}\,m)(0.16)}{0.9\times10^{-6}\,m} = 33.5.$$

Example 2:

A step index fibre of diameter 50 μm has a numerical aperture of 0,23 . If the wavelength of input light energy is 0.82 μm, determine the number of modes in the cable.

Solution:

The number of modes,

$$M_N = \left[\frac{2.22d(NA)}{\lambda}\right]^2$$

$$= \left[\frac{2.22 \times 5010^{-6} m \times 0.23}{0.82 \times 10^{-6} m}\right]^2 = 9.69.$$

Example 3:

Optical power of 1 mW is launched into an optical fibre of length 100 m. If the power emerging from the other end is 0.3.mW, calculate the fibre attenuation.

Solution:

$$\text{Attenuation, } \alpha = \frac{10}{L}\log\frac{P_i}{P_o}$$

$$= \frac{10}{0.1km}\log\frac{1mW}{0.3mW} = 52.\ dB/km.]$$

Example 4:

In an optical fibre, the core material has refractive index 1.6 and refractive index of land material is 1.3. What is the value of critical angle? Also calculate the value of angle of acceptance cone.

Solution:

Critical angle is given by $\sin\phi_c = \frac{n_2}{n_1} = \frac{1.3}{1.61} = 0.8125$

Acceptance angle $\theta_0 = \sin^{-1}\left[\sqrt{{}_1^2 - n_2^2}\right] = \sin^{-1}[\sqrt{1.6 - 1.3^2}]$

$= \sin^{-1}(0.87)$

$= 60.5°$

Angle of acceptance cone $= 2\theta_0 = 121°$

Example 5:

The numerical aperture of an optical fibre is 0.5 and the core refractive index is 1.54. Find the refractive index of the cladding.

Solution:

Cladding refractive index, $n_2 = \sqrt{n_1^2 - (NA)^2}$

$$= \sqrt{1.54^2 - 0.5^2} = 1.46.$$

Example 6:

What is the numerical aperture of an optical fibre cable with a clad index of 1.378 and a core index of 1.546?

Solution:

$$NA = \sqrt{n_1^2 - n_2^2} = \sqrt{1.546^2 - 1.378^2} = \sqrt{0.491} = 0.70.$$

Example 7:

A fibre cable has an acceptance angle of 30° and a core index of refraction of 1.4. Calculate the refractive index of the cladding.

Solution:

$$\sin\theta_0 = \sqrt{n_1^2 - n_2^2}$$

$$\therefore \sin^2\theta_0 = n_1^2 - n_2^2$$

$$n_2^2 = n_1^2 - \sin^2\theta_0 = (1.4)^2 - \sin^2 30° = 1.96 - 0.25 = 1.71$$

$$\therefore n_2 = 1.308.$$

EXERCISES

1. Explain what is step-index, graded index, mono-mode, multimode fibre.
2. Explain the difference between the step-index fibre and graded-index fibre.
3. Describe various mechanisms of dispersion in optical fibres. Explain the effect of dispersion on the bandwidth of optical communication channel.
4. What are the different types of attenuation losses in an optical fibre? Discuss the absorption losses.
5. Draw the diagram for an optical fibre link and explain the function of each block.
6. A glass fibre optic cable has a relative dielectric constant of 2.30. Calculate the index of refraction for this cable.
7. Find the critical angle for a ray travelling from glass with refractive index 1.7 to water (n = 1.333).
8. An optical fibre is 2 m long and has a diameter of 20 μm. If a ray of light is incident on one end of the fibre at an angle of

40°, how many reflections does it undergo before emerging from the other end? Refractive index of fibre is 1.3.

9. What is the numerical aperture of a cable whose critical angle is 26.1°?
10. Discuss the advantages and disadvantages of optical fibres over conventional communication transmission media.
11. Using ray theory, describe the mechanism of transmission of light within an optical fibre.
12. With the help of a ray diagram, show how optical fibres can guide light waves. Explain what you understand by modes and numerical aperture.
13. Derive an expression for N.A. for S.I. fibre in terms of refractive index of the core and relative refractive index difference between the core and the cladding.
14. Explain with diagram step index and graded index fibre. What is multimode index fibre?
15. A SI fibre made with a core index 1.52 and a fractional difference index of 0.0007 has a diameter of 29 μm. It is operated at a wavelength of 1.3 μm. Find :

    (a) the fibre V-number, and

    (b) the number of modes the fibre will support.
16. An optical signal has lost 85% of its initial power after traversing 500 m of fibre. What is the loss in dB/km of this fibre?
17. A certain optical fibre has an attenuation of 3.5 dB/km at 850 nm. If 0.5 nm of optical power is launched into the fibre, what is power level in μW after 4 km?
18. Compare a monomode step index fibre with a multimode step index fibre.
19. List the main components of optical communication system. Describe the basic optical communication system.
20. Explain optical communication through block diagram. For long distance communication whether :

    (i) monomode or multimode, and

    (ii) step index or graded index fibre, which are preferable and why?

8

NON-LINEAR OPTICS

INTRODUCTION

Lasers generate coherent radiation at many wavelengths ranging from meter wavelength region to the soft x-rays region. However, it is not possible to produce light covering all wavelengths of interest in spite of the fact that a large number of active materials are available and lasers can be built using them. It becomes, therefore, necessary to transform the frequency of light generated by lasers into light of desired frequency. Nonlinear optical media help us generate frequencies that were not available, through frequency conversion techniques. Harmonic generation, sum and difference frequency generation and parametric oscillations are some of the important nonlinear processes utilized in laser light frequency transformations. Stimulated Raman scattering is another important process that is used in generating new wavelengths. The processes such as second harmonic generation, sum and difference frequency conversion, parametric oscillation are associated with *passive media*, *i.e.*, media that do not make evident their own characteristic frequencies. Stimulated Raman scattering arises in active media that impose their characteristic frequencies on the light wave. In this chapter we acquaint ourselves with the methods of nonlinear processes.

WAVE PROPAGATION AND MOMENTUM CONSERVATION

A plane wave propagating through a vacuum in a general direction is represented by the equation (1).

$$E = E_o e^{i(\omega t - k.r)} \quad ...(1)$$

where k is the vector that defines the direction of propagation of the wave, k can be written as

$$k = ik_x + j\, k_y + k\, k_z \quad ...(2)$$

When a light wave interacts with an isotropic material, it will be affected in the same way whatever may be the direction of the beam with respect to the material. Hence, the refractive index of the material is the same in every direction and the velocity of the beam passing through the material is also the same in every direction. Since the magnitude of k is related to the velocity, the components k_x, k_y and k_z would all be the same in case of isotropic crystals. In anisotropic crystals, the light beam experiences different refractive indices in different directions and propagates at different velocities in different directions. Therefore, the components of k will have different values in different directions.

If we consider the photon description of light beam, we know that the momentum and wavelength of a photon are related through the following expression.

$$p = \frac{h}{\lambda} = \frac{2\pi h}{2\pi\lambda} = \hbar k$$

Momentum p is a vector quantity and hence we write the above equation in vector form as

$$p = \hbar\ k \qquad ...(3)$$

where k is the vector which defines the propagation direction of the wave associated with the photon. Thus, the significance of k in eqn. (1) and (3) is the same.

When a beam of light passes from air into a crystal, the momentum of the wave must be conserved at the interface of the two media. It requires that the normal component and parallel component of the momentum must be separately conserved at the boundary. Therefore, the transverse component of the momentum must be continuous across the boundary. There cannot be an abrupt jump or change in the transverse momentum of the wave as the beam enters the medium. In view of the relation (3), it means that the transverse component of k should not change abruptly at the boundary.

LINEAR MEDIUM

The optical media are basically dielectric materials. They do not allow electric current to pass through when an electric field is applied across them. Instead, they get polarized. The electric field exerts forces on the valence electrons. These forces are quite small, and induce *electric dipoles* in the medium. These dipoles orient in the direction of the electric field and the dielectric is said to be *polarized*. Dielectrics are polarized, for example, when they are placed between the charged plates of a capacitor. The polarization vector, P, denotes the extent to which the dielectric is polarized. The electric polarization is parallel with and directly proportional to the applied field, E. It is given by

$$P = \varepsilon_0 \chi E \quad ...(4)$$

where ε_0 is the permitivity of free space and χ [= $(\varepsilon/\varepsilon_0) - 1$] is a dimensionless constant known as the *electric susceptibility* of the medium. The stronger the electric field, the greater will be the polarization and a plot of P versus E is a straight line. Materials, in which such kind of linear relationship holds, are known as *linear dielectrics*. Light waves are electromagnetic waves and when they propagate through a dielectric, the electric field of the waves polarizes the dielectric. Hence the optical parameters of the dielectric are closely related to the dielectric polarization. The refractive index of the medium is given by,

$$\mu = \sqrt{\varepsilon} = \sqrt{1+4\pi\chi} \quad ...(5)$$

As long as the intensity of the light propagating in the dielectric medium is small, the parameters χ and μ are constant quantities and are independent of the intensity of light. Ordinary light sources generate light of field strengths of the order of 10^5 V/m which are very small compared to atomic fields, and therefore cannot affect the optical parameters of the medium.

NONLINEAR POLARIZATION

When high electric field strength is used, it is expected that P cannot increase linearly indefinitely with E and will become saturated. Therefore, we may anticipate nonlinear behaviour of Pat very high field strengths. Lasers produce light of field strengths of the order of 10^7 to 10^{11} V/m, which are of the order of the atomic field strengths. Therefore, the intense light of lasers is in a position to cause nonlinearity of P and influence the optical parameters of the medium. When the electric field E in the light is very large, the parameters χ, ε and μ become the functions of E. Since the directions of P and E coincide in an isotropic medium, we can express χ as a power series in the field strength as,

$$\chi(E) = \chi_1 + \chi_2 E + \chi_3 E^2 + ... \quad ...(6)$$

and as $P = \varepsilon_0\, \chi(E)\, E$

$$\therefore \quad P = \varepsilon_0(\chi_1 E + \chi_2 E^2 + \chi_3 E^3 + ...) \quad ...(7)$$

where χ_1 is linear susceptibility and is much greater than the coefficients of the nonlinear terms χ_2, χ_3 and so on. The nonlinear terms contribute noticeably only at very high-amplitude electric fields.

The second order nonlinear polarization is given by

$$P_2 = \varepsilon_o \chi_2 E^2 \quad ...(8)$$

and third order nonlinear polarization by

$$P_3 = \varepsilon_o \chi_3 E^3. \quad ...(9)$$

Nonlinear polarization leads to nonlinear optical effects. Materials, in which polarization exhibits nonlinear dependence on the field strengths, are called *nonlinear media*. Fig. 8.1 shows the nonlinear variation of electric polarization with the electric field strength in a nonlinear medium.

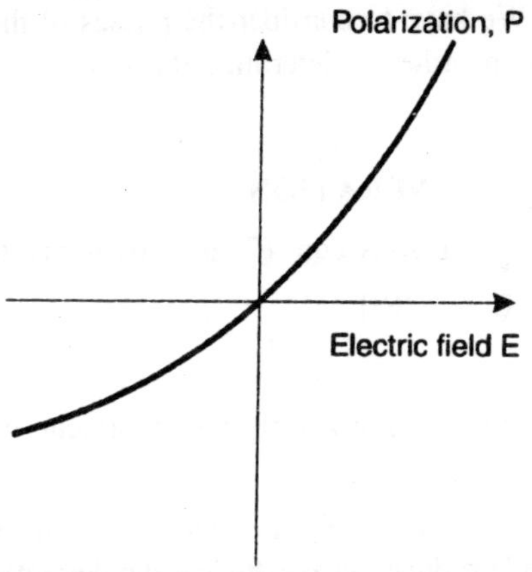

Fig. 8.1 : A typical plot of variation of electric polarization with electric field in a nonlinear medium.

In optically isotropic materials, the coefficients of even powers of E in Eqn. (7) are zero. But in case of anisotropic materials, coefficients of both odd and even powers of E exist. Strictly speaking, any medium becomes nonlinear provided the electric field of the incident radiation is very high. With the advent of intense lasers, the *nonlinear* polarization has assumed importance and made it possible the phenomena of frequency conversion and other non-linear effects.

Now if we take into the consideration of electric field oscillations of a light wave, we can write for the electric field of the wave as

$$E = E_o \sin \omega f$$

The polarization caused by this wave can be expressed as

$$P = \varepsilon_o (\chi_1 E_o \sin \omega t + \chi_2 E_o^2 \sin^2 \omega t + \chi_3 E_o^3 \sin^3 \omega t + ...) \quad ...(10)$$

The above equation can be re-written as

$$P = \varepsilon_o [\chi_1 E_o \sin \omega t + 1/2\ \chi_2 e\ E_o^2 (l - \cos 2\omega t) + 1/4\ \chi_3 E_o^3 (3 \sin \omega t - \sin 3 \omega t) + ...] \quad ...(11)$$

Thus, as the electric field in the incident wave oscillates, the array of dipoles produces an electromagnetic wave, which may be called a *polarization wave*. When the incident light beam is not intense, the polarization wave will be in phase and have the same frequency as the incident light wave. The higher order terms in eqn. (11) indicate that for higher beam intensities, polarization waves of higher frequencies also will be produced. We have to consider the phases of the waves at these other frequencies in order to determine the frequencies that get enhanced.

SECOND HARMONIC GENERATION

In case of anisotropic crystals eqn. (7) may be rewritten as,

$$P = \varepsilon_0[\chi_1 E + \chi_2 E^2] \quad ...(12)$$

$$= P_1 + P_{nl}$$

where P_1 and P_{nl} are linear polarization and nonlinear polarization components respectively.

Fig. 8.2 shows the response of the nonlinear medium, in which a symmetrical electric field produces an asymmetrical polarization. Fourier analysis of the polarization shows that it consists of components having frequencies v and 2v as well as a d.c. component. Let us consider a plane monochromatic light wave of frequency v, travelling at the velocity υ_1 in a nonlinear medium. The electric field of the wave may be of the form,

$$E = E_o \sin [2\pi v (t - x/\upsilon_1),)] + \chi E_o^2 \sin^2 [2\pi v (t - x/\upsilon_1)]$$

$$= E_o \sin[2\pi v(t - x/\upsilon_1),) + 1/2\chi_1 R_o^2 - E_o^2 \cos[4\pi v (t - x/\upsilon_1)] \quad ...(13)$$

Eqn. (13) contains a term in $4\pi v$ which corresponds to a wave of polarization propagating in the same direction and at the same speed of light wave with twice higher frequency. The magnitude of the term in $4\pi v$ {= 2 (2 π v)} approaches the magnitude of the first term $\chi_1 E_o$ at high vames of E_o. Secondly, under certain conditions, this polarization wave may cause emission of a new light wave. It means that a strong light of frequency v propagating in a nonlinear medium can generate a new light wave at frequency 2v. It is a second harmonic and hence the phenomenon is known as *second harmonic generation.* At some initial point the incident wave generates the second harmonic wave and at that point the two waves are coherent. As the incident wave propagates through the crystal, it continues to generate second harmonic waves. All these waves combine constructively only if they satisfy the phase matching condition (14).

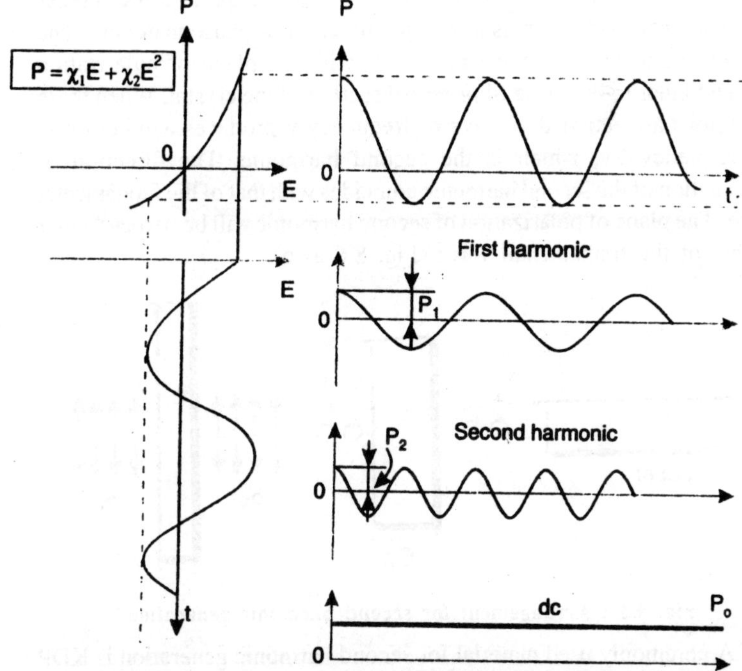

Fig. 8.2 : Fourier analysis of the asymmetrical polarization wave into fundamental, second harmonic and d.c. components

The input wave of frequency v is normally called the *pump wave.* The ratio of the intensity of the generated second harmonic wave to that of the pump wave is dependent on the *phase mismatch factor*, Δk, which is given by

$$\Delta k = k\ (2v) - 2k(v) = 2v\ [\mu\ (2v) - \mu\ (v)]$$

In order to obtain an intense second harmonic wave, which is directed like the pump laser beam, one has to achieve the condition $\Delta k = 0$. This implies that

$$\mu\,(2v) = \mu(v) \qquad ...(14)$$

Eqn. (14) is known as the *phase matching condition.* It is also known as *index matching condition.*

This condition is satisfied only when the pump and generated waves are of orthogonal polarization. Therefore, the process of second harmonic generation requires a unaxial double refracting crystal exhibiting high nonlinear susceptibility χ_2. It must be cut in the shape of a rectangular parallelepiped with the axis along the direction of phase matching. The incident light must be plane polarized, with the plane of polarization perpendicular to the plane of principal section of the crystal. When these conditions are satisfied, a wave of frequency v produces another wave of frequency 2v, which is the second harmonic. The direction of propagation of the second harmonic coincides with that of the fundamental wave. The plane of polarization of second harmonic will be perpendicular to that of the fundamental wave (Fig. 8.6 also).

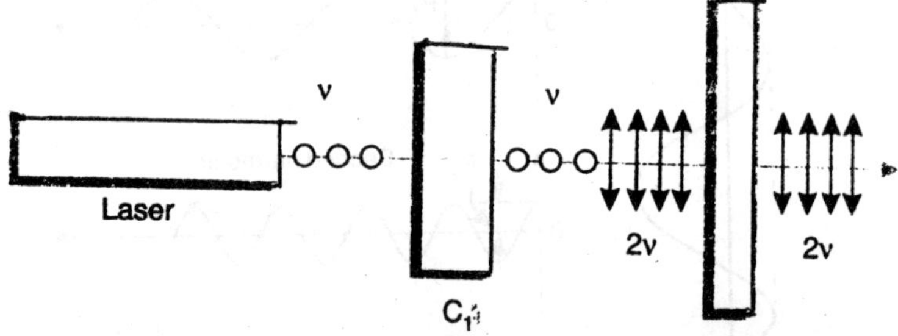

Fig. 8.3 : Arrangement for second harmonic generation.

A commonly used material for second harmonic generation is KDP which is a negative double refracting crystal. In negative crystals, if the fundamental light is a linearly polarized *ordinary wave*, the resulting second harmonic will be an *extraordinary wave*. The reverse is the situation in case of positive crystals. If the fundamental light is a linearly polarized extraordinary wave, the resulting second harmonic will be an *ordinary wave.*

The second harmonic generation may also be described from the standpoint of photon concept. According to this picture (Fig. 8.4), two photons of the fundamental frequency v, each having energy hv combine to produce an energy of 2hv. The energy levels located at hv and 2 hv are known as virtual levels and shown by dashed lines, as they are not allowed states of the material and as such are not populated. In this process, two photons of frequency v are destroyed and one photon of frequency 2v is simultaneously created.

Second harmonic generation was first demonstrated by Franken and his co-workers in 1961. They focused a 3kW ruby laser pulse (λ = 6943Å) onto a quartz crystal and obtained a very low intensity output at a wavelength of 3471.5Å. Second harmonic generation helps in extending the range of laser wavelengths into the blue and UV part of the spectrum, which are not rich in naturally occurring laser lines. Second harmonic generators are now available in market for use with visible and IR lasers.

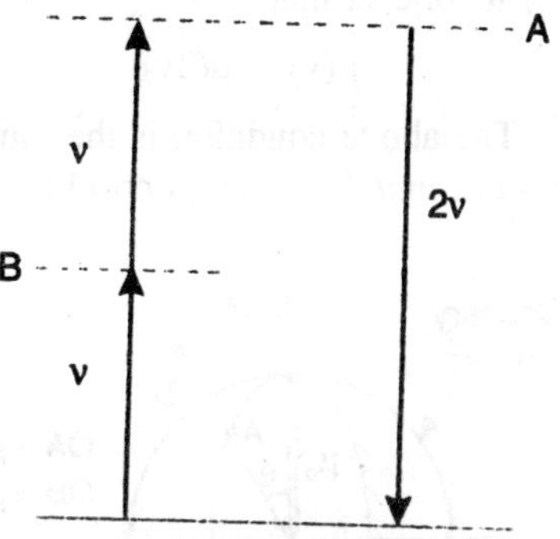

Fig. 8.4 : Second harmonic generation.

PHASE MATCHING

It was noted earlier that light waves are electromagnetic waves and the electric field of the waves polarizes the medium through which they propagate. The polarization response of the medium consists in the orientation of the electric dipoles in the direction of the externally applied electric field. As the electric field in the incident wave oscillates, the array of dipoles produces *polarization wave*. For higher beam intensities, polarization waves of higher frequencies also will be produced. To understand the condition under which the enhancement of a particular frequency occurs, let us take the case of second harmonic generation. In this case, the polarization wave propagates at the velocity υ_1, the second harmonic propagates at the velocity v^ in the medium. The velocities are given by,

$$\upsilon_1 = \frac{c}{\mu(v)} \quad ...(15a)$$

and $$\upsilon_2 = \frac{c}{\mu(2v)} \quad ...(15b)$$

The velocity Vy of the generated wave differs from that of the incident wave, υ_1 because the refractive index is a function of the frequency. Efficient transfer of energy from the polarization wave to the second harmonic requires that the two waves must be matched in momentum, that is velocity. It requires that the waves should experience identical refractive indices in the medium. Thus, from the set of eqn. (15) it follows that,

$$\mu(v) = \mu(2v) \quad ...(16)$$

The above condition is the same as the *phase matching condition or index-matching* condition (14).

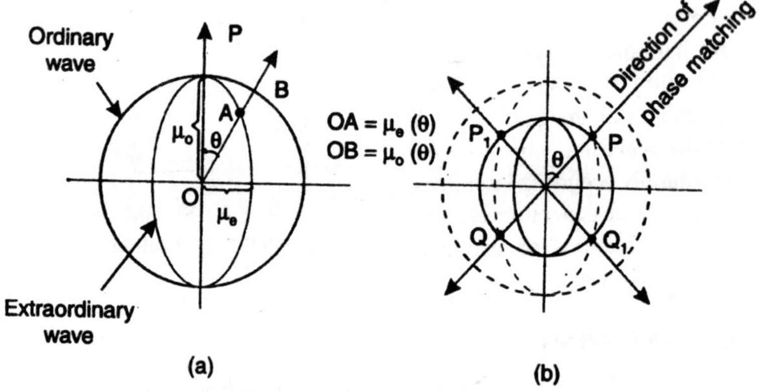

Fig. 8.5 : (a) Index ellipsoid ana spnere corresponding to a e-ray and o-ray in a negative uniaxial crystal. (b) Index matching in a negative crystal.

This condition is not satisfied in general cases. It is satisfied *only* in anisotropic crystals known as uniaxial *crystals.* When a light ray is incident on a *uniaxial crystal*, it splits up into two rays, namely, an *ordinary ray* (o-ray) and an *extra ordinary ray* (e-ray). The refractive index of an e-ray depends upon the direction of propagation. The refractive index surface is a *sphere* in case of o-ray and an *ellipsoid* in case of e-ray. The two refractive index surfaces in case of a negative *uniaxial crystal* are shown in Fig. 8.5(a). Fig. 8.5(b) shows the refractive index surfaces for the frequency v and for the frequency 2v. It is seen that the refractive index surface of the o-ray with frequency v and that of the e-ray with frequency 2v intersect; one of the points of intersection is at point P. At the point of intersection the velocities υ_1 and υ_2 are equal.

It is obvious that the phase-matching condition is satisfied for light waves propagating in the direction PQ. Thus, the direction PQ is *the direction of phase- matching.* Along PQ,

$$\mu_o(\nu) = \mu_e(2\nu) \qquad ...(17)$$

Angle Tuning for Phase Matching

There are two methods for achieving phase matching in double refracting crystals. They are known as angle *tuning* and *temperature tuning*. The refractive index of extraordinary ray 4 varies with the angle between the direction of propagation and the optic axis. The variation is given by

$$\mu_e(\theta) = \left[\frac{\cos^2\theta}{\mu_o^2} + \frac{\sin^2\theta}{\mu_e^2}\right]^{-1/2} \qquad ...(18)$$

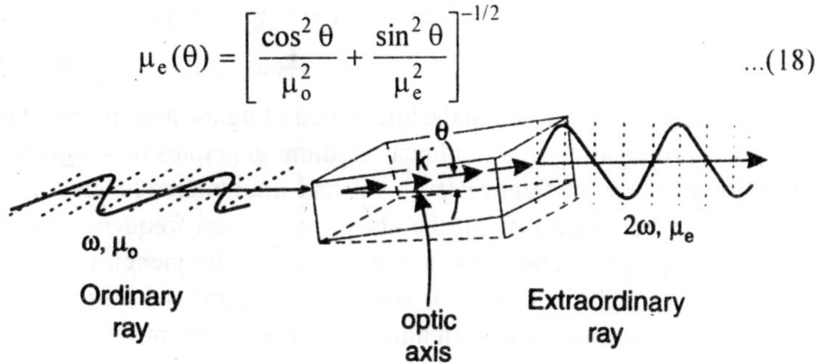

Fig. 8.6 : Phase matching using angle tuning.

By rotating the crystal in the plane of the optic axis and the incident beam (Fig. 8.6), it is possible to select the value of μ_e that will satisfy the phase matching condition (16).

SUM AND DIFFERENCE FREQUENCY GENERATION

Nonlinear crystals make it also possible to generate beams of sum and difference frequencies when two light waves of different frequencies interact within them. The interaction of waves with frequencies ν_1 and ν_2 generate light waves with frequencies $\nu_1 + \nu_2$ and $\nu_1 - \nu_2$.

Let us consider two coherent waves of frequencies ν_1 and ν_2 propagating simultaneously in a nonlinear medium. Let the waves be represented by $E_1 \sin 2\pi\nu_1 t$ and $E_2 \sin 2\pi\nu_2 t$, respectively. The superposition of the waves leads to,

$$E = E_1 \sin 2\pi\nu_1 t + E_2 \sin 2\pi\nu_2 t \qquad ...(19)$$

Using Eqn. (19) in Eqn. (12), we obtain the following expression for the nonlinear term in the polarization of the medium.

$$P_{nl} = \varepsilon_o\chi_2 E^2$$
$$= \varepsilon_o\ \chi_2\ [E_1 \sin 2\pi v_1 t + E_2 \sin 2\pi v_2 t]^2$$
$$= \varepsilon_o\chi_2\ E_1{}^2 \sin^2 2\pi v_1 t + \varepsilon_o\chi_2 E_2{}^2 \sin^2 2\pi v_2 t$$
$$+ 2\varepsilon_o\chi_2 E_1\ E_2 \sin(2\pi v_1 t) \sin(2\pi v_2 t)$$
$$= \varepsilon_o\chi_2\ E_1{}^2\ (1 - \cos 4\pi v_1 t) + \varepsilon_o\chi_2 E_2{}^2 (1 - \cos 4\ 2\pi v_2 t)$$
$$+ \varepsilon_o\chi_{2\ E} 1E_2\ \{\cos 2\pi(v_1 - v_2)t - \cos 2\pi(v_1 + v_2)t\}$$

$$\text{or } P_{nl} = \varepsilon_o\chi_2 + \left(E_1^2 + E_2^2\right) - \varepsilon_o\chi_2 E_1^2\ \cos 2\pi(2v_1)\ t - \varepsilon_o\chi_2\ E_1^2$$
$$\cos 2\pi(2v_2)\ t + \varepsilon_o\chi_2\ E_1E_2\{\cos 2\pi(v_1 - v_2)\ t$$
$$- \cos 2\pi\ (v_1 + v_2)\ t\} \quad \text{...(20)}$$

The Eqn. (20) shows that the interaction of light waves of frequencies v_1 and v_2 propagating in a nonlinear medium generates new light waves of sum $(v_1 + v_2)$, difference $(v_1 - v_2)$ and double frequencies $2v_1$ and $2v_2$. It is to be noted that all the above mentioned frequencies are not generated simultaneously when two waves of frequencies v_1 and v_2 propagate through the crystal. If any of these additional frequencies are to be generated, the phase matching must be implemented and it can be implemented only to one of the frequencies at a time.

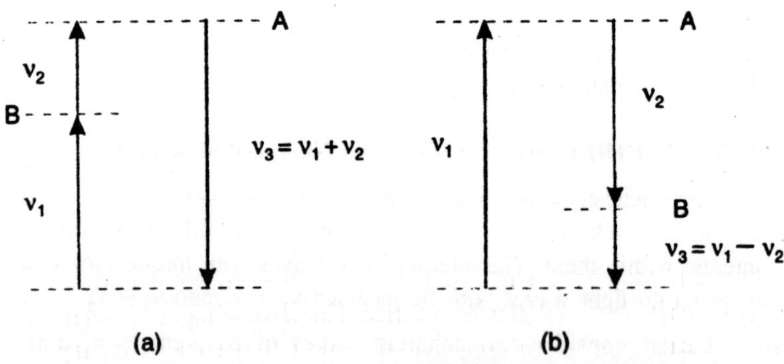

Fig. 8.7 · (a) sum frequency generation (b) difference frequency generation

The processes of sum and difference frequency generation are depicted in Fig. 8.7. In sum frequency generation the frequencies v_1 and v_2 are destroyed and a new photon $v_3 = (v_1 + v_2)$ is created, as shown in Fig. 8.7 (a). In difference frequency generation, the photon of higher frequency v_1 is destroyed and two new photons of frequencies v_2 and

$v_3 = (v_1 - v_2)$ are created, as shown in Fig. 8.7 (b). Note that v, is already present as one of the input beams; it indicates that v_2 is amplified in the process.

In sum frequency generation, v_1 is taken as a weak infrared wave that sums with a strong pump wave v_2 to generate a high frequency wave at $v_3 = (v_1 - v_2)$. Thus, the light of wavelength of 10.6 μm produced by a CO_2 laser is made to interact with a laser beam of 1.06 μm wavelength produced by Nd: YAG laser to yield a light beam of wavelength of 0.964 μm. Such upconversion is done because it is difficult to detect the radiation at 10 μm while the radiation at 0.964 μm can be easily detected.

PARAMETRIC OSCILLATION

The production of coherent radiation at frequencies in the UV region is very much essential. Apart from the technique of the harmonic generation, tuning of frequency by parametric oscillation is another technique available for this purpose. The method of parametric excitation of oscillations is widely used in electronics, where use of nonlinear capacitors is made. The parametric generation of light involves use of nonlinear medium and thus bears close similarity with the electronic process.

In sum frequency generation, light at frequencies v_1 and v_2 add together to produce the frequency $(v_1 + v_2)$. On the other hand, if a strong beam at frequency $v_1 = (v_2 + y_3)$ alone is applied to a suitable nonlinear material such as lithium niobate, two beams at lower frequencies v_1 and v_3 can be generated. For exciting parametric oscillations, the nonlinear medium is placed between two mirrors, which form a resonant cavity comparable to a laser cavity. The wave of frequency $v_1 = (v_2 + v_3)$ is the *pump wave* and it should be highly intense such that it can induce nonlinear behaviour of the crystal. Secondly the process requires that the phase-matching condition be satisfied. The wave of higher frequency, say v_2 is called the *signal wave* and the lower frequency wave, say at v_3, is called the idler wave. Both the *signal and idler waves* are weak and always present in the crystal in the form of noise, which arises due to spontaneous photons. If the pump frequency $(v_2 + v_3)$ is fixed, then the two frequencies v_2 and v_3 are free to spread over a wide range of values. This effect is known as *parametric amplification.* By varying the orientation of the crystal, the values of v_2 and v_3 can be varied. If the mirrors of the resonant cavity are reflective at v_2 and v_3 and not at the frequency ($v_2 + v_3$), resonance of these waves in the cavity enhances the interaction in the nonlinear crystal and builds up the waves. Thus the system acts as a parametric oscillator. Tuning can be achieved by varying the phase-matching conditions through mechanical or temperature control of the cavity.

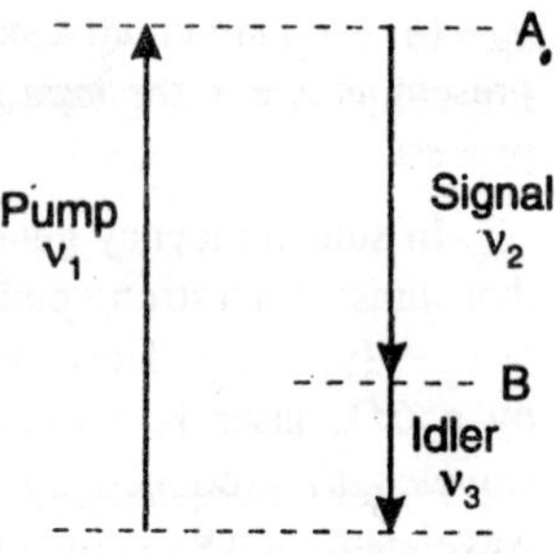

Fig. 8.8 : Optical parametric oscillation

Parametric oscillation was first observed by Giordmaine and Miller in 1965. The schematic of the set up used by them is shown in Fig. 8.9. The output was tuned by changing the temperature of the lithium niobate crystal. A temperature range of about 11°C produced wavelengths in the range 9680Å to 11540Å. The conversion efficiency was low as 1 %. Currently, conversion efficiency is improved from 40% to 50%.

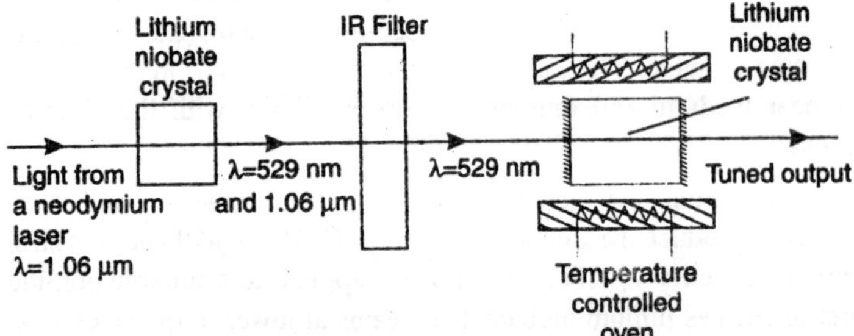

Fig. 8.9 : A typical arrangement for observing parametric oscillation in a nonlinear medium

SELF-FOCUSSING OF LIGHT

When a light beam of very high intensity passes through an optical medium, an internal force will result. It in turn alters the density, changes permittivity and thereby the refractive index. If the beam has a Gaussian cross-section, the intensity would be the greatest on the axis of the rod and hence the index of refraction would be greater on the axis than off the axis.

These induced refractive index variations will cause the medium to function as a lens, as illustrated in Fig. 8.10. Consequently, the beam contracts and results in the process known as self-focussing of the beam.

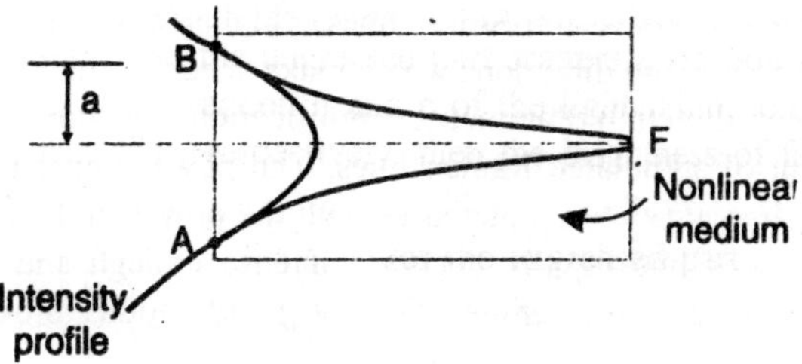

Fig. 8.10 : Self-focussing

STIMULATED RAMAN SCATTERING

Stimulated Raman effect is a process in which an optical medium absorbs a higher energy photon and emits a lower energy photon and simultaneously absorbs the balance of the original photon energy and goes to an excited state. The process is illustrated in Fig. 8.11. A photon of frequency v_1 is incident upon the material. A molecule in the material absorbs the photon and goes to a virtual level A, from where it emits a photon of lower frequency v_2 and simultaneously absorbs the energy $hv_\upsilon = h\,(v_1 - v_2)$. As a result the molecule is left at the excited level A B. The normal Raman scattering is incoherent whereas the stimulated Raman scattering is coherent.

The main advantage of stimulated Raman scattering is that it provides a mechanism for generating coherent radiation in infrared region.

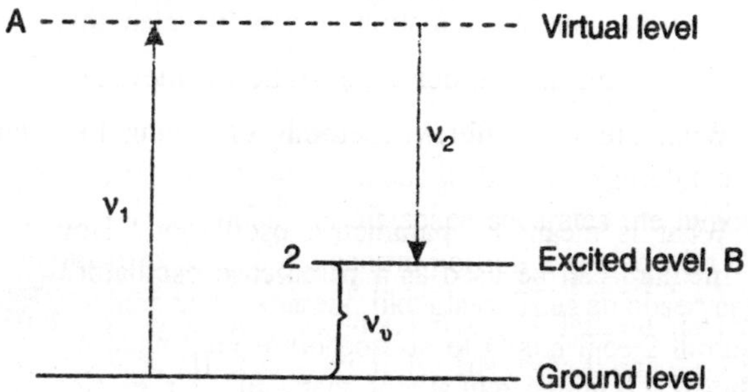

Fig. 8.11 : Stimulated Raman scattering

Now if the scattered radiation is photographed on a colour sensitive film, a striking pattern of concentric coloured rings are seen. It is found that the additional wavelengths are as coherent and well collimated as the main beam. The Stokes bands, which lie at lower wave numbers are emitted essentially along the laser beam direction. The coloured rings correspond to successive anti-Stokes lines at higher wave numbers which are emitted only along directions which make specific small angles with the laser beam direction. Parametric amplification is responsible for the enhancement of stimulated Raman lines. The power transferred to the first Stokes line at $(v_o - v_M)$ increases with the power in the laser at v_o. The first Stokes line rapidly becomes intense enough and acts as a powerful source at wave numbers $(v_o - v_M)$ and another Stokes line at

$$(v_o - v_M) - v_M = (v_o - 2v_M)$$

is generated. As this line gains in intensity, it acts as another source giving rise to a third line and so on. The essential difference between stimulated Raman scattering and stimulated emission in lasers is that for producing coherent radiation in stimulated Raman scattering there is no need for inverting the population of the states.

EXERCISES

1. Explain what is meant by self-focussing?
2. Explain stimulated Raman scattering? How can it be explained using quantum concept?
3. What is a nonlinear medium?
4. What is nonlinear polarization and why it assumed importance after the advent of lasers?
5. Explain why phase matching condition must be satisfied to generate coherent radiation at new wavelengths?
6. How is second harmonic generation achieved?
7. What are the different methods of tuning to obtain index-matching?
8. What is meant by parametric oscillation? How a nonlinear medium can be used as a parametric oscillator?

9

ATOM LASER

INTRODUCTION

An atom laser is analogous to an optical laser. Optical laser generates a coherent beam of electromagnetic waves whereas an atom laser produces a coherent beam of matter waves. Each atom is associated with a de Broglie wave. When a dilute gas is cooled to temperatures near to zero, the de Broglie waves merge together to form a coherent matter wave. Therefore, an atom laser requires a thermal cloud of ultra-cold atoms which constitutes active medium of the laser. Hence, the main challenge in making an atom laser is the cooling of atoms to near zero temperature. In 1975 T.W. Hansch and A.L. Schawlow, USA, proposed first the technique of laser cooling. They suggested that neutral atoms might be cooled to millikelvin temperatures by the action of a near resonant laser beam that excites the atoms followed by spontaneous emission. Steven Chu, Claude Cohen-Tannoudji and William D.Phillips succeeded in cooling and trapping atoms with laser light and were awarded in 1997 the Nobel prize in Physics for their work. Cooling to millikelvin temperatures is not enough to create coherent matter waves. Atoms are to be cooled to nanokelvin temperatures, when they form a Bose-Einstein condensate and create the state of coherent matter waves. In 1995 Comell, Wieman and their colleagues succeeded for the first time in cooling a dilute gas of rubidium atoms (Rb^{87}) to nanokelvin temperatures. They were awarded the Nobel prize in physics in the year 2001 for this work. The discovery of Bose-Einstein condensation has heralded the invention of atom laser. An atom laser will have a major impact on the fields of atom optics, atom lithography and precision measurements.

BOSE-EINSTEIN CONDENSATION

It is necessary to understand what a Bose-Einstein condensate is and how it is produced, before we go into the details of atom laser.

In 1924 the Indian physicist Satyendra Nath Bose derived the Planck law for black-body radiation by treating the photons as a gas of identical particles. Photons have an intrinsic angular momentum, or "spin", of the Planck constant h divided by 2π. Einstein generalized Bose's theory to other particles of integral spin. The theory is now known as *Bose-Einstein statistics*. Particles that have a spin that is an integer multiple of $h/2\pi$ obey Bose-Einstein statistics and are called bosons. It was shown that more than one boson can occupy the same quantum state. Einstein predicted that at sufficiently low temperatures all the atoms in an ideal gas of identical atoms might be condensed to a single lowest quantum state of the system. This large number of atoms locked together in the same state is called *Bose-Einstein condensate* (BEC). Bose-Einstein condensation is an exotic quantum phenomenon that was observed in dilute atomic gases. Bose-Einstein condensation happens only for "bosons". The low temperature necessary for obtaining BEC is of the order of nanokelvin.

METHODS OF COOLING ATOMS

Let us consider a gas inside a container at room temperature. The atoms of gas move at random in all directions and attain thermal equilibrium with the container. They move with velocities, υ, which are of the order of a few hundred meters/second. According to the kinetic theory the absolute temperature is proportional to $\left\langle \upsilon^2 \right\rangle$. Temperature is an expression of kinetic motion of the particles. Hence the kinetic energy of a particle may define the kinetic temperature.

The successful route to form Bose-Einstein condensate of the gas involves two steps. The first step consists in cooling down the atoms at low density to millikelvin temperatures by the action of a laser beam of an appropriate frequency. In the second step, the cold atoms are further cooled by using the technique of evaporative cooling. These techniques provided a new route to ultracold temperatures that does not involve cryogenics.

LASER DOPPLER COOLING

The basic principle of this method is as follows. If a photon falls on an atom, which has a resonance frequency equal to that of the photon, the atom absorbs the photon. If an atom absorbs a photon coming from the opposite direction, the momentum of the photon is transferred to the atom and the atom is pushed back and loses its velocity. In order to ensure deceleration of the atom, the atoms should absorb only the oppositely moving radiation. This is achieved by using Doppler effect.

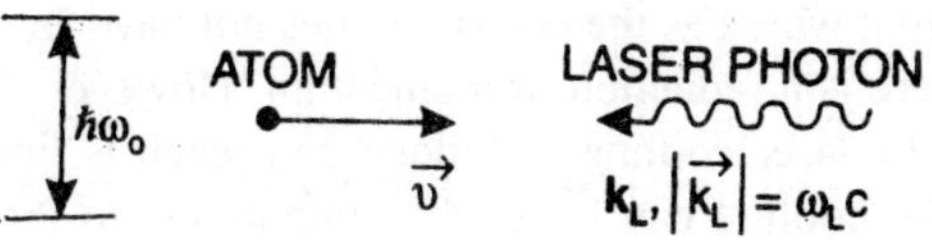

Fig. 9.1 : Doppler-resonant absorption of a red-detuned laser beam by a moving two-leve) atom

An atom with a lower energy level E_1 and an upper energy level E_2 loses or gains energy by emitting or absorbing light of frequency v_L such that $E_2 - E_1 = hv_L$ (Fig. 9.1). The photons associated with a plane light wave of frequency v_L, propagating along a direction with unit vector k_L have an energy $E = hv_L$ and a momentum $P_L = \left(\frac{hv_L}{c}\right)$ each. Absorption of a single photon by an atom results in a momentum transfer of P_L to the atom. In one second, the atom can absorb and emit a large number of times and hence-the net momentum transfer rate to the tiny atom become substantial. After about ten thousand absorption and subsequent emission cycles, an atom which was originally moving at about 700 m/s can be slowed down to near zero speed.

Fig. 9.2 shows a beam of atoms, each of mass m traveling with a velocity υ, colliding with a counter directed beam of laser photons having a propagation vector k_L. The laser frequency v_L is selected such that it is just beneath the resonant frequency v_0 of the atoms. Because of its motion, any particular atom sees an oncoming photon with a frequency that is Doppler-shifted upward by an amount $v_L \upsilon/c$. When the laser frequency is tuned so that $v_0 = v_L (1 + \upsilon/c)$, collisions with photons will resonate the atoms. In the process, each photon transfers its momentum of $\hbar k_L$ to the absorbing atom, whose speed there upon reduced by an amount $\Delta\upsilon$ where $m \Delta\upsilon = \hbar k_L$. The cloud of atoms is not very dense, and each excited atom can drop back to its ground state with spontaneous emission of a photon of energy hv_0. The emission is randomly directed and so although the atom recoils, the average amount of momentum regained by it over thousands of cycles tends to zero. The change in momentum of the atom per photon absorption-emission cycle is therefore effectively $\hbar k_L$ and it slows down. By contrast, an atom moving in the opposite direction, away from the light source sees photons to have a frequency v_L $(1 - \upsilon/c)$, far enough away from v_0 that there can be little or no absorption and therefore no momentum gain. If there are two counter-propagating waves, they can be used to decelerate the atoms coming from both directions. The process is repeated several times and slows down the atoms. Hence, the gas is cooled. Thus, the absorption slows down the atom where as the emission does not have an effect and on the average there is a reduction in momentum. However, the lowest temperature that the laser cooling technique can reach is limited. The process of Doppler cooling can lower the temperature to the order of 100 μK for alkali atoms. Alkali atoms are well suited to laser- based methods because their optical transitions can be excited by available lasers and because they have a favourable internal energy-level structure for cooling to low temperatures. Once the atoms are cooled to uX temperature, we have to confine them. If the cooling laser is turned off, the atoms will start moving apart with their residual velocities, and will hit the walls and eventually fall down. The cold atoms usually have a velocity of a few centimeters per second.

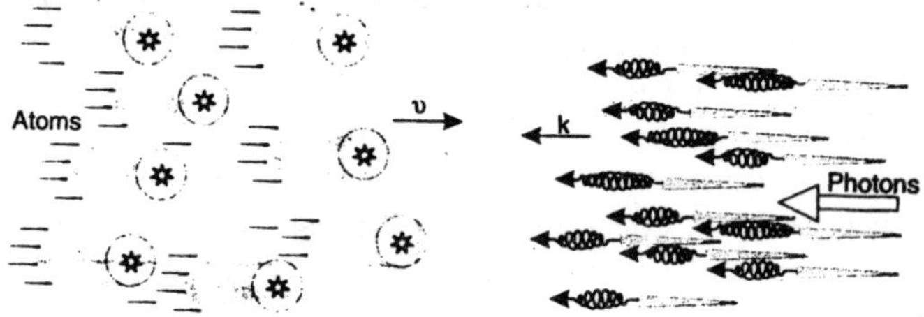

Fig. 9.2 : A stream of atoms colliding with a laserbeam in the process of laser cooling.

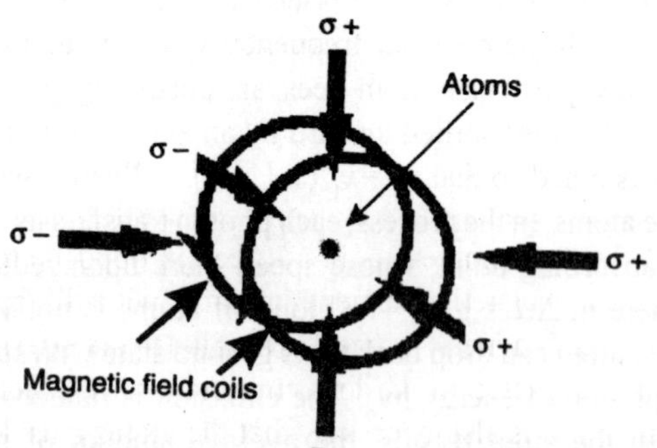

Fig. 9.3 : Six laser beams and a pair of magnetic field coils forming a magneto-optical trap can cool more than a billion atoms to micro-kelvin temperatures.

In practice six laser beams are arranged-two counter-propagating beams each, (one along the positive direction and the other along the negative direction) along each of the three Cartesian axes, as shown in Fig. 9.3. An atom, in whichever direction it moves, always encounters abeam in opposite direction and the net effect is that the atom experiences a strong viscous force in the three dimensional space in the field of photons. This six laser beam configuration used to cool a sample of atoms from a vapour is given the name *optical molasses*, since the light bath always opposes the motion of atoms as if they were submerged in molasses and causes viscous effect on the atomic motion. The optical molasses could produce a temperature much lower than Doppler cooling limit. However there is no position dependent force and since the atoms are not slowed to complete standstill, they diffuse out of the laser beams and fall away under gravity.

Trapping of Atoms

The frictional force arising from the Doppler effect cools the atoms, but does not capture them; the atoms fall out of the molasses in a few seconds. It becomes therefore necessary to confine the atoms in a small region of space. Position dependence is introduced by using ($\sigma^+/\sigma-$) polarized laser beams and applying a weak magnetic field. This arrangement is known as a *magneto-optical trap* (MOT). The trap may be a spherical quadrupole magnetic field, produced by two anti-Helmholtz coils carrying currents in opposite directions (Fig. 9.3). Such an arrangement gives rise to a zero-field region between the two coils, at the intersection of the six laser beams and the field increases in all directions away from this point. This field minimum attract molecules in the low-field seeking states, that is molecules with their magnetic moments oriented anti-parallel to the magnetic fields, and repel strong field seeking molecules (Fig. 9.4).

Alkali atoms have a magnetic moment because they have an unpaired electron. The magnetic moment is in a direction opposite to that of the electron spin. The magnetic moment interacts with the applied weak magnetic field. If the magnetic moment is parallel to the external magnetic field, the atom is attracted to the local minimum of the field (Fig. 9.4) and can be trapped. It is also necessary that the atoms must be thermally isolated from their surroundings, since at ultracold temperatures atoms stick to all surfaces. After the atoms are trapped and cooled with lasers, all light is extinguished and a potential is built up around the atoms with an inhomogeneous magnetic field. Thus, the magnetic field can act as a little bowl and which confines the atoms to a small region of space, as shown in Fig. 9.5 (a). In addition to holding the atoms at a point in space, the trapping force compresses them into a dense cold cloud.

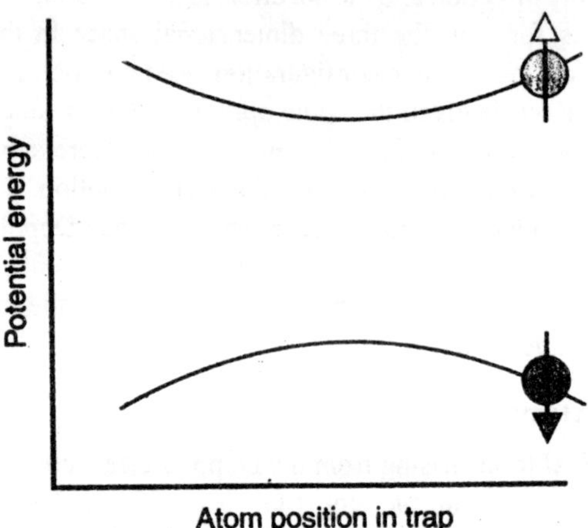

Fig. 9.4 : An atom with spin parallel to the magnetic field is attracted to the energy minimum and an atom with anti-parallel spin is repelled.

EVAPORATIVE COOLING

The atoms are further cooled to ultracold temperatures using the technique of evaporative cooling. The basic physics of evaporating cooling is very simple. It is known that a cup of hot milk or coffee cools down by evaporation. In the same way, high-energy atoms are allowed to escape from the sample so that the average energy of the remaining atoms is reduced.

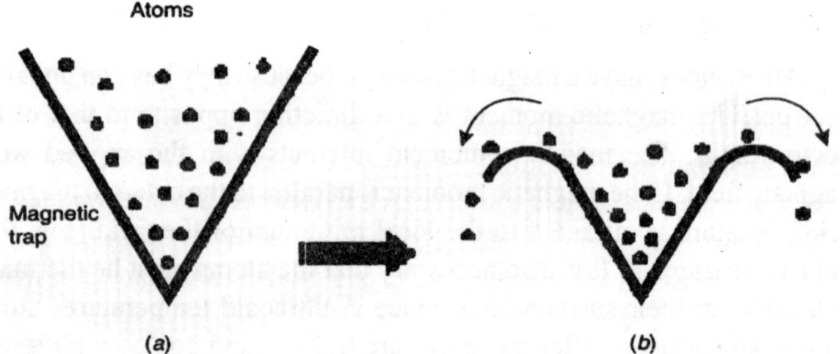

Fig. 9.5 : Evaporative cooling, (a) Atoms held in a magnetic trap. (b) Cooling works by slowly lowering the sides to release the hot atoms at the top.

It was shown by H.F. Hess that the idea of evaporative cooling could be applied to the atoms confined in a magnetic bowl. In this method the atoms are cooled first in the presence of the magneto-optic trap. Then the laser beams are turned off. Then a magnetic field, of the same type but much stronger than that used in MOT, is applied to perform the rol of a bowl and confine the cold atoms. Now, a radio frequency oscilla field is applied to induce transitions in the atom between spin u (attracted to the magnetic trap) and the spin down state (repell y the magnetic trap), so that the atoms will undergo a spin-flip transition. If the rf field is tuned to the higher side of the magnetic bowl the atoms having higher energy will jump the well and fall out of the bowl (Fig. 9.6). Lowering the rf frequency will induce some more atoms to leave the bowl. When the atoms with higher energy leave the bowl, the average energy of the remaining atoms becomes lower and atoms get colder in the process.

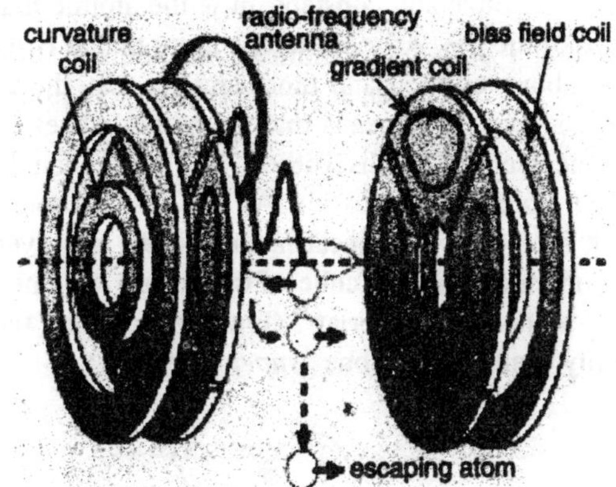

Fig. 9.6

Fig. 9.7 : Cooling to progressively colder temperatures resulting in the formation of a Bose- Einstein condensate (far right).

Continuing to cool involves continuously lowering the sides of the magnetic trap (Fig. 9.5 b). The process is halted when the temperature goes down to nearly 100 nK. and only a few atoms are left to form a condensate, with the only drawback being that the number of trapped atoms is reduced. The process of Bose-Einstein condensate formation is shown in Fig. 9.7.

Bose-Einstein Condensate

Each atom has an associated de Broglie wavelength λ_B. In accordance with the Heisenberg uncertainty principle, the position of an atom is smeared out over a distance given by the thermal de Broglie wavelength,

$$\lambda_B = \left[\frac{h^2}{2kmT}\right]^{1/2}$$

where k is the Boltzmann constant, m is the atomic mass and T is the temperature of the gas. At room temperature the de Broglie wavelength is typically about ten thousand times smaller than the average distance between the atoms. This means that the matter waves of the individual atoms are uncorrelated or "disordered" and the gas can thus be described by *classical Boltzmann statistics*. As the gas is cooled, however, the wavelength increases, the smearing increases, and eventually there is more than one atom in each cube of dimension λ_B. The wave functions of adjacent atoms then "overlap" (Fig. 9.8), causing the atoms to lose their identity, and become one *'super-atom'*.

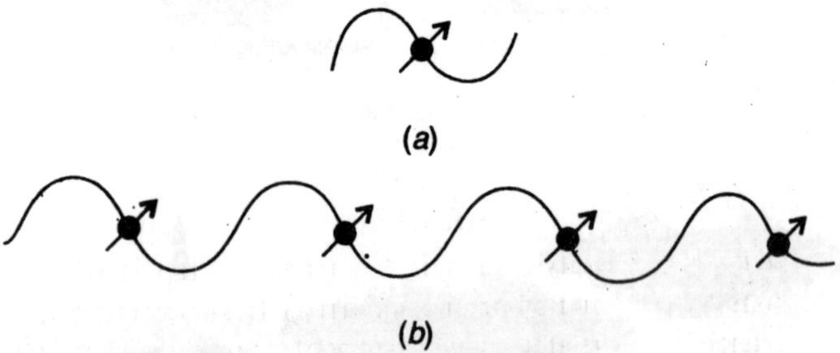

Fig. 9.8 : (a) An atom behaves as a de Broglie wave; (b) at critical temperature Bose-Einstein condensate or a coherent matter wave is formed.

Bose-Einstein statistics dramatically increase the chances of finding more than one atom in the same state. Thus, at very low temperatures, all the atoms fall down to the lowest state, as they do not have enough energy to go to higher energy states. The result is *Bose-Einstein condensation*, a macroscopic occupation of the ground state of the gas. Although small, typically 0.1 mm across, a Bose-Einstein condensate can be seen with weakly magnifying lenses and a video camera. Thus, it is a microscopic object. The transition to BEC corresponds to a transition from a set of disordered atoms to coherent matter waves. Indeed, the transition from disordered to coherent matter waves can be compared to the change from incoherent to laser light. In a Bose condensate all the atoms occupy the same quantum state and can be described by the same wave function. The condensate therefore has many unusual properties not found in other states of matter.

BASIC ATOM LASER

An optical laser consists of an active medium held in a cavity resonator (Fig. 9.9). When the active medium is pumped, say with an optical source, into the state of population inversion, stimulated emission is triggered and an intense beam of light emerges out of the partially reflecting mirror of the cavity. An atom laser also should consist of similar principal parts.

1. *Active medium* : In the case of atom laser, the active medium is a thermal cloud of ultracold atoms.

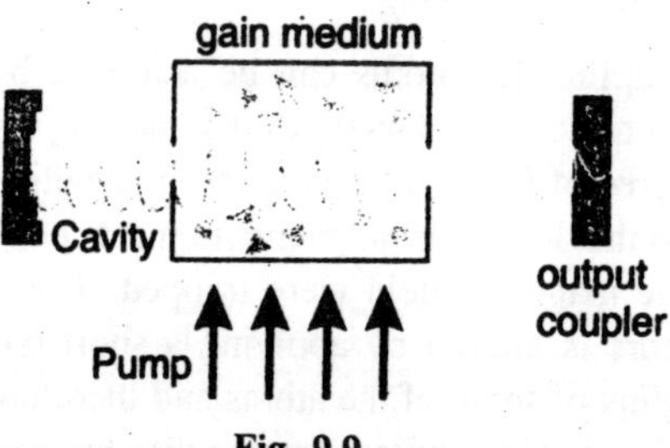

Fig. 9.9

2. *Cavity* : The cavity resonator is a magnetic trap in which the atoms are confined by magnetic mirrors. In a magnetic trap, for instance, once the atoms have been cooled and trapped by lasers, the light is switched off and an inhomogeneous magnetic field provides a confining potential around the atoms. The trap is analogous to the optical cavity formed by the mirrors in a conventional laser.
3. *Pumping* : Pumping is done by the evaporative cooling. The evaporation process creates a cloud which is not in thermal

equilibrium and relaxes toward colder temperatures. This results in growth of the condensate.

4. *Stimulated emission :* In an atom laser, the presence of a Bose-Einstein condensate with N atoms enhances the probability that an atom joins the condensate increasing the size of the condensate to N+l atoms. The process of condensing atoms into the ground state of a magnetic trap is analogous to stimulated emission into a single mode of an optical laser.
5. *Laser threshold :* An optical laser starts working when the losses in the cavity are balanced by the gain in the medium. In atom laser the critical temperature for Bose-Einstein condensation resembles to the laser threshold. When the critical temperature for BEC is reached, atoms predominantly go into the lowest energy state of the system.
6. *Output-coupler :* An important feature of a laser is an output coupler to extract a fraction of the coherent field in a controlled way. In the case of a conventional laser the output coupler is a partially transmitting mirror.

We may think of the Bose condensate as being held in a container much like water in a bowl and to get a beam of coherent matter we just puncture the container and Bose condensate leaks out. The extraction process is called *out-coupling.*

Output coupling for atoms can be achieved by transferring them from states that are confined to states that are not, typically by changing an internal degree of freedom, such as the magnetic states of the atoms. Only the atoms that had their magnetic moments pointing in the opposite direction to the magnetic field were trapped. The "reflectivity" of the magnetic mirrors is altered by applying a short radio-frequency pulse to "flip" the spins of some of the atoms and therefore release them from the trap (Fig. 9.10). The extracted atoms then are accelerated away from the trap under the force of gravity. By changing the amplitude of the radio-frequency field, the extracted fraction could be varied between 0% and 100%. A pulsed output beam of atoms.

7. *Output of laser :* The output of an optical laser is a well collimated beam of light (electromagnetic waves). For an atom laser, the output is a beam of atoms (matter waves).
8. *Nature of output :* Optical lasers can work either in a pulsed mode or continuous wave (cw) mode. So far, the atom laser has

been realized in the pulsed mode only. Most continuous-wave optical lasers are truly continuous in the-sense that they are continually fed energy or "pumped" so that they can supply photons indefinitely. A truly continuous source of coherent matter waves could be similarly achieved only if thê condensate could be replenished continually. Schemes for steady- state condensate formation are being explored currently.

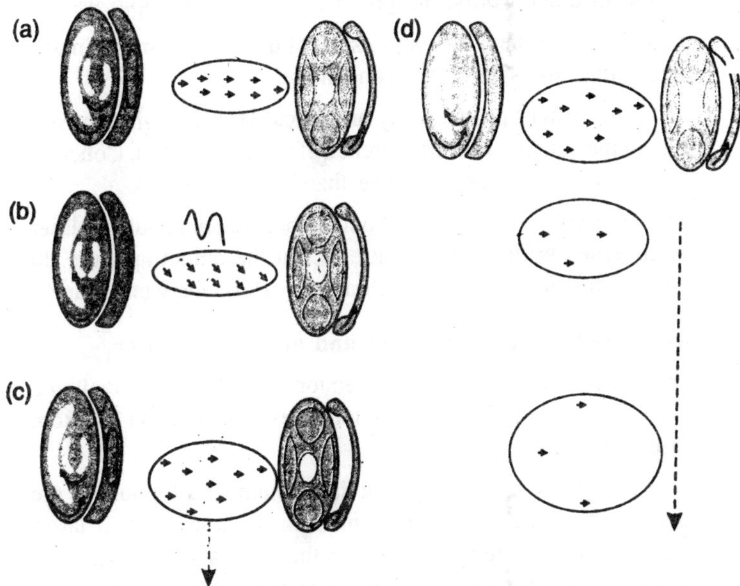

Fig. 9.10 : The rf output coupler (a) shows a Bose condensate trapped in a magnetic trap (b) A short pulse of rf radiation tilts the spins of me atoms (c) Quantum-mechanically, a tilted spin is a superposition of spin up and down. The cloud is split into a trapped cloud and an out-coupled cloud, (d) Several output pulses can be extracted, which spread out and are accelerated by gravity.

9. *Modes :* Optical lasers radiate in several modes, *i.e.*, at several nearby frequencies. The cavity in an optical laser is typically many wavelengths long and as a result can support several different frequencies or modes. Atom laser operates in a single mode, since the Bose condensate actually involves mode competition and the atoms typically occupy the lowest energy state of the trap.

Important Features of a Laser Beam

1. *Monochromaticity :* An important property of laser light is that it is monochromatic. By analogy, in a Bose-Einstein condensate all the atoms have the same energy and hence the same de Broglie wavelength. If this property can be maintained when the atoms are released from the condensate, we will have a highly *monochromatic* source of matter waves.
2. *Coherence :* A crucial feature of a laser is the coherence of its output-in other words, the presence of a macroscopic wave.

   It is likely that the *coherence length* of the lasers is in fact much larger than the size of the condensate.
3. *Intensity :* Light produced by an optical laser is highly intense due to the constructive interference of coherent waves. Coherent matter waves are more intense than ordinary atoms.
4. *Directionality :* The spread of optical beam issuing out of a laser is governed by the diffraction effect at the output aperture. In case of atom laser, the beam is limited by Heisenberg uncertainty.

Differences Between an Atom Laser and an Optical Laser

1. In optical lasers population inversion is essential for realizing laser action. Population inversion does not occur in an atom laser.
2. In case of an optical laser, a large number of photons are generated through stimulated emission. In case of atom laser, atoms are not created. The atoms in the lower state are increased while the number of atoms in the upper states decreases.
3. Atoms are heavier than photons. They are therefore accelerated by gravity and a matter wave beam will fall like a beam of ordinary atoms.
4. A light beam from an optical laser can travel very far and does not require a vacuum for its operation. Unlike photons, atoms will not travel very far in air, so that the atom laser must be used in a vacuum.

ATOM LASER APPLICATIONS

The possibility of producing a coherent beam of atoms that could be collimated to travel long distances, or brought to a tiny focus like an optical laser, opens up a whole host of applications. Atom lasers may have a major impact on the fields of atom optics, atom lithography, precision atomic clocks and other measurements of fundamental standards. We study here three typical applications of atom laser.

Holography

One application for which the coherence of an atom laser is critical is atom holography. Just as conventional holography uses the diffraction of a photon beam to reconstruct a 3-D image, atom holography uses the diffraction of atoms. As the de Broglie wavelength of the atoms is much smaller than the wavelength of light, an atom laser could create much higher resolution holographic images. Atom holography might be used to project complex integrated-circuit patterns, just a few nanometres in scale, onto semiconductors.

The first atom holograms were demonstrated in 1996 by Fujio Shimuzu and colleagues at the University of Tokyo, using laser-cooled atoms. In the case of laser-cooled gases, the level of coherence needed to create a hologram is achieved by selecting a small portion of the atoms. The problem would be simplified if a source, in which most of the atoms are in the same quantum state, is used. An atom laser is such a source and could provide a much more intense and fully coherent beam of atoms.

Holography is a two-step process. First, a hologram—a sort of diffraction grating containing information about the object—is produced. Then a beam of light (or atoms) is diffracted by the hologram to form the image. In optical holography, the hologram is often made by interfering a laser beam with light that has been reflected from an object. The resulting diffraction pattern is recorded on photographic film.

In the atom-holography experiments, an image has been created by diffracting a coherent beam of atoms through a grating that was manufactured using electron-beam lithography. The image may be recorded on a "microchannel plate"—a detector that is sensitive to atoms. So far, atom holography has been able to produce 2-D images.

Atom Interferometry

Another important application is atom interferometry. In an atom interferometer an atomic wave packet is coherently split into two wave packets that follow different paths before recombining. The interference pattern created when the two wave packets recombine tells us something about the phase difference between the two paths. Atom interferometers that are more sensitive than optical interferometers could be used to test quantum theory, and may even be able to detect changes in space-time. This is because the de Broglie wavelength of the atoms is smaller than the wavelength of light, and the atoms have mass. Atom lasers would allow the use of devices with unequal path lengths, such as Michelson interferometers. Such devices may provide a way to measure lengths over large distances with unprecedented precision.

Nonlinear Atom Optics

Just as the invention of the laser enabled the field of nonlinear optics to flourish, intense sources of coherent matter waves have opened up a similar field in atom optics. Until recently, most atom-optics experiments could be thought of as single-particle phenomena where the interactions between particles could be neglected. In conventional nonlinear optics, photons interact with each other through some mediating material, such as a transparent crystal. A common nonlinear optical phenomenon is "four-wave mixing". Typically three waves, of frequency ω_1, ω_2 and ω_3, are sent into a nonlinear crystal. The exchange of energy and momentum between the waves, mediated by the nonlinear crystal, results in the production of a fourth wave with frequency $\omega_4 = \omega_1 + \omega_2 - \omega_3$ (Fig. 9.13a). A quantum mechanical description of this process shows that two photons from separate beams annihilate in the crystal and produce two new photons. The energy and momentum of one of these photons adds to the third beam, while the other photon corresponds to a new, fourth beam.

In 1998 an analogous process with matter waves was predicted. It is predicted that if three condensates of appropriate momenta collided, the term in the nonlinear Schrodinger equation that describes the interactions between the atoms would give rise to a fourth. At the atomic level, this process can be described as a collision between two atoms from separate matter-wave beams. One of the atoms is stimulated so that it scatters in the direction of the third incident matter-wave beam. By the conservation of momentum, the other atom goes off to make a fourth, separate beam.

The actual experiment did not use three separate condensates. Instead, lasers were used to divide H single condensiale into three different momentum states via a process called Bragg diffraction. Starting with a condensiale at rest, two separate pulses of interfering laser beams were applied to create the Bragg "diffraction grating" that divided the atoms roughly equally into three different momentum states, including the state of the initial condensate. When these pulses were applied fast enough-that is Before the different momentum states had a chance to separate-atoms in a fourth momentum state were produced.

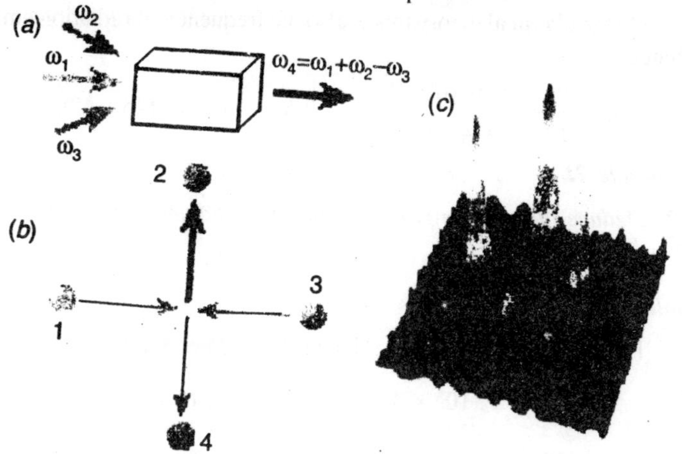

Fig. 9.11

The four-wave mixing process arises from collisions between pairs of atoms from two matter- wave beams (1 and 2 in Fig. 9.11b). One pair of atoms scatters in the direction of the third, incident Biatter-waye beam and-amplifies (3). By the conservation of energy and momentum, the other pair of atoms produces a fourth, separate beam (4). Fig. 9.11(c) shows an image of the experimental atomic distribution showing the fourth (small) wave packet generated by the matter-wave mixing process.

SOLVED EXAMPLES

Example 1:

At what frequency should an electron revolve around a proton if orbit radius is 0.53Å? Estimate the wavelength of radiations emitted on the classical theory.

Solution:

Equating centripetal force with the Coulomb force,

$$\frac{mv^2}{r} = \frac{e^2}{4\pi \epsilon_0 r^2} = \Rightarrow v = \left(\frac{e}{4\pi \epsilon_0 nr}\right)^{1/2}$$

Frequency is given by $v/2\pi r$. Hence,

$$f = v/2\pi r = [e/2\pi]\ (4\pi\epsilon_0 m)^{-12} r^{-3/2}$$

with $r = 0.53 \times 10^{-8}$ cm, substitution gives

$$f = 6.6 \times 10^{15}\ s^{-1}$$

On the classical theory this is also the frequency of radiation emitted. Hence

$$\lambda = \frac{c}{f} = \frac{3.0\times10^{10}\,\text{cms}^{-1}}{6.6\times10^{15}\,\text{s}^{-1}} = 4.5 \times 10^{-6}\ \text{cm (450 A°)}.$$

Example 2:

Deduce the wavelength of the first two Balmer lines in He+, given $R = 1.097 \times 10^5\ cm^{-1}$.

Solution:

For He^+ we have Z = 2. Hence the Balmer series is

$$\bar{v} = 1.097 \times 10^5 \times 4 \left(\frac{1}{2^2} - \frac{1}{n^2}\right) \text{cm}^{-1};\ n = 3, 4, ...$$

Substituting n = 3 and 4 in turns and taking the reciprocal gives:

$$\lambda = 1.64 \times 10^{-5}\ \text{cm};\ 1.21 \times 10^{-5}\ \text{cm}.$$

Example 3:

The ns and np states of sodium atom have quantum defects 1.37 and 0.88 respectively. Calculate the energy values for the lowest three levels of each set, draw an energy level diagram, and deduce the wavelength of the spectral line for the transition 3p – 3s.[13] ($R = 109670\ cm^{-1}$).

Solution:

In Na the lowest n is 3. Using the a values for s and p states.

We have,

$$E_s = -\frac{109670}{(1.63)^2}, -\frac{109670}{(2.63)^2}, -\frac{10970}{(3.63)^2}\text{cm}^{-1}$$

$$= -41290,\ -15850,\ -8326\ \text{cm}^{-1}$$

$$E_p = -\frac{109670}{(2.12)^2}, -\frac{109670}{(3.12)^2}, -\frac{109670}{(4.12)^2}\text{cm}^{-1}$$

$$= -24410,\ -11270,\ -6463\ \text{cm}^{-1}$$

The transition m = 3 to n = 3 is also shown. The wavelength is given by:

$$\lambda = \frac{1}{v} = \frac{1}{41290-24410}\text{cm}$$

$$= \frac{1}{16880} = 5.9 \times 10^{-5} \text{ cm.}$$

Example 4:

Deduce the Duane and Hunt limit for an X-ray tube working at 40 kV.

Solution:

We have

$$v_{max} = \frac{40000\times1.6\times10^{-19}}{6.62\times10^{-34}}\text{Hz} = 0.97 \times 10^{19}\text{Hz}$$

$$\lambda_{min} = \frac{c}{v_{max}} = 0.31 \times 10^{-8} \text{ cm.}$$

Example 5:

With α-particles from polonium ($v = 1.6 \times 10^9$ cm/sec) calculate the nearest approach r_o with silver nuclei (Z = 4.7).

Solution:

Nearest approach is for head-on collision, and all kinetic energy then becomes potential energy.

$$\frac{1}{2}m_\alpha v^2 = \frac{2e.Ze}{4\pi\varepsilon_0 r_0}$$

$$\therefore r_o = \frac{1}{4\pi\varepsilon_0}\cdot\frac{4Ze^2}{m\alpha v^2} = (9 \times 10^9)\ \frac{4\times47\times(1.6\times10^{-19})^2}{(4\times1.67\times10^{-27}(1.6\times10^{7})^2}\text{m}$$

$$= 2.5 \times 10^{-13} \text{ cm.}$$

EXERCISES

1. Explain the principle of an atom laser and describe its important components.
2. Describe the similarities and differences between an optical laser and an atom laser.

3. Describe any two important applications of atom laser.
4. Explain how atom laser is used in holography?
5. Explain the four-wave mixing of matter waves.
6. Explain in brief the method of cooling a dilute gas to nanokelvin temperatures.
7. What is meant by Bose-:Einstein condensation? What is its importance?
8. Explain the technique of Doppler cooling? What is the temperature limit that can be attained through this method?
9. What is an optical molasses? What is its role in cooling atoms?
10. What is a magneto-optic trap? Why is it required in cooling of atoms?
11. Describe the technique of evaporative cooling.

10

THE SPECIAL THEORY OF RELATIVITY

If the same event is observed by two observers who are in *relative motion*, then their measurements would differ. In Galileo's idea if x', y', z', t' are the measurements by observer S', who has velocity V relative to observer S then the measurements x, y, z, t as made by S (Fig. 10.1) are given by:

$$\left.\begin{aligned} x &= x' + Vt' \\ y &= y' \\ z &= z' \\ t &= t' \end{aligned}\right\} \text{ or } \left.\begin{aligned} x' &= x - Vt' \\ y' &= y \\ z' &= z \\ t' &= t \end{aligned}\right\} \quad ...(1)$$

where axes x, y, z and x', y', z' are taken parallel, the motion of S' is along the x' axis, and the origin 0' crosses 0 at time ' = t = 0. The set of Eqn. (1) is called the *Galilean transformation* of co-ordinates between observers in uniform relative motion.

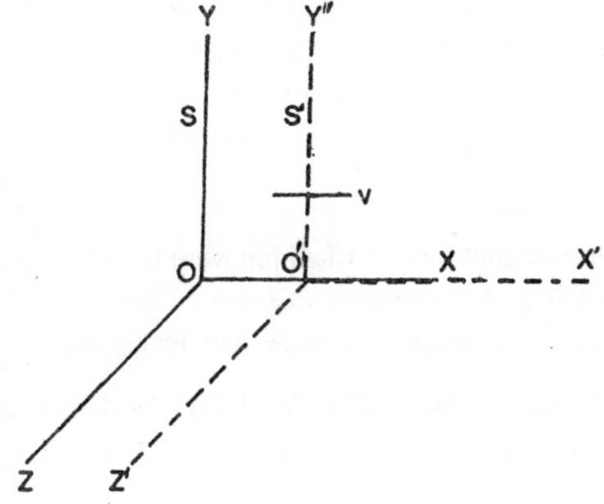

Fig. 10.1 : Two frames S,S' in relative motion along x-axis.

Newtonian mechanics accepted Galilean transformation, which leads to:

$$\left.\begin{aligned}\Delta x &= \Delta x' \\ \Delta y &= \Delta y' \\ \Delta z &= \Delta z' \\ \Delta t &= \Delta t'\end{aligned}\right\} \quad \text{...(2)}$$

For speed measurements we get

$$\upsilon = \frac{dx}{dt} = \frac{dx'}{dt}\ V = \upsilon' + V \text{ or } \upsilon' = \upsilon - V \quad \text{...(3)}$$

These give the Galilean transformation for intervals of space (lengths) and time, and the law of addition of velocities. Eq. (3) leads to

$$\frac{d^2x}{dt^2} = \frac{d^2x'}{dt'^2} \quad \text{...(4)}$$

which is independent of V. The Newtonian mechanics will apply identically in the measurements made by S or S'. For this reason Eqs. (1) to (4) are said to represent *Newtonian relativity*. Measurements x and $\frac{dx}{dt}$ differ from x' and $\frac{dx'}{dt'}$ respectively, but intervals Δx and Δt and acceleration $\frac{d^2x}{dt^2}$ are identical with $\Delta x'$, $\Delta t'$, and $\frac{d^2x'}{dt^2}$ respectively.

These transformations are simple and elegant, and we use them widely in everyday life. But with the growth of electromagnetism, and with highspeed particles available at the beginning of this century, these transformations failed to explain experimental results. *Einstein* came up with the *Special Theory of Relativity*, and it came up with new transformations which not only explained the existing experimental data, but made several predictions, all of which were found to be quantitatively verified. This Chapter deals with this most remarkable theory, which deeply influenced philosophy, science and technology.

RELATIVITY AND THE PROPAGATION OF LIGHT

If ether is accepted as the medium for the propagation of waves of light, then the waves have a speed c *relative to ether*. Now consider a telescope on earth viewing a star (Fig. 10.2). The telescope moves with earth's orbital speed V, which is about 10^{-4}c. If we ignore V the image will be formed at P, the true focus. But if we take the telescope moving *towards* the star, then by the time the waves travels length L to P, we will find P at P' where PP' = (V/c)PL.

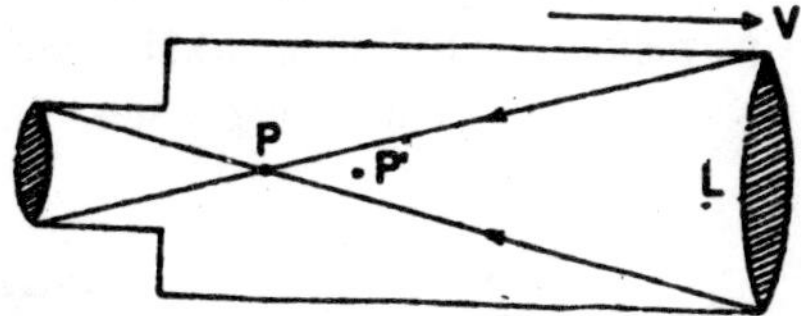

Fig. 10.2 : Effect of motion of a telescope through the ethers.

If the telescope is moving away from the star, the shift of P will be away from L by the same amount, so that by observations in two situations the difference should be observed. This effect was looked for by Arago long ago, but was not observed. Bradley looked for a similar angular deflection when V owns perpendicular to the line of sight. Many such results, involving differences of the first order in V/c were expected, but were not observed.

Did it mean that each observer (like the telescope) carried ether along with itself? Fresnel in 1818 produced arguments showing that a material medium moving with speed ^dragged ether with itself to the extent of $V(1 - n^{-2})$, where n is its refractive index. This is called *the ether drag*. Fizeau in 1851 showed by direct experiment that a column of water does drag the waves of light to this extent. In 1895 Lorentz deduced Fresnel's formula for ether drag, using the electromagnetic theory. Ether drag would mean that relative to the moving water, for example, ether would have speed Vn^{-2}. This will be the ether wind.

Michelson examined the possibility of using interference fringes to measure the *ether wind.* He found that any device used would only measure effects proportional to $(V/c)^2$, i.e., the *second order* effect. Since earth's orbital speed is $\sim 10^{-4}c$, the effect in (say) 1 m length will amount to $\sim 10^{-8}$m, which is $\sim \lambda/50$ for visible light. Yet, he devised his famous interferometer for just this experiment 'In 1887 he performed the crucial experiment along with Morley in which this effect, if present, could have been measured.

THE MICHELSON AND MORLEY EXPERIMENT

The Michelson interferometer has a significant feature that a beam of light splits into two parts, which travel along *mutually perpendicular* directions, and then return to form interference fringes. Michelson and Morley mounted the interferometer on a rock floated on mercury, and keeping one arm OA (Fig. 10.3a) parallel to earth's orbital velocity V, they got the interference fringes. Then the rock was turned by 90°, making arm OA perpendicular to F (Fig. 10.3b), and a close watch was kept on the fringes. No shift of the fringes was observed.

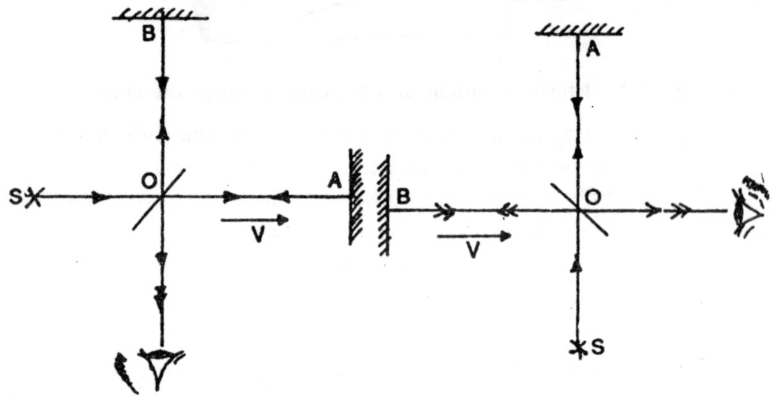

Fig. 10.3(a) Michelson interferometer with OA parallel V.

Fig. 10.3(b) Michelson interferometer with OB parallel to V.

The theory, based on Galilean velocity addition, indicated that a change equal to path change (OA + QB)V^2/c^2 should have occurred. Making each distance ~10m, they expected a shift of n fringes, where with ($\lambda = 5 \times 10^{-7}$m)

$$n = \frac{2 \times 10\text{m}}{5 \times 10^{-7}\,\text{m}} \times (10^{-4})^2 \approx 0.4$$

But Michelson and Morley state that the *observed shift* was "certainly less than one-twentieth part of the computed shift and possibly less than one fortieth part." Quite reasonably, this was within the error of the observations. So it was a *null result, i.e.*, there is no ether wind at all.

Theory of the Michelson-Morley Experiment : In Fig. 10.3 the pulse travelling from OO to A approaches A at speed c-V, and the return pulse from A to O approaches O at a speed c + V, according to Galilean transformation. So the two-way time is

$$t_A = \frac{OA}{c-V} + \frac{OA}{c+V} = \frac{2OA}{c} \cdot \frac{1}{1-\beta^2}, \text{ where } \beta = \frac{V}{c}$$

For the pulse travelling towards B we note that by the time it travels there B will move to B', where magnitudewise BB'/OB' = V/c. Thus the pulse approaches B at speed $(c^2 - V^2)$. The same speed applies for the return pulse from B approaching O. So for this path the return time is

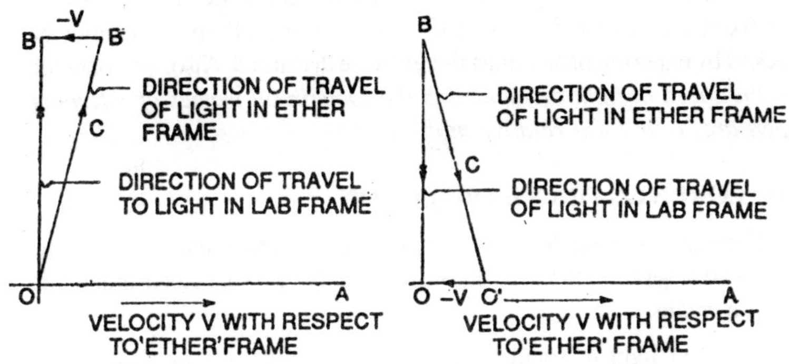

Fig. 10.4(a) : Pulse from O reaching B'

Fig. 10.4(b) : Pulse from B reaching O'

$$t_B = \frac{OB}{(c^2 - V^2)^{1/2}} + \frac{OB}{(c^2 - V^2)^{1/2}} = \frac{2OB}{c}\frac{1}{\sqrt{1-\beta^2}}$$

The time difference is thus given by

$$t_B - t_B = \frac{2OB}{c}\frac{1}{1-\beta^2} - \frac{2OB}{c}\frac{1}{\sqrt{1-\beta^2}}$$

Now comes the crucial step. Turning the two path by 90° leads to the new time difference

$$t_A' - t_B' = \frac{2OA}{c}\frac{1}{\sqrt{1-\beta^2}} - \frac{2OB}{c}\frac{1}{1-\beta^2}$$

because OB is now parallel to V and OA is perpendicular to V. Thus the result of 90° turning is to change the time difference by

$$\Delta t \equiv (t_A' - t_B') - (t_B - t_B) = \frac{2(OA + OB)}{c}\left[\frac{1}{\sqrt{1-\beta^2}} - \frac{1}{1-\beta^2}\right]$$

$$= \frac{OA + OB}{c}\beta^2$$

using binominal expansion, since $\beta << 1$. The equivalent change in path difference is

$$\Delta c = c\,\Delta t = (OA + OB)\beta^2 \qquad ...(5)$$

The Michelson-Morley experiment and other variations of it have been repeated under many different circumstances, and the null result has been established fairly firmly. We therefore conclude that the concept of a fixed frame of reference (like 'ether' filling all space) cannot be checked by experiment and must therefore be discarded. Also, we conclude that *the speed of light in vacuum is the same in all frames of reference which are in uniform relative motion.*

THE LORENTZ TRANSFORMATION

Lorentz searched for that transformation which may replace the Galilean one (eqn. 1) to keep Maxwell's equations of the electromagnetic field *invariant.*

Speed of light in vacuum is one of the results of e.m. theory, and we will deduce Lorentz transformation from the consideration that speed of light be the same in all directions for any observer in uniform motion.

Consider a reference frame S in which a pulse generated at the origin at time t = 0 grows at time t into a sphere given by

$$x^2 + y^2 + z^2 - c^2t^2 = 0 \qquad ...(6)$$

If the pulse is observed from a frame S', moving relative to S along the x-axis with speed V, then Galilean transformation (24.1) would give the pulse for an observer in S' as

$$(x' + Vt')^2 + y'^2 + z'^2 - c^2t^2 = 0$$

or $$x'^2 + 2x'Vt' + V^2t'^2 + y'^2 + z'^2 - c^2t'^2 = 0$$

This is not a sphere. The terms in y' and z' are of the proper form to give us a sphere, but the x' and t' terms are not. To get rid of the term 2x' Vt' let us change the time transformation from t = t' to

$$t = t' + \frac{Vx'}{c^2} \qquad ...(7)$$

Substitution of this in eqn. (6) now gives us

$$x'^2 + 2x'Vt' + V^2t'^2 + y'^2 + z'^2 - c^2t^2 - 2x' - Vt' - \frac{V^2x'^2}{c^2} = 0$$

or $$x'^2\left(1-\frac{V^2}{c^2}\right) + y'^2 + z'^2 - c^2t'^2\left(1 - \frac{V^2}{c^2}\right) = 0$$

Now we find that a common factor $\left(1-\frac{V^2}{c^2}\right)$ occurs with bolts x'^2 and t'^2 terms. Hence the transformation for x as well as t needs a further factor $\left(1-\frac{V^2}{c^2}\right)^{1/2}$ Thus set of eq. (1) is replaced by

$$x = \frac{x' + Vt'}{\sqrt{1-\frac{V^2}{c^2}}} = \gamma\,(x' + Vt')$$

$$y = y' \qquad \qquad ...(8)$$

$$z = z'$$

$$t = \frac{t' + \frac{Vx'}{c^2}}{\sqrt{1-\frac{V^2}{c^2}}} = \gamma\left(t' + \frac{Vx'}{c^2}\right)$$

where $\gamma = 1\sqrt{1-\beta^2}$. The set of Eqn. (8) is called the *Lorentz transformation for space and time co-ordinates* for frames in uniform relative motion. Substitution of these into Eqn. (6) gives us

$$x'^2 + y'^2 + z'^2 - c^2t'^2 = 0$$

which is a sphere, as desired. In other words, the pulse, as observed from frame S', also grows into a sphere at the rate if the measurements are related by Eqn. (8) and not by Eqn. (1).

One notable point about Eqns. (8) is that the space transformation involves time and the time transformation involves the space co-ordinate. Thus the transformation is not a "space and time" transformation but "space-time" transformation.

Invariance of Maxwell's Equations of the Electromagnetic Field : Transformation (8) has significance beyond just keeping the velocity of light invariant with the observer's motion. All Maxwell's equations of the *e.m.* field remain invariant under this transformation (we shall not prove it here). Under Galilean transformation these equations would vary with the observer (though Newton's law of mechanics did not vary), *i.e.*, the taws of electrodynamics would vary. Therefore, between two observers A and B in uniform relative motion, one could decide not only the relative velocity V_{AB}, but also the absolute velocities V_A and V_B with reference to some absolute reference frame. Lorentz transformation means that all frames in uniform relative motion show the same laws of electrodynamics, so that no experiment on electrodynamics can measure V_A and V_B only V_{AB} is measurable.

THE SPECIAL THEORY OF RELATIVITY

Einstein took the fundamental view that what cannot be measured by an experiment has no meaning in science. Hence the concept of an *absolute* reference frame, in space is meaningless, and only *relative* motions are measurable and hence meaningful. Thus the concept of ether or any other 'fixed' frame of reference was discarded by him altogether.

The null result of the Michelson-Morley experiment on electromagnetic waves was extended by Einstein to include all conceivable experiments, including those in mechanics, and given the status of a fundamental postulate. In 1905 he put forth the following postulates of what is now called the Special Theory of Relativity:

1. The laws of electrodynamics (including the propagation of light with a speed c in free space) and the laws of mechanics are exactly the same in all frames in uniform relative motion.
2. It is impossible to devise an experiment whioh may detect a state of absolute motion.

The first postulate is equivalent to the null result of the Michelson-Morley, experiment, but extends it to all conceivable experiments. The second says that not only the laws are identical, but also the physical constants like h, c, e, k would be the same for measurements in all frames in uniform relative motion.

Sometimes the postulates are put in three parts as follows:

(A) For frames of reference in uniform relative motion, the speed of light is independent of the reference frame in which the measurement is made.

(B) The space is isotropic and uniform.

(C) The fundamental laws of physics arc identical for any two observers in uniform relative motion.

In this form the null result of Michelson-Morley experiment is put separately in (A) and its extension to *all laws* of physics is stated separately in (C). Postulate (B) is really an additional statement included in postulate 2 given earlier but often not clearly appreciated. It is obvious that if space were not isotropic or not uniform, one could use the varying values of the constants as a device for determining absolute motion.

The Operational View of Relativity : The theory of relativity of essentially a theory of measurement. Consider, for instance, a railway train in uniform motion relative to the earth. Events occurring in the train are observed by observers in the train and also by observers on the earth; similarly events occurring on the earth are observed by observers on the earth and also by observers in the train. So far we have not carefully examined *how* an observer in the train would measure the length of a radon the ground, or how an observer on the ground would measure the time interval between two events occurring in the train? The *operation* for such measurements has to be clearly specified. Einstein stated the operation as follows:

(a) The instruments in the two frames of references cannot communicate directly except when they coincide in position.

(b) The only other communication can be through pulses of light.

If we proceed through this operational point of view we find that the transformation is not of the Galilean form, but of the Lorentz form. Thus the Lorentz transformation comes out as a consequence of the operational point of view about measurements of events in one reference frame from another frame in uniform relative motion. To emphasize this we shall use the operational procedure in one although the equivalent results are more easily obtained from the Lorentz transformation.

THE OPERATIONAL VIEW OF TIME DILATION

Fig. 10.5 shows a reference frame S' moving relative to another frame S' with speed V along x' axis. A pair of events at the same place in S' is the emission of a pulse of light from O and its return to O' from a mirror M placed at transverse distances. Let this time interval be $\Delta t'$. We have to see the *operation* by which an observer in frame S would measure this time interval. Recall rules (a) and (b) laid down by Einstein that a communication to S can occur only when the instruments coincide in position, or through pubes of light. So we consider a pair of events involving pulses of light. Now, frame S' contains a clock at O' and frame S contains a chain of clocks along the x-axis; these will start when a pulse of light reaches there, and stop when the next pulse reaches there. The pulse of light is created just as O' coincides with O during the motion of S'.

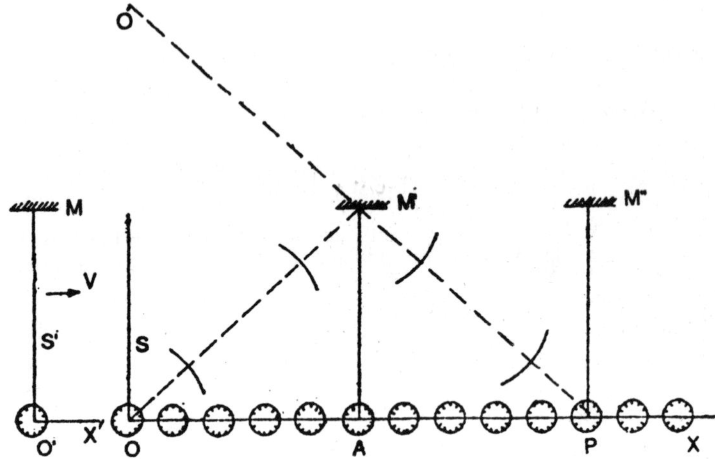

Fig. 10.5 : The Operational view of lime dilation.

Now, for frame S', the mirror M is all the lime along perpendicular to x' axis at O'. The pulse starting from O' and returning to it after reflection from M takes time

$$\Delta t' = \frac{2y}{c} \qquad ...(9)$$

For frame S the pulse starts from O, and as it grows on, the mirror M (fixed in S' frame) moves along the-c-axis at speed V. So the pulse contacts M at a position like M' and the reflected pulse reaches a particular clock P to stop it. (The S' frame has its MO' at position M" P just at this instant.)

What interval Δt does the observer in S frame record? The event started at O and ended at P in this frame. The "start' signal came to S frame by method (a), *i.e.*, clock O' coinciding with O. The 'stop' signal came to S frame by method (b), *i.e.*, the pulse light. Clock P also is started by the direct pulse, which reached there at time OP/c later than the start of the event.

We presume that this delay is taken care of by synchronizing the clocks in the chain appropriately. With this understanding, the time Δt is measured by clock P in frame S.

By Einstein's postulate of constancy of the speed of light, OM' + M'P = $c\Delta t$. Simple geometry gives now the relation between Δt arid $\Delta t'$.

$$AM'^2 = OM'^2 - OA^2 = \left(\frac{1}{2}c\Delta t\right)^2 - \left(\frac{1}{2}V\Delta t\right)^2 = \left(\frac{1}{2}c\ \Delta t\right)\left(1 - \frac{V^2}{c^2}\right)$$

But Eqn. (12) gives $\frac{1}{2}\ c\Delta t' = y = AM'$. Hence

$$\left(\frac{1}{2}\ c\Delta t'\right)^2 = \left(\frac{1}{2}c\Delta t\right)^2\left(1 - \frac{V^2}{c^2}\right)$$

This is the same as Eqn. (11), deduced from Lorentz equations.

EXPERIMENTAL VERIFICATION OF TIME DILATION

Time dilation is best verified in experiments on decay half-life of fast moving nuclear particles. π^+ meson is one such which can travel io the laboratory at quite high speeds, close to c, and which decays with time.

Now, we measure the time of travel Δt of a π^+ meson beam across length L in the laboratory. We also determine the π^+ flux at the start and end of this length, and from this we get the half-life $\tau_{1/2}$ for their decay. Observations show that the $\tau_{1/2}$ thus measured varies with velocity $V(=L/\Delta t)$ of the particles. The larger the V value, the greater is the measured $\tau_{1/2}$. Nuclear decay lime should be a characteristic constant. So how do we understand this varying $\tau_{1/2}$. Convert $\tau_{1/2}$ (measured) to $\tau'_{1/2}$ (proper half-life) using

$$\tau'_{1/2} = \tau_{1/2}\sqrt{1 - \frac{V^2}{c^2}}$$

Then you find that for all the varying $\tau_{1/2}$ for π^+ mesons the $\tau'_{1/2}$ value comes out the same 1.8×10^{-8}s.

As one example, a π^+ meson beam produced with speed 0.990c travelling a 30 m distance decreased to 58% of the original flux. Using

$$2^{-\Delta t/\tau_{1/2}} = 0.58$$

gives $\tau_{1/2} = 12.8 \times 10^{-8}$s. But $\tau'_{1/2}$ comes to 1.8×10^{-8}s.

EQUIVALENCE OF MASS AND ENERGY

We may consider m_0c^2 as an energy associated with mass m_0. This is called the *'rest energy'*, and is in addition to all other earlier known forms of energy. The sum of rest energy and kinetic energy then becomes

$$E = mc^2 \quad ...(10)$$

An alternative way of showing that such a hypothesis is a necessity is as follows:

Consider a *'massless box'* containing a number of particles of different masses and momenta. A solid body fits this description if the geometrical form alone is' taken as the massless box and the atoms and molecules in it are taken as the contained particles with various masses and momenta. In a reference frame fixed with the 'box' the total momentum of the particles is zero. If the ith particle has mass m, and kinetic energy T_i we would say

$$\text{total mass } m_o = \Sigma m_i$$

$$\text{total kinetic energy } T = \Sigma m_i$$

where the summation extends over all the particles.

In the *same frame of reference*, we can have an alternative description of the same system. We may consider the 'box with the particles' (the solid as such in our example) as a collective whole. Now in this description the system has no kinetic energy. Yet, if we take the total mass as just Σm_i (that is m_o) there will be discrepancy-about energy. Hence to keep the law of conservation of energy intact we must take the mass m, in the second description, such that

$$m - m_o = \frac{T}{c^2}$$

or $$c^2(m - m_o) = T = \Sigma T_i \quad ...(11)$$

The right-hand side expresses the sums of kinetic energies of the separate particles in the first description. When we switch over to the collective description, this appears in the form of increase of mass

$$\Delta m = m - m_o = \frac{T}{c^2}$$

However, energy can occur in various forms, the equivalence of which has been well established. Hence the statement may be generalised. We write Eqn. (10) now as

$$E = mc^2 = m_oc^2 + T + E_1 + E_3 + ...$$

where E_1, E_2, etc., are energies in other forms like electrical energy, gravitational energy, etc., and E is the 'total energy* including m_oc^2. From this we put most generally

$$c^2\Delta m = \Delta T + \Delta E_1 + \Delta E_2 + \ldots \qquad ...(12)$$

which means that a change of mass Δm may appear in the form of energies of various kinds, whose total is $c^2 \Delta m$,

The most spectacular examples of mass-energy equivalence are the *pair production* of electron and positron from y-rays and the *annihilation*, of a particle and its anti-particle to produce γ-rays photons. In the first case, new material particles are generated from energy of an electromagnetic field, while in the second case, a pair of material particles is totally destroyed and equivalent energy appears as photons.

An electron and its anti-particle (the positron) each has mass 9×10^{-31} kg whose mc^2 value converted to electron-volts is 0.51 Mev. Hence the minimum energy of a γ-ray photon to produce an electron-positron pair is $2 \times 0.51 = 1.02$ Mev.

SOLVED EXAMPLES

Example 1:

A rod placed along the length of a railway train measures 2 metres for an observer standing on the ground. How much would it measure for an observer in the train itself? Speed of the train in 30 m/sec.

Solution:

The given length is the non-proper length to be designated by Δx. To deduce the proper length $\Delta x'$ we have,

$$\Delta x' = \frac{\Delta x}{\sqrt{1 - V^2/c^2}}$$

$$= \frac{2}{\sqrt{1 - 10^{-14}}} = \left(2 + \frac{1}{2} \times 10^{-14}\right) \text{ metre}$$

[The change is negligible, but the point is that $\Delta x'$ is greater than Δx].

Example 2:

A laboratory observation shows that a length 2 metre is crossed by a beam of radioactive particles in time 1.0×10^{-8} second and in this process half of the particles disintegrate. Deduce the proper half-life of these particles.

Solution:

The event (of disintegration) occurs *in the particle,* whose speed in the laboratory is, from the given data, $V = 2 \times 10^8$ m/s.

The half-life as measured *in the lab* is 1.0×10^{-8} second. This is the non-proper measurement, to be designated Δt. The proper half-life $\Delta t'$ is therefore given,

$$\Delta t' = \Delta t \sqrt{1 - V^2/c^2} \doteq 10^{-8} \sqrt{1-(2/3)^2}$$

$$= 0.78 \times 10^{-8} s.$$

Example 3:

In Example 24.4 deduce the traversed length as seen by an observer fixed relative to the particle. Use it to deduce the half-life.

Solution:

the path in the lab as measured by a lab observer is 2 metres. So this is the proper length $\Delta x'$. The length of this as seen by the moving observer would be the non-proper length Δx. Hence,

$$\Delta x = 200 \sqrt{1-(2/3)^2} = 1.56 m.$$

Example 4:

Calculate the relative values of kinetic energies of a particle accelerated to a speed of 0.500c, 0.900c, and 0.980c.

Solution:

$$T = m_o c^2 \left[\frac{1}{(1 - V^2/c^2)^{1/2}} - 1 \right]$$

Since $m_0 c^2$ is common for all three cases, the bracketted term alone determines the relative values of kinetic energies. The values of this term give

$$T_1 : T_2 : T_3 = 0.16 : 1.27 : 4.0 \text{ (check it).}$$

Example 5:

The temperature of a body of specific heat 0.2 is raised by 200°C. Calculate the fractional increase of its mass.

Solution:

$$\Delta E = ms\, \Delta\theta = m \times (0.2 \times 10^3 \times 4.2) \times 200$$

$$= 1.7 \times 10^5 \text{ m joule}$$

$$\frac{\Delta m}{m} = \frac{\Delta m}{m} = \frac{1.7 \times 10^5 m}{9 \times 10^{16} m} = 2 \times 10^{-12}$$

Thus mass of a 1 tonne body will increase by 2 microgram, which is too small for all practical purposes.

Example 6:

A nucleus of mass 435.124 a.m.u. breaks into two nuclei of masses 97.936 and 135.951 a.m.u. and a neutron of mass 1.009 a.m.u. Calculate the energy released when 1 milligram of such nuclei distintegrate.

Solution:

Loss of mass per fission = 0.228 a.m.u. (check)

$\therefore$ fractional loss of mass $= \dfrac{.228}{235} = 1.0 \times 10^{-3}$

$\therefore$ loss of mass in 1 mg $= 1.0 \times 10^{-9}$ kg

Energy released $= \Delta mc^2 = 1.0 \times 10^{-9} \times 9 \times 10^{16}$

$= 9 \times 10^7$ joule.

Example 7:

In the laboratory one particle A moves with speed + 0.9c along the x-axis, while another particle B moves with speed –0.8c along the x-axis. Deduce the velocity of A relative to B.

Solution:

Consider a frame moving with particle B. Relative to this B-frame the laboratory frame moves with speed 0.8c. Then we have

$$v_x' = 0.9c \text{ and } V = 0.8c.$$

Substitution we gives

$$v_x = \frac{(0.9+0.8)c}{\left(1+\dfrac{0.9\times0.8}{1}\right)} = \frac{1.7}{1.72}c = 0.998c.$$

Example 8:

Deduce the velocity at which the mass of a particle becomes 10 times its rest mass.

Solution:

$$\frac{m}{m_o} = \frac{1}{(1-V^2/c^2)^{1/2}} = 10$$

$$\therefore \quad 1 - V^2/c^2 = 0.01$$

or $V/c = (1 - .01)^{1/2} = (1 - .005)$

$\therefore$ $V = 0.995c.$

Example 9:

Deduce the speed of an electron accelerated to a potential of 1 million volts.

Solution:

Non-relativistic equation $T = 1/2nv^2$ cannot be used here,, when re-arranged, gives

$$\frac{1}{1 - V^2/c^2} = \left(1 + \frac{T}{n_o c^2}\right)^2$$

We have

$$m_o c^2 = 9.1 \times 10^{-31} \times (3.0 \times 10^8)^2 \text{ joule}$$

$$= \frac{8.2 \times 10^{-14}}{1.6 \times 10^{-19}} \text{ev} = 0.51 \text{ Mev}^5$$

With T = 1.00 Mev (given), we have

$$\frac{1}{1 - V^2 c^2} = \left(1 + \frac{1.00}{0.51}\right)^2 \Rightarrow V = 0.94c.$$

Example 10:

A reference frame S' moves along the x-axis with velocity 300 m/sec relative to frame S' their origins coinciding at t = t' = 0. If an event occurring at x' = 10m, y' = 5 m,z' = 2m, t' = 5 sec, is observed from frame S deduce x, z, t.

Solution:

The data given

$Vi' = 1500$ m; $Vx' = 3000$ m^2s^{-1};

$$\lambda = \left[1 - \left(\frac{300}{3 \times 10^8}\right)^2\right]^{-1/2} = 1 + \frac{10^{-12}}{2}$$

We have

$$x = (x' + Vt')\,\lambda = (1510 + 755 \times 10^{-12}) \text{ m} = 1510\text{m};$$

$$y = y' = 5\text{m};\ z = z' = 2\text{m}$$

$$t = \left(t' + \frac{Vx'}{c^2}\right)\lambda$$

$$= \left(5 + \frac{10^{-13}}{3}\right) \times \left(1 + \frac{10^{-12}}{2}\right) s = 5s$$

[Thus x and x' differ by larger amounts, y and z do not differ at all from y' and z' respectively, while t and t' differ but by negligible amounts].

Example 11:

Repeat $x' = 10^{18}$ m, rest of the data remaining the same. (Star distances are of this order).

Solution:

$$x = (x' + Vt')\lambda = (10^{18} + 1500)\lambda \text{ m};$$

$$\lambda = 1 + \frac{10^{-12}}{2}$$

$$t = t' + Vx'/c^2 = (5 + 333)\lambda s$$

[In this case due to large x' we find t to be several times t' even though λ is still quite close to unity].

EXERCISES

1. Maxwell's equations of the electromagnetic field are given show that these are invariant in the Lorentz transformation but not in the Galilean one.
2. A rod travels parallel to its length with a speed 2.5×10^7m/sec in a frame S. If the length as measured by a person in frame S is 25.0 cm, deduce the length as measured by a person riding on the rod.
3. Deduce that kinetic energy T is given by $(m - m_0)c^2$, and that $E^2 = m_0c^4p^2c^2$
4. Compute the speed of an electron when it is accelerated to (i) 10^5 volt, (ii) 2.0×10^6 volt, (iii) 8.0×10^6 volt.
5. Deduce the momentum of a proton with kinetic energy 2×10^3 Mev. (proton rest mass = 1.673×10^{-27} kg).,
6. Show that at relativistic speed v, the force vs. acceleration relation becomes $F = m_0a/(1 - \upsilon^2/c^2)^{3/2}$.

7. Prove that any velocity added to c gives c itself. Prove also that any velocity subtracted from c also gives c itself.
8. In a laboratory one particle A travels with velocity $+ 2.5 \times 10^8$ m/sec, another particle B travels with velocity -2.9×10 m/sec, both along x-axis. Deduce the velocity of A relative to B.
9. A square frame 1.0m × 1.0m travels parallel to one of its sides with speed 2.8×1.0^8m/sec. Deduce how the frame would appear to a lab observer.
10. The half-life of a radioactive particle is 2.0×10^{-8}sec. Deduce the fraction of such particles which can survive after travelling 2.0m distance in the laboratory when their speed is (i) 2.0×10^5m/sec, (ii) 2.0×10^8m/sec, (iii) 2.96 × 10sm/sec.
11. As a beam of particles crosses 12.0 m length of a lab in 4.14×10^{-8}s, half of these particles decay. Deduce the proper half-life of the particles. What would an agency moving with the particles get as a measure for the length of the lab?
12. Without relativistic transformations the measurements of elm of fundamental particles and life-times of decaying particles would be in a mess. Explain.
13. Define *proper and non-proper* intervals of space and time, using the example of events on the ground and a jet plane in uniform motion. A carrom board measuring 1.00m × 1.00m when at rest on the ground is placed in the jet plane. How would it appear to players in the plane when the plane has velocity 1 km/s?
14. In the theory of the Michelson-Morley experiment one does not need OA and OB to be equal. Examine why the difference of OA and OB must not be large.
15. Examine the postulates A, B, C of what is the meaning of postulate B, and why is it needed? Comment on the statement that laws of physics are not the same in different frames of reference.
16. Prove that total energy E and momentum are related by $E^2 = (pc)^2 + (m_0c^2)^2$. Show that for $E_2 << m_0c^2$ this reduces to $p = E/c$, and for kinetic energy $K << m_0c^2$ it leads to $K=p^2/2m_0$.
17. If a reference frame A moves relative to another frame B with velocity $V_xi + Vj + V_zk_i$ write down the Lorentz transformation

for x_a, y_a, t_a, z_a in terms of x_b, a_b, z_b, t_b and the velocity of frame A. The axes in the two frames are orthogonal add parallel, and the origins coincide at time $t_a = t_b = 0$.

18. Show that for $V/c << 1$, the Lorentz transformation reduces to the Newtonian (or Galilean) one.

19. If $\beta = V/c$ and $y = \frac{1}{\sqrt{1-\beta^2}}$ obtain a plot of γ against β. Also, if we define $\alpha = 1 - \beta$ express y in terms of a for $\alpha << 1$. Deduce the values of α, β and V when γ takes a value 10.

20. Consider the four-dimensional volume element $\Delta c\ \Delta y\ \Delta z\ \Delta t$. Show that this remains invariant under Galilean transformation as well as Lorentz transformation.